Advances in Building Energy Research

Advances in Building Energy Research

Volume 3

earthscan
from Routledge

First published 2019 by Earthscan

2 Park Square, Milton Park, Abingdon, Oxfordshire OX14 4RN
52 Vanderbilt Avenue, New York, NY 10017

Routledge is an imprint of the Taylor & Francis Group, an informa business

First issued in hardback 2019

Notices
Practitioners and researchers must always rely on their own experience
and knowledge in evaluating and using any information, methods, compounds,
or experiments described herein. In using such information or methods they should be
mindful of their own safety and the safety of others, including parties for whom they
have a professional responsibility.

Product or corporate names may be trademarks or registered trademarks,and are
used only for identification and explanation without intent to infringe.

ISBN 978-1-84407-752-6 (hbk)

Typeset by Domex e-Data, India
Cover design by Giles Smith

A catalogue record for this book is available from the British Library

Library of Congress Cataloging-in-Publication Data

Advances in building energy research / editor-in-chief, Mat Santamouris.
    v. cm.
Includes bibliographical references.
1. Buildings–Energy conservation. I. Santamouris, M. (Matheos),
1956-TJ163.5.B84A285 2007
696–dc22

                            2007004087

Abstracting services which cover this title: Elsevier Scopus

Contents

List of Acronyms and Abbreviations

AC	architecture and construction
ACH	air changes per hour
ACM	absorption cooling machine
ADF	Approved Document F
ADINAT	a general finite element thermal analysis model
AHP	analytic hierarchy process
AHU	air-handling unit
AICVF	Association des Ingenieurs de Climatisation et de Ventilation de France
AIJ	Architectural Institute of Japan
AIST-CM	Advanced Industrial Science & Technology – Canopy Model
ANN	artificial neural network
ANP	analytic network process
AP	atmospheric pressure
ARD	automatic relevance determination
ASHRAE	American Society of Heating, Refrigerating and Air-Conditioning Engineers
BCR	bio-climatic roof
BEMS	building energy management systems
BMT	Buildings Market Transformation
BP	back-propagation
BRE	Building Research Establishment
°C	degrees Centigrade
CABE	Commission for Architecture and the Built Environment
CCE	cost of conserved energy
CEN	European Committee for Standardization (Comité Européen de Normalisation)
CFC	chlorofluorocarbon
CFD	computational fluid dynamics
CHP	combined heat and power
CIBSE	Chartered Institution of Building Services Engineers
cm	centimetre
cm^2	square centimetre
CO_2	carbon dioxide
COP	coefficient of performance
CSUMM	Colorado State University Mesoscale Model
CTF	conduction transfer function
CTTC	cluster thermal time constant
CWE	computational wind engineering
DAGT	decrement average ground temperature

DAT	discharge air temperature
DB	dry-bulb temperature
DE	double effect
DEC	desiccant evaporative cooling
DEROB	dynamic energy response of buildings
DES	detached-eddy simulation
DF	daylight factor
DL	astronomical day length
DL	double lift
DNS	direct numerical simulation
DQI	Design Quality Indicator
DSF	double-skin facade strategy
DSM	Differential Stress Model
DSS	decision support systems
EB	Energy Band
EC	European Community/Commission
ECMWF	European Centre for Medium-Range Weather Forecasts
EDEM	ESRU Domestic Energy Model
EEP	Energy and Environmental Prediction
EERS	energy efficient retrofit score
EHCS	*English House Condition Survey*
EI	Environmental Index
ELECTRE	Elimination et Choix Traduisant la Realité
EoS	end of setback
EPBD	Energy Performance of Buildings Directive
EPIQR	Energy Performance Indoor Environmental Quality Retrofit Method for Apartment Building Refurbishment
EST	Energy Saving Trust
ETI	energy and time consumption index
EU	European Union
FDD	fault detection and diagnosis
FDS	fault diagnosis system
FLL	Forschungsgesellschaft Landschaftsentwicklung Landschaftbau (German Landscape Development Research Society)
FPE	final prediction error
GA	genetic algorithm
GAEIEM	Genetic Algorithm Energy Input Estimation Model
GDP	gross domestic product
GIS	geographical information system
g/kg	grams per kilogram
GRNN	general regression neural network
GSCW	glazed solar chimney wall
GWP	global warming potential
H_2O	water

ha	hectare
HOTMAC	Higher Order Turbulence Model for Atmospheric Circulation
HVAC	heating, ventilating and air-conditioning
IAQ	indoor air quality
IEA	International Energy Agency
IEQ	indoor environment quality
IRR	internal rate of return
ISO	International Organization for Standardization
ITPE	interzone temperature profile estimation
K	degrees Kelvin
$kgCO_2/kWh$	kilograms of carbon dioxide per kilowatt hour
kg/kWh	kilograms per kilowatt hour
kg/s	kilograms per second
kg/yr	kilograms per year
km^2	square kilometre
kW	kilowatt
kWh	kilowatt hour
kWh/day	kilowatt hours per day
kWh/m^2	kilowatt hours per square metre
$kWhm^{-2}a^{-1}$	kilowatt hours per square metre per annum
kWh/yr	kilowatt hours per year
LAI	leaf area index
LAWEPS	Local Area Wind Energy Prediction System
LC	low cloud amount
LCA	life cycle analysis
LCC	life cycle cost
LEED	Leadership in Energy and Environmental Design
LES	large-eddy simulation
LM	Levenberg–Marquardt
m	metre
m^2	square metre
m^2kW^{-1}	square metres per kilowatt
m^2/m^3	square metres per cubic metre
m^3	cubic metre
m^3/s	cubic metres per second
$m^3/s/m^2$	cubic metres per second per square metre
MCDA	multi-criteria decision analysis
MI	mould index
MIMO	multi-input multi-output
MJ	megajoule
MJ/m^2	megajoules per square metre
MKEP	Milton Keynes Energy Park
mm	millimetre
MM5	5th Generation Mesoscale Model

MOGA	multi-objective genetic algorithm
MOP	multi-objective programming
MRT	mean radiant temperature
ms^{-1}	metres per second
MSI	mould severity index
MSW	metallic solar wall
MT	megatonne
MTW	modified Trombe wall
MW	megawatt
MWh	megawatt hour
NCAR	National Center for Atmospheric Research
NCEF	National Clearinghouse for Educational Facilities
NH_3	ammonia
NHER	National Home Energy Rating
NNARX	neural network autoregressive with exogenous input
NPV	net present value
OBS	optimal brain surgeon
OECD	Organisation for Economic Co-operation and Development
ORME	Office Rating Methodology
PE	primary energy
PET	physiological equivalent temperature
PHS	predicted heat strain
PM	observed minimum temperature of the previous night
PMV	predicted mean vote
POE	post-occupancy evaluation
PV	photovoltaic
PW	present weather condition
QUICK	Quadratic Upstream Interpolation for Convective Kinematics
RAMS	Regional Atmospheric Modelling System
RANS	Reynolds averaged Navier-Stokes
RBF	radial basis function
RCD	robust construction detail
RCMAC	recurrent cerebellar model articulation controller
RCS	residential–commercial sector
REFLEX	Effective Retrofitting of an Existing Building Tool
RH	relative humidity
RNG	renormalization group
RSC	roof solar collector
SAP	Standard Assessment Procedure
SBS	sick building syndrome
SCG	scaled conjugate gradient
SDWH	solar domestic water heating
SE	single effect
SET*	standard effective temperature
SISO	single-input single-output

SSE	sum of the squared error
STA	stabilization strategy
SUB	substitution strategy
SVD	singular value decomposition
SVPX	standardized vapour pressure excess
SWH	solar water heater
TARBASE	Technology Assessment for Radically Improving the Built Asset Base
TC	thermodynamic class
TC	total cloud amount
TE	triple effect
TOBUS	Tool for Selecting Office Building Upgrading Solutions
TSP	travelling salesman problem
TMY	typical meteorological year
TOW	time of wetness
TUT	Tampere University of Technology
TVOC	total volatile organic compounds
TW	Trombe wall
TWh	terawatt hour
UCL	University College London
UHI	urban heat island
USGBC	United States Green Building Council
VI	visibility
VL	vertical landscaping
VOC	volatile organic compounds
VTT	Technical Research Centre of Finland
W	watt
WB	wet-bulb temperature
WBGT	wet bulb globe temperature
W/m^2	watts per square metre
Wm^{-2}	watts per square metre
$Wm^{-1}K^{-1}$	watts per metre per degrees Kelvin
W/m^2K	watts per square metre degrees Kelvin
WRF	Weather Research and Forecasting model

earthscan

publishing for a sustainable future

1

Energy, Carbon and Cost Performance of Building Stocks: Upgrade Analysis, Energy Labelling and National Policy Development

J. A. Clarke, C. M. Johnstone, J. M. Kim and P. G. Tuohy

Abstract

The area of policy formulation for the energy and carbon performance of buildings is coming under increasing focus. A major challenge is to account for the large variation within building stocks relative to factors such as location, climate, age, construction, previous upgrades, appliance usage and type of heating/cooling/lighting system. Existing policy-related tools that rely on simple calculation methods have a limited ability to represent the dynamic interconnectedness of technology options and the impact of possible future changes in climate and occupant behaviour. The use of detailed simulation tools to address these limitations in the context of policy development has hitherto been focused on the modelling of a number of representative designs rather than dealing with the spread inherent in large building stocks. Further, these tools have been research-oriented and largely unsuitable for direct use by policy-makers, practitioners and, ultimately, building owners/occupiers. This chapter summarizes recent initiatives that have applied advanced modelling and simulation in the context of policy formulation for large building stocks. To exemplify the stages of the process, aspects of the ESRU Domestic Energy Model (EDEM) are described. EDEM is a policy support tool built on detailed simulation models aligned with the outcomes of national surveys and future projections for the housing stock. On the basis of pragmatic inputs, the tool is able to determine energy use, carbon emissions and upgrade/running costs for any national building stock or subset. The tool has been used at the behest of the Scottish Building Standards Agency and South Ayrshire Council to determine the impact of housing upgrades, including the deployment of new and renewable energy systems, and to rate the energy/carbon performance of individual dwellings as required by the European Commission's Directive on the Energy Performance of Buildings (EC, 2002).

ADVANCES IN BUILDING ENERGY RESEARCH ■ 2009 ■ VOLUME 3 ■ PAGES 1–20

doi:10.3763/aber.2009.0301 ■ © 2009 Earthscan ■ ISSN 1751-2549 (Print), 1756-2201 (Online) ■ www.earthscanjournals.com

■ *Keywords* – national building stock; performance simulation; upgrade policy; energy performance certificate

INTRODUCTION

An essential element in promoting the rational use of energy is that decision makers be given access to relevant sources of information. These include energy demand profiles and characteristics of potential sources of supply; and the outputs from studies to assess the impact of possible alternative strategies. However, indications are that at present comprehensive information is rarely in the hands of those who require it and the use of modelling in strategy formulation is virtually unknown. This chapter describes projects that imply an attempt to change this situation.

The worldwide building stock (residential and commercial) is responsible for more than 33 per cent of global CO_2 emissions (de la Rue du Can and Price, 2008) and a large mitigation potential has been identified (Urge-Vorsatz and Novikova, 2008). To attain aspirational CO_2 reduction targets, such as the EU's 20 per cent reduction by 2020 or the UK's 60 per cent reduction by 2050, will require the implementation of radical construction upgrades, new technology deployments and, contentiously, lifestyle change.

The simplified energy/carbon calculators presently in use, such as defined in CEN Standard 13790 (ISO, 2007) or the UK's Standard Assessment Procedure (BRE, 2005), are based on energy balance methods that do not account fully for the dynamic characteristics of buildings; Figure 1.1 contrasts graphically the energy balance underlying a simplified and simulation-based approach. Simplified methods cannot adequately represent the performance of the myriad upgrade options that may be applied individually or in combination. Also, as buildings have extended lifetimes, it is important to assess performance under likely future contexts, such as occupant behavioural change, climate change and the emergence of new technologies.

There are many available building simulation tools (Crawley et al, 2008) and these offer considerable advantages over their simplified counterparts, particularly in the areas of dynamic building response, adaptive occupant comfort, ventilation, indoor air quality, novel control and renewable energy systems integration. The simulation approach is increasingly being mandated in building performance legislation and applied in studies to inform policy.

Crawley (2007a, 2007b) reported on a simulation-based study to assess the impact of climate change and urban heat island effects on future building performance. A method was developed and applied to generate future weather collections for 25 locations in 20 climate zones around the world. This resulted in 525 weather collections encapsulating recent urban heat island data and the four economic scenarios and general climate assumptions of the Intergovernmental Panel on Climate Change. The intention is to use these collections to establish future climate impacts for a range of building types after various energy performance enhancements have been applied. To date, a model representing 25 per cent of the US office stock has been simulated for US locations, with three levels of enhancement applied corresponding to current practice, best practice and future practice scenarios. The results give the impact on heating and cooling energy use for combinations of location, climate change and selected efficiency measures.

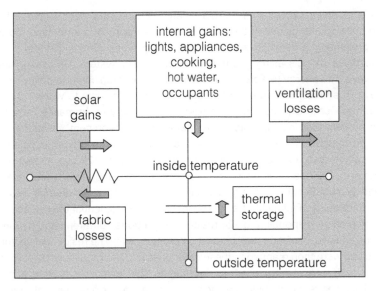

a) simplified

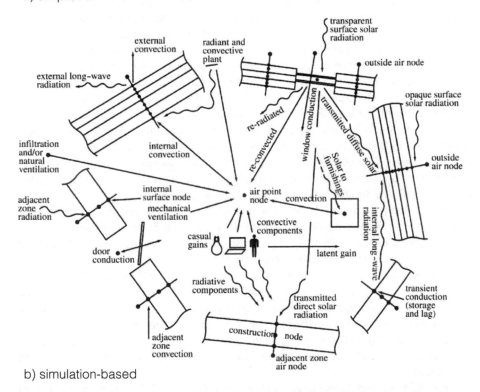

b) simulation-based

FIGURE 1.1 Visual comparison of simple and simulation-based room energy models

Heiple and Sailor (2008) investigated energy supply and heat island effects by simulating building energy use at the urban scale. Their approach employed prototypical models representing 8 dwelling and 22 non-dwelling types. Entire districts were then mapped to these models using geographical information system (GIS) techniques and simple surveys. The energy use predictions were then communicated to policy-makers via GIS overlays. The researchers reported agreement with utility data to within 10 per cent.

Research at Osaka University characterized urban energy performance via the use of simulation models that included stochastic algorithms to represent variations in usage patterns. The method was initially demonstrated for two large offices supplied by co-generation plant and absorption chillers (Yamaguchi et al, 2003). A second study (Hashimoto et al, 2007) focused on two office districts of different density and investigated various energy supply solutions (e.g. district heating/cooling and water/air source heat pumps). A third study (Taniguchi et al, 2007) investigated residential energy use. Models were constructed corresponding to 228 residential categories differentiated by family and building type and five insulation levels based on property age. The residential sectors of 20 Japanese cities were then mapped to these categories and the models, after calibration by comparison with utility data, used to investigate the efficacy of energy efficiency measures, including reduced appliance use based on Japan's 'top-runner' policy, increased insulation, improved air conditioner efficiency and modified temperature setpoints.

In the UK, the Carbon Vision project (Carbon Trust, 2008) set out to provide the strategy and evidence base to allow the UK to meet its 60 per cent carbon reduction goal by 2050. The Technology Assessment for Radically Improving the Built Asset Base (TARBASE) sub-project carried out dynamic simulation modelling for representative building types in order to determine the applicability and likely impact of specific upgrade measures. Preliminary results indicate that a 70 per cent reduction in emissions is possible at a cost of between £7000 and £21,000 for the selected dwellings (Staunton, 2008).

A number of simulation-based policy studies have been undertaken in China. Li et al (2007) proposed a method to investigate building upgrade scenarios for housing in Xi'an City, with results displayed on GIS maps. Hu et al (2007) carried out a study of upgrade options based on apartments in a typical multi-dwelling building. Xie et al (2007) evaluated office cooling loads for three different cities in Hunan province.

Studies such as those above have generally set out to answer specific policy questions, with the tools developed and applied by building performance researchers in order to explore policy implications. In contrast, some recent projects in the UK have set out to develop tools for direct use by policy-makers.

Jones et al (2001) developed the Energy and Environmental Prediction (EEP) planning support tool, which is able to quantify energy use and associated emissions for cities and regions. EEP is based on GIS techniques, with representative models for the building, industry and transport sectors. The tool has been applied to the Neath Port Talbot County Borough comprising 60,000 dwellings and 4000 commercial properties. Figure 1.2, for example, shows a typical outcome from a cost-performance assessment of houses at the district level.

The EnTrak system (Clarke et al, 1997; Kim and Clarke, 2004) enables the tracking of regional/city energy use over time, with a range of interrogations provided to support

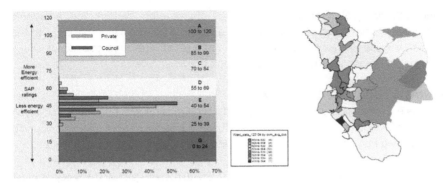

a) Percentage of homes having SAP ratings   b) Average domestic heating cost per ward
   in the indicated ranges

FIGURE 1.2 Example of home energy/cost rating using the EEP system

energy action planning. Embedded simulation models are used to augment sparse data sets and to quantify the impact of upgrades or the deployment of renewable energy systems both locally and nationally. Figure 1.3 presents two example outputs: a) corresponds to a combined heat and power (CHP) feasibility study and presents heat-to-power ratios on a street-by-street basis when classified into excellent, good, fair, poor and bad; b) corresponds to wind farm development control in Caithness, Scotland (Bamborough et al, 1996) with the two GIS maps representing policy and technical ratings to a 1 km² resolution using appropriate scoring criteria. In the latter case, a comparison shows that policy rather than technical issues are likely to be the constraint on wind farm development within the region.

Clarke et al (2004) have developed the ESRU Domestic Energy Model (EDEM, 2007) as a generalized housing stock modeller, which encapsulates the results of simulations

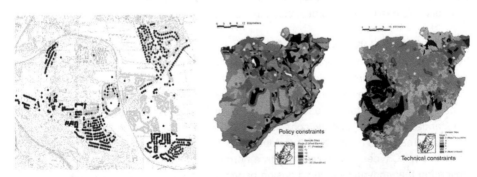

a) city CHP feasibility study   b) policy/technical rating of wind farm proposals

FIGURE 1.3 Example analyses using the EnTrak system

together with financial information to predict the energy, carbon and economic implications of stock upgrading over time. As a result of its simulation basis, EDEM is equipped to analyse all possible future upgrade options, while taking into account issues such as climate change and user behaviour adaptation.

BUILDING STOCK DESCRIPTION

All the above approaches depend on a sound knowledge of the building stock to be analysed and, although complex, this can usually be gleaned from national housing surveys, Building Regulation change histories, landlord property inventories, maintenance records and/or site surveys.

In Scotland, for example, there are around 2,278,000 dwellings of which 4 per cent are vacant and 2.5 per cent are due for demolition. The dwellings comprised houses (62 per cent) and flats (38 per cent). More than 40 per cent of all dwellings were built since 1965, with 24 per cent constructed between 1945 and 1965. The 2002 *Scottish House Condition Survey* (Scottish Homes, 2002) identified seven predominant house types (detached, semi-detached, terraced, tenement flat, four-in-a-block, conversion and tower/slab block) and established a mean National Home Energy Rating (NHER) of 4.5 (on a scale of 0/poor to 10/good), with an associated mean Standard Assessment Procedure (SAP) rating of 46.5 (on a scale of 0/poor to 100/good). CO_2 emissions were estimated at 16.2 million tonnes per year. By comparison, the 1996 *Scottish House Condition Survey* identified a mean NHER rating of 4.1 and a mean SAP rating of 43, indicating a 10 per cent improvement over the intervening period, with 12 per cent of all dwellings achieving an NHER rating between 7 and 9, but no dwellings attaining a rating of 10.

From the 2002 survey, around 86 per cent of dwellings were identified as having whole house central heating, with a further 8 per cent having partial central heating. This represents a 6 per cent improvement since 1996, with the number of dwellings with no central heating down from 13 per cent to 5.5 per cent. This small but significant figure gives rise to concerns about fuel poverty and the related health problems associated with hypothermia, condensation and mould growth. Although around 90 per cent of houses have loft insulation, in only 27 per cent of cases does this meet or surpass the 1991 building standard. In the Scottish context, therefore, the need for housing stock improvement is palpable: this need is likely to be echoed in many countries throughout the world.

STOCK MODELLING APPROACH

While it is a straightforward task to identify house types from an architecture and construction viewpoint (hereinafter referred to as an AC type), the task becomes semi-intractable when viewed thermodynamically. Setting aside the effect of occupant behaviour, two separate houses, each belonging to the same AC type, may have substantially different energy consumption patterns as a result of dissimilar energy efficiency measures having been previously applied. Likewise, two houses corresponding to different AC types may have the same normalized energy consumption because the governing thermodynamic-related design parameters are essentially the same.

One approach to stock modelling is to establish prototype models for each AC type and to then apply design parameter variations to each model to represent all possible upgrade combinations, while accepting that many of the resulting permutations will give rise to identical performance outcomes. Another approach, and the one adopted within the EDEM tool, is to operate in terms of unique thermodynamic classes (TCs) so that different AC types may belong to the same TC. A representative model may then be formed for each TC and its energy performance determined by simulation using real, representative weather data. Any actual house may then be related to a TC via the present level of its governing design parameters. Should any of these parameters be changed as part of an upgrade then that house would be deemed to have moved to another TC.

The simulation results for the set of representative TC models, scaled by the appropriate factors representing their proportion of the overall population, then define the possible performance of the entire housing stock, present *and* future, for the climate, exposure, occupancy and system control assumptions made within the simulations. By varying these assumptions and resimulating, scenarios such as future climate change and improved living standards may be readily incorporated.

Within the EDEM project, the ESP-r system (ESP-r, 2007) was used to determine dwelling performance by subjecting the TC models to long-term weather sequences that typify the range of possibilities for the region in question. The time series performance predictions for all TC models, when re-expressed as regression equations defining energy use as a function of prevailing weather parameters, are then encapsulated within the EDEM tool for use by relevant user groups: policy-makers engaged in the development of building regulations in response to national policy drivers; building stock owners/ managers to appraise the impact of candidate improvement measures; and local authorities in a performance rating context.

The evaluation of any given upgrading scenario is quantified by assigning the dwellings in question to a TC based on principal design parameters. The energy reduction brought about by relocation to another TC may then be simply 'read off' as depicted in Figure 1.4, which shows the main EDEM control screen. Because each TC corresponds to a unique design parameter combination, the required upgrade is immediately apparent from the TC relocation. The impact of technologies that may be considered independent of house type, such as district heating or community CHP, are then modelled separately based on dwelling energy demands. Specific upgrade scenarios are accepted or discarded as a consequence of the resulting performance benefit and cost. The calculation of the domestic energy rating band and associated indexes as defined within the regulatory standards, are output for both the unimproved and improved case.

By ensuring that the TC models encapsulate the assumptions underpinning the regulatory simplified calculation methods presently used for energy ratings (e.g. SAP in the UK), EDEM can be used in emulation mode when applied to the restricted cases for which these regulatory tools are valid.

TC FORMULATION

TC models cover construction- and technology-related design parameters to allow an impact assessment of deploying different construction/technology upgrade combinations over time.

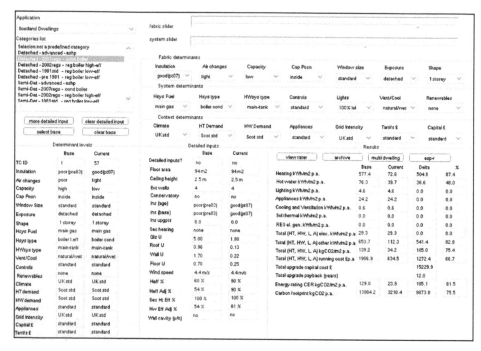

FIGURE 1.4 EDEM control screen

CONSTRUCTION ASPECTS

The range of TC models to be processed are established as unique combinations of design parameters that may be considered as the main determinants of energy use and can be adjusted as part of any upgrade: insulation (six levels), thermal capacity (two levels), capacity position (three levels), air permeability (three levels), window size (three levels), exposure (five levels) and wall-to-floor ratio (two levels). If each of these parameters can exist at the level indicated in parentheses then there will be 3240 (6 x 2 x 3 x 3 x 3 x 5 x 2) TC models representing the universe of possibilities. That is, any dwelling, existing or planned, will correspond to a unique combination of these parameters and therefore belong to one, and only one, TC. Significantly, most TC models will not correspond to existing dwellings because the stock presently comprises designs that may be regarded as poor in terms of energy use and carbon footprint. Rather, the majority represent future possibilities that will result from the application of upgrades.

Long-term ESP-r simulations were conducted for the 3240 TC models and the predicted energy demands normalized by floor area to render the results independent of dwelling size and so facilitate inter-comparison. The models were then resimulated for each of 24 context combinations relating to climate (two), occupancy (two), temperature setpoint (three) and appliance efficiency (two).

To facilitate the simulations, a standard house model was constructed comprising living, eating and sleeping areas, with appropriate parametric modifications applied to realize the individual TCs. While the assumptions underlying this standard model correspond to the UK situation as determined from appropriate publications (Bartholomew and Robinson, 1998; Scottish Homes, 2002; Shorrock and Utley, 2003; BRE, 2005; CIBSE, 2006), these can be readily changed to reflect any other situation.

TECHNOLOGY ASPECTS

Dwelling energy demands, as extracted from the TC model simulations, are applied to technology models corresponding to the range of possible environmental control systems.

For heating/cooling systems, type, age, control and fuel type are used to set an efficiency value in line with CIBSE and SAP defaults (BRE, 2005), the BRE Domestic Energy Fact File (Shorrock and Utley, 2003) and the Carbon Trust's Building Market Transformation project database (MTP, 2006). Hot water loads are determined in relation to standard domestic system capacities and water usage rates (BRE, 2005), whereas lighting energy use is calculated using a standard model. New and renewable energy systems are also selectable: currently mono-/poly-crystalline and amorphous photovoltaic components, micro wind turbines, solar thermal collectors, air/ground source heat pumps, heat recovery ventilation, and combined heat and power.

A number of user-replaceable defaults are set within the tool: the default mapping of energy use to CO_2 emission is based on data published by the UK Carbon Trust (0.42 kg/kWh for grid electricity and 0.19 kg/kWh for gas); fuel unit costs and standing charge values are based on the standards set for the UK's SAP; and the capital cost of construction and technology upgrades are based on current market information.

EDEM VERIFICATION

Detailed models of five actual houses were subjected to simulation, energy efficiency improvements applied and the simulations rerun. The houses and their variants were then related to TCs based on the level of their principal design parameters. The predicted heating energy demands resulting from the detailed simulations were then compared with the value associated with the matched TC model. The results indicated disagreements averaging around 5 per cent (3 per cent to −13 per cent range), indicating that the TC approach is a reasonable proxy for the simulated situation.

A second study compared EDEM output with energy performance as determined using the UK's NHER methodology. A local authority energy officer carried out detailed surveys of dwellings with electric and gas heating systems and computed their energy performance using the NHER Surveyor tool (NHER, 2008). EDEM was used to calculate carbon and energy performance data and the results compared with those from NHER; Figure 1.5 shows the good agreement obtained.

EDEM APPLICATION

EDEM is designed to be flexible in its application. The control screen (Figure 1.4) enables analysis at scales of integration from individual dwellings to entire housing stocks, with input data accepted at various levels of detail. The tool is configured to allow user customization of the control screen labels, drop-down menus and data tables. A scripting language is used to adapt the underlying models and contexts, to rerun simulations and

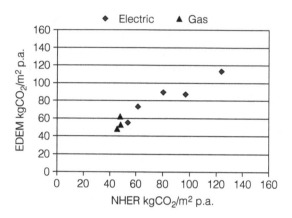

FIGURE 1.5 EDEM versus NHER Surveyor – CO_2 emissions for dwellings with gas or electric heating systems

to extract new values for the associated data tables. Alternative values for cost, carbon and system parameters can be directly entered.

Once the context is defined and pragmatic input data gathered, construction- and technology-related parameters may be selected either directly (Figure 1.6) or by predefined category (Figure 1.7) and the outputs expressed in terms of energy/carbon/cost (Figure 1.8)

Fabric determinants						
Insulation	Air changes	Capacity	Cap Posn	Window size	Exposure	Shape
good (pt07)	tight	low	inside	standard	detached	1 storey
System determinants						
Hsys Fuel	Hsys type	HWsys type	Controls	Lights	Vent/Cool	Renewables
main gas	boiler cond	main-tank	standard	100% lel	natural/wet	none
Context determinants						
Climate	HT Demand	HW Demand	Appliances	Grid Intensity	Tarrifs £	Capital £
UK std	Scot std	Scot std	standard	UK std	standard	standard

FIGURE 1.6 Selection of input parameter levels in EDEM

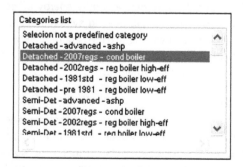

Categories list

Selecion not a predefined category
Detached - advanced - ashp
Detached - 2007regs – cond boiler
Detached - 2002regs - reg boiler high-eff
Detached - 1981std - reg boiler low-eff
Detached - pre 1981 - reg boiler low-eff
Semi-Det - advanced - ashp
Semi-Det - 2007regs - cond boiler
Semi-Det - 2002regs - reg boiler high-eff
Semi-Det - 1981std - reg boiler low-eff

FIGURE 1.7 EDEM user configurable predefined categories provide a short-cut input data entry method for frequently used building types

	Base	Current	Delta	%
Heating kWh/m2 p.a.	577.4	72.6	504.8	87.4
Hot water kWh/m2 p.a.	76.3	39.7	36.6	48.0
Lighting kWh/m2 p.a.	4.6	4.6	0.0	0.0
Appliances kWh/m2 p.a.	24.2	24.2	0.0	0.0
Cooling and Ventilation kWh/m2 p.a.	0.6	0.6	0.0	0.0
Sol thermal kWh/m2 p.a.	0.0	0.0	0.0	0.0
RES el. gen. kWh/m2 p.a.	0.0	0.0	0.0	0.0
Total (HT, HW, L, A) elec. kWh/m2 p.a	29.3	29.3	0.0	0.0
Total (HT, HW, L, A) other kWh/m2 p.a	653.7	112.3	541.4	82.8
Total (HT, HW, L, A) kgCO2/m2 p.a.	139.2	34.2	105.0	75.4
Total (HT, HW, L, A) running cost £p.a	1906.9	634.5	1272.4	66.7
Total upgrade capital cost £			15229.9	
Total upgrade payback (years)			12.0	
Energy rating CER kgCO2/m2 p.a.	129.0	23.9	105.1	81.5
Carbon footprint kgCO2 p.a.	13084.2	3210.4	9873.8	75.5

FIGURE 1.8 EDEM results summary illustrating a case where poor fabric and heating systems are upgraded to current standards

and an Energy Performance Certificate (Figure 1.9). Improvements in construction- and technology-related parameters may then be evaluated by selecting parameters as above or by using sliders to rapidly evaluate upgrade options.

An application will typically proceed as follows. The context of the analysis is set, e.g. 'current UK standard', '2050 climate with high indoor comfort' or '2050 climate with low carbon intensity electricity grid', and then for each dwelling type to be analysed the construction and technology input data are entered. The governing design parameters are then inferred from these data and the corresponding TC model automatically selected. The energy demand for the given model is then determined from the embedded database, adjusted for building specifics, and the delivered energy and associated costs

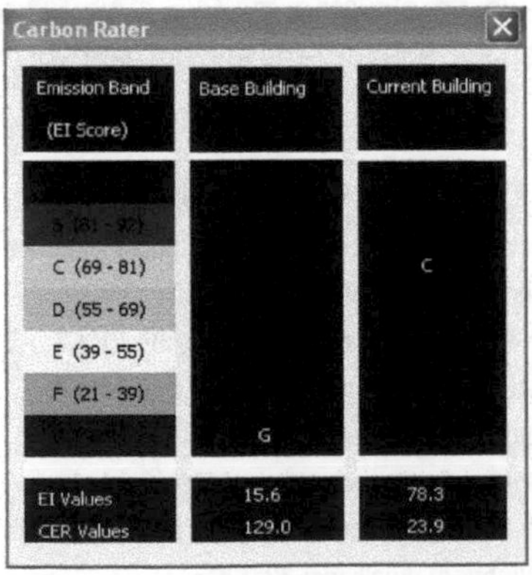

FIGURE 1.9 An EDEM Energy Performance Certificate for the case described in Figure 1.8

and carbon emissions established. Upgrades of interest are then applied, resulting in a new TC model for processing to give new energy, carbon and financial results. Finally, combinations of interest are saved for comparison and presentation.

CASE STUDY: NATIONAL STOCK UPGRADE

A digest of the 2002 *Scottish House Condition Survey* data has shown that the 2,278,000 dwellings in Scotland give rise to a total annual space heating demand of 14.5 TWh and CO_2 emission of 5.5 MT (domestic space heating accounts for 17 per cent of the total Scottish energy demand). The entire national housing stock can be classified into three groups (or TCs) as listed in Table 1.1; the largest housing sector is contained within Group 1, which includes TCs associated with unimproved dwellings constructed prior to 1981. This grouping accounts for 11.1 TWh of annual space heating energy.

Practical considerations dictate that any upgrading strategy should focus on low-cost technologies initially to maximize the return on investment, and be phased over time thereafter to accommodate technical advances. Reducing ventilation and fabric heat loss are the most effective measures to improve dwelling thermal performance and these were assessed at the outset of the study.

A preliminary EDEM analysis indicated that the most cost-effective upgrade strategy should be to target an appropriate subset of the Group 1 dwellings by improving their air tightness to standard (through the application of basic draught proofing) and applying lower-cost insulation measures where appropriate to improve insulation to standard (e.g. cavity, internal or external wall insulation, double glazing and loft insulation). These actions would shift these properties to a Group 2 TC, with an associated annual saving of around 40 kWh/m². Further analysis indicated that the remaining Group 1 dwellings along with the Group 2 dwellings (comprising original and previously upgraded Group 1 members) could be cost-effectively improved to a Group 3 TC by more aggressively improving insulation and infiltration to achieve compliance with all elements of the 2002 UK regulations.

The implementation of the first phase of improvement measures was predicted to result in savings in the annual space heating energy demand of 4.7 TWh (or 33 per cent of the national energy demand). In the second phase of the programme, the annual space heating energy savings could be elevated to 7.36 TWh. Overall, a phased programme would reduce the annual space heating energy demand of the Scottish housing stock

TABLE 1.1 Digest of Scottish dwellings
1: high thermal mass, poor insulation, high air change rate
   Number of dwellings: 1,594,600
   Average heating demand (kWh/m².yr): 87
2: standard insulation, high air change rate
   Number of dwellings: 660,620
   Average heating demand (kWh/m².yr): 47
3: high insulation, standard air change rate
   Number of dwellings: 22,780
   Average heating demand (kWh/m².yr): 26

from 14.5 to 7.14 TWh, i.e. a 52 per cent reduction of the space heating energy demand. Further details on the outcome of the project are given elsewhere (Clarke et al, 2004).

CASE STUDY: REGIONAL HOUSING UPGRADE

A local authority housing stock comprising 7876 dwellings was evaluated using EDEM to determine the impact on the carbon footprint of a range of upgrades (Tuohy et al, 2006). The stock was decomposed into TCs using the local authority's available property data, while a range of possible upgrades were identified from the Energy Saving Trust's practical help publications (EST, 2007):

0. Current stock – no upgrades applied.
1. Low-cost fabric improvement – where there is a pitched roof and a suspended wooden floor then loft insulation is increased and the suspended timber floors insulated; all dwellings to have basic double glazing and be brought up to a tight infiltration standard.
2. Major fabric upgrade – in addition to the low-cost measures, flat roofs are upgraded to a U-value of 0.16 W/m²K, cavity wall properties have insulation added to give a U-value of 0.35, solid wall properties are improved to a U-value of 0.6, and windows improved to a U-value of 1.5.
3. 2007 heating systems – gas, electricity and solid fuel heating systems are upgraded to meet the 2007 Building Regulation standards by the installation of a condensing boiler with instantaneous water heating, an air source heat pump with radiators and a wood boiler respectively.
4. Upgrades 1 + 2 + 3.
5. Upgrade 4 plus solar hot water heating (delivering 920 kWh/yr useful energy, applied to properties with an exposed roof).
6. Upgrade 5 plus local renewable energy generation (650 kWh/yr) in the form of either photovoltaic (PV) or small-scale wind turbines at appropriate locations.
7. Upgrade 5 with gas boilers replaced with Stirling engine CHP.
8. Upgrade 5 with heating through individual or community wood boiler systems.

Figure 1.10 shows the impact of each upgrade option on the carbon footprint. These results show the current carbon footprint per dwelling to be 4.9 tonnes of CO_2 per year, while future scenarios are presented with emissions below 1 tonne. Further details on the selected upgrade options and a breakdown of the study results by dwelling type are reported elsewhere (ibid).

CASE STUDY: DWELLING ENERGY LABELLING

EDEM can be used to provide energy performance ratings. In this case, the Environmental Index (EI) and Energy Band (EB) are calculated from the generated energy demands in accordance with a standard UK method (BRE, 2005). Table 1.2 shows EDEM output when applied to an electrically heated 1980s dwelling, which had previously been upgraded with cavity wall insulation, double glazing and 200 mm of loft insulation. A number of further improvements were explored commencing with fabric improvements to 2002 standards, followed by alternative system replacement options: gas-fired condensing combination

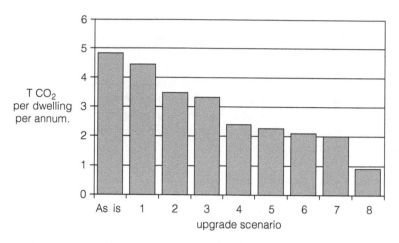

FIGURE 1.10 Impact of upgrade options on the carbon footprint of a local authority housing stock

TABLE 1.2 CO_2 emission (kg/yr), Environmental Index (EI) and Energy Band (EB) for alternative upgrade options

UPGRADE	EMISSION	EI	EB
0. As is	3391	57	D
1. 2002 fabric	2778	66	D
2. 1 + gas condensing combi-boiler	1679	81	B
3. 1 + ground source heat pump	1515	83	B
4. 1 + community biomass heating	817	93	A
5. 1 + community gas-fired CHP	1000	98	A
6. 2 + PV + solar thermal	1454	84	B

boiler; ground source heat pump; community biomass heating; community gas-fired combined heat and power; and a hybrid system comprising a condensing combination boiler, solar water heating and a PV panel producing 920 kWh and 650 kWh of thermal and electrical energy, respectively, annually. From the results of Table 1.2, it can be seen that two upgrade options were able to raise the initial 'D' rating to 'A' – upgraded fabric with either community biomass heating or community gas-fired combined heat and power.

CASE STUDY: IMPACT OF GRID ELECTRICITY GENERATION MIX

EDEM allows investigations across a range of system types, dwelling types and contexts (climates, behaviour patterns, grid carbon intensity, etc.) to inform policy. One study analysed the 2050 scenarios proposed by the UK Carbon Trust's Buildings Market Transformation (BMT) project (Carbon Trust, 2008) in order to establish the impact of the electricity grid generation mix on technology performance. An assumption used in some scenarios is that while imported grid electricity has associated overall carbon emissions, electricity generated locally (CHP or renewable generation) displaces only the carbon-fuelled portion of grid generation plant. Multiple grid generation mixes were included in the EDEM study, including a current UK grid (0.54 kgCO$_2$/kWh overall, 0.73 kgCO$_2$/kWh for the carbon-fuelled portion), a projected 2020 grid (0.42 kgCO$_2$/kWh and 0.57 kgCO$_2$/kWh, respectively) and a projected 2050 grid (0.3 kgCO$_2$/kWh and 0.4 kgCO$_2$/kWh, respectively). The BMT scenarios include a range of gas-fired CHP systems (with various overall and electrical efficiencies) and electric heat pump options (with various coefficient of performance [COP] values).

EDEM quantified the carbon performance of various technologies applied to dwellings with poor, average or 2002 standards of insulation/infiltration for each grid scenario. Figures 1.11 and 1.12 show the results for the 2020 and 2050 grids, while Figure 1.13 shows the carbon benefit relative to a condensing boiler for selected CHP and heat pump systems across the three grid scenarios. Decarbonizing the grid reduces the calculated carbon performance of CHP and other local generation technologies, while grid electricity-fuelled systems such as heat pumps benefit from lower associated grid emissions.

CASE STUDY: FINANCIAL APPRAISAL OF UPGRADE OPTIONS
EDEM can be used for financial appraisal in support of policy or building upgrade strategy. The BMT 2050 scenarios were analysed for both a medium feed-in tariff (locally generated electricity is exported to the grid at a tariff equal to half the electricity import price) and a high feed-in tariff (locally generated electricity is consumed locally or exported at a tariff equal to the import price). While more detailed financial analyses may be carried out, the

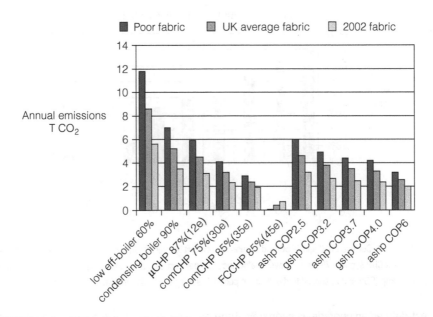

FIGURE 1.11 2020 Grid – annual emissions associated with dwellings of poor, UK average and 2002 regulation insulation/infiltration and a range of heating systems including gas boilers, CHP (micro (µ), community (com) and fuel cell (FC)) and heat pumps (air and ground source)

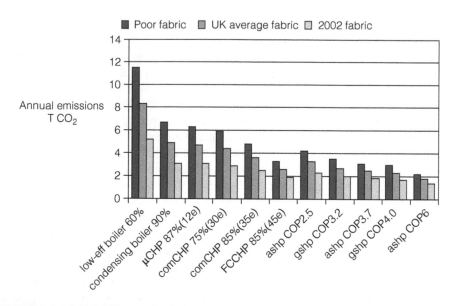

FIGURE 1.12 2050 Grid – annual emissions associated with dwellings of poor, UK average and 2002 regulation insulation/infiltration and a range of heating systems including gas boilers, CHP (micro, community and fuel cell) and heat pumps (air and ground source)

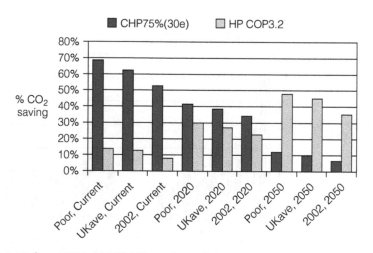

FIGURE 1.13 Current, 2020 and 2050 Grid scenarios – calculated carbon savings (relative to a gas condensing boiler) for a gas CHP system (75 per cent overall efficiency, 30 per cent electrical, supplying 70 per cent of demand) and a heat pump (COP of 3.2) applied to dwellings of poor, UK average and 2002 insulation/infiltration

project defined an upgrade as economic if the payback period was less than the expected lifetime (20 years for a technical system, 40 years for fabric upgrades and 30 years if combined). From the results (Figure 1.14) the upgrades applied to the poor dwelling were shown to be economic while the upgrades were marginal or uneconomic for a UK average

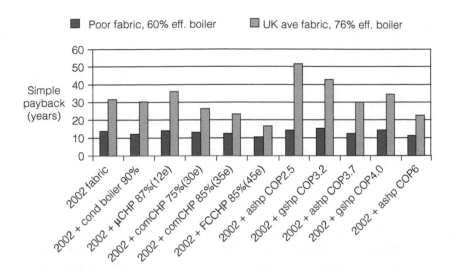

FIGURE 1.14 Simple payback for a range of upgrades applied either to a dwelling with poor insulation/ infiltration and 60 per cent efficient gas boiler or a dwelling with UK average insulation/infiltration and 76 per cent efficient gas boiler where a high electricity price is associated with the local electricity generation (high feed-in tariff)

dwelling except for the 2002 fabric upgrade, low-cost/high-efficiency heat pump or efficient CHP systems where there is a high feed-in tariff.

CASE STUDY: FINANCIAL APPRAISAL OF INDIVIDUAL DWELLING UPGRADE
Specific measures applied to individual dwellings can be assessed using EDEM. For example (Table 1.3), a three-bedroom mid-terrace house built in 1929 with electric storage heating was upgraded with: ground floor insulation, external wall insulation, loft insulation, timber-framed double glazing, low-energy lighting, efficient A-rated appliances, ground

TABLE 1.3 Analysis of an upgrade package applied to an individual dwelling

	ORIGINAL	UPGRADED	DELTA
Heating kWh/m² pa.	204	25	179
Hot water kWh/m² pa.	36	8	28
Lights kWh/m² pa.	9	5	4.5
Appliances kWh/m² pa.	25	15	10
Total KWh/m² pa.	274	53	221
Running cost £ pa.	£2245	£472	£1773
Total T CO₂ pa.	9.26	1.78	7.48
EPC Rating	F	B	
Capital cost £			£13,492
Simple payback years.			7.6

source heat pump, controls and a solar water heater. The calculated cost of this upgrade package was £13,492 and the calculated fuel cost saving was estimated as £1773 per year giving a simple payback of 7.6 years. This upgrade produced a calculated reduction in carbon footprint from 9.3 to 1.8 tonnes of CO_2 per year, i.e. a predicted saving of 80 per cent.

CONCLUSIONS

Building performance simulation is now being used in support of policy development. The goal is to provide decision makers with the means to compare energy supply and demand, at a local or regional level, in terms of the match at present or as it may exist under some future scenario. By enabling informed decisions, opportunities for cost-effective upgrades and the exploitation of renewable energy systems at local (autonomous) and strategic (grid-connected) levels can be fully explored. The approach allows the accuracy and applicability advantages of dynamic simulation to be made accessible to those who are concerned with the management and adaptation of large estates. Typically, an energy simulation programme is applied to a set of building models that represent the spectrum of upgrade possibilities and the output predictions embedded in another, easy-to-use policy support tool. In this chapter, the use of the approach has been demonstrated at a number of levels from national stock to individual dwellings. The methodology is general and can be applied to any building stock and climate region.

ACKNOWLEDGEMENTS

The authors thank Dr Linda Sheridan and Mr Gavin Peart of the Scottish Building Standards Agency for supporting EDEM development. Recent developments of EDEM were performed within the Building Market Transformation project sponsored by the UK Carbon Trust: particular thanks to Sabeeta Ghauri of ESRU for her inputs to the EDEM interface.

AUTHOR CONTACT DETAILS

J. A. Clarke, C. M. Johnstone, J. M. Kim and **P. G. Tuohy**: Energy Systems Research Unit, University of Strathclyde; esru@strath.ac.uk

REFERENCES

Bamborough, K., Brown, A., Clarke, J. A., Evans, M., Grant, A. D., Lindsay, M. and Morgan, J. (1996) *Integration of Renewable Energies in European Regions*, Final Report to the European Commission for Project RENA-CT94-0064, ESRU, University of Strathclyde, Glasgow

Bartholomew, D. and Robinson, D. (eds) (1998) *Building Energy and Environmental Modelling: Applications Manual 11*, CIBSE, London

BRE (Building Research Establishment) (2005) *Standard Assessment Procedure for Energy Rating of Dwellings*, Building Research Establishment, Garston, UK

Carbon Trust (2008) www.carbontrust.co.uk/technology/carbonvision/buildings.htm

CIBSE (Chartered Institution of Building Services Engineers) (2006) *Environmental Design*, Guide A, CIBSE, London

Clarke, J. A., Evans, M. S., Grant, A. D. and Kelly, N. (1997) 'Simulation tools for the exploitation of renewable energy in the built environment: The EnTrak System', in *Proceedings of Building Simulation '97 Conference*, Prague, vol 1, pp9–17

Clarke, J. A., Johnstone, C. M., Kondratenko, I., Lever, M., McElroy, L. B., Prazeres, L., Strachan, P. A., McKenzie, F. and Peart, G. (2004) 'Using simulation to formulate domestic sector upgrading strategies for Scotland', *Energy and Buildings*, vol 36, pp759–770

Crawley, D. (2007a) 'Creating weather files for climate change and urbanisation impacts analysis', in *Proceedings of Building Simulation '07 Conference*, Beijing, pp1075–1082

Crawley, D. (2007b) 'Estimating the effects of climate change and urbanisation on building performance', in *Proceedings of Building Simulation '07 Conference*, Beijing, pp1115–1122

Crawley, D., Hand, J. W., Kummert, M. and Griffith, B. T. (2008) 'Contrasting the capabilities of building energy performance simulation programs', *Building and Environment*, vol 43, pp661–673

de la Rue du Can, S. and Price, L. (2008) 'Sectoral trends in global energy use and greenhouse gas emissions', *Energy Policy*, vol 36, pp1386–1403

EC (European Commission) (2002) *On the Energy Performance of Buildings*, Directive 2002/91/EC of the European Parliament

EDEM (ESRU Domestic Energy Model) (2007) www.esru.strath.ac.uk/Programs/EDEM.htm

ESP-r (2007) www.esru.strath.ac.uk/Programs/ESP-r.htm

EST (Energy Saving Trust) (2007) 'UK Government Energy Efficiency Best Practice Programme', www.est.org.uk/bestpractice

Hashimoto, S., Yamaguchi, Y., Shimoda, Y. and Mizuno, M. (2007) 'Simulation model for multi-purpose evaluation of urban energy system', in *Proceedings of Building Simulation '07 Conference*, Beijing, pp554–561

Heiple, S. and Sailor, D. J. (2008),'Using building energy simulation and geospatial modeling techniques to determine high resolution building sector energy consumption profiles', *Energy and Buildings*, vol 40, pp1426–1436

Hu, J., Yang, C. and Yu, C. (2007) 'Building energy analysis and simulation of Changsha area', in *Proceedings of Building Simulation '07 Conference*, Beijing, pp1871–1876

ISO (International Organization for Standardization) (2007) *Energy Performance of Buildings – Calculation of Energy Use for Space Heating and Cooling*, Standard 13790, ISO, Geneva

Jones, P. J., Lannon, S. and Williams, J. (2001) 'Modelling building energy use at the urban scale', in *Proceedings of Building Simulation '01 Conference*, Rio de Janeiro, pp185–172

Kim, J. and Clarke, J. A. (2004) 'The EnTrak System: Supporting energy action planning via the Internet', in *Proceedings of CTBUH '04 Conference*, Seoul, pp441–448

Li, Q., Jones, P. and Lannon, S. (2007) 'Planning sustainable in Chinese cities: Dwelling types as a means to accessing potential improvements in energy efficiency', in *Proceedings of Building Simulation '07 Conference*, Beijing, pp101–108

MTP (Market Transformation Project) (2006) 'UK Market Transformation Project', www.mtprog.com

NHER (National Home Energy Rating) (2008) 'NHER: Software for existing dwellings, NHER Surveyor', www.nher.co.uk/pages/software/existing.php

Scottish Homes (2002) *Scottish House Condition Survey 2002*, www.shcs.gov.uk

Shorrock, L. D. and Utley, J. I. (2003) *Domestic Fact File*, BRE Publications, Garston, UK

Staunton, G. (2008) 'The Carbon Vision Buildings Programme', Presentation at *CIBSE/ASHRAE Conference: Sustainability – Niche to Norm, 29–30 April 2008*, Gateshead (available from www.cibse.org/content/Natconf2008presentations/Garry%20Staunton.pdf

Taniguchi, A., Shimoda, Y., Asahi, T., Yamaguchi, Y. and Mizuno, M. (2007) 'Effectiveness of energy conservation measures in residential sectors of Japanese cities', in *Proceedings of Building Simulation '07 Conference*, Beijing, pp1645–1652

Tuohy, P. G., Strachan, P. A. and Marnie, A. (2006) 'Carbon and energy performance of housing: A model and toolset for policy development applied to a local authority housing stock', in *Proceedings of Eurosun '06 Conference*, Glasgow, paper no 197, www.eurosun2006.org/

Urge-Vorsatz, D. and Novikova, A. (2008) 'Potentials and costs of carbon dioxide mitigation in the world's buildings', *Energy Policy*, vol 36, pp642–661

Xie, D., Zhang, G. and Zhou, J. (2007) 'Analysis of building energy efficiency strategies for the hot summer and cold winter zone in China', in *Proceedings of Building Simulation '07 Conference*, Beijing, pp1877–1882

Yamaguchi, Y., Shimoda, Y. and Mizuno, M. (2003) 'Development of district energy system simulation model based on detailed energy demand', in *Proceedings of Building Simulation '03 Conference*, Eindhoven, pp1443–1450

earthscan
publishing for a sustainable future

2

Solar Chimneys in Buildings – The State of the Art

A. Dimoudi

Abstract

Improvement of the energy performance of buildings and achievement of indoor comfort conditions can be attained by passive cooling techniques and energy efficient measures. Natural ventilation is one of the simplest and most widely applied cooling strategies, achieved by permitting air to flow through the building. The solar chimney is a construction on the building's envelope that exploits solar energy to promote ventilation through the building. The typical configurations of a solar chimney and the principles of its operation are presented in this chapter. An overview is given of past and ongoing research on experimental work and modelling of the performance of different solar chimney configurations. The work is concluded by identifying the main advantages and disadvantages of solar chimneys.

■ *Keywords* – solar chimney; natural ventilation; stack effect

INTRODUCTION

Buildings are a major energy consumer as they account for almost 40 per cent of the EU's energy requirements. Energy consumption for space cooling shows an increasing trend in recent years. The economic growth of many countries and increased living standards are expected to boost electricity consumption for space cooling in the future. In Organisation for Economic Co-operation and Development (OECD) countries, electricity demand for residential space cooling increased by 13 per cent between 1990 and 2000 (IEA, 2003). Furthermore, the total demand for air conditioners, at an international level, increased by 6.8 per cent during the period 1999–2002 (JRAIA, 2008).

The challenge of reducing greenhouse gas emissions at local and global levels requires not only the use of more energy efficient production, processing and distribution technologies but also behavioural changes in lifestyle and energy consumption. Passive cooling techniques (Santamouris et al, 1996; Santamouris, 2007) and energy efficient measures can considerably contribute to the improvement of the energy performance of buildings and achievement of interior comfort and air quality conditions.

doi:10.3763/aber.2009.0302 ■ © 2009 Earthscan ■ ISSN 1751-2549 (Print), 1756-2201 (Online) ■ www.earthscanjournals.com

Natural ventilation is one of the simplest and most widely applied cooling strategies, achieved by permitting air to flow through the building.

TYPOLOGIES – OPERATION MODES

A solar chimney is a construction used to promote air movement throughout the building by using solar energy to promote ventilation. The principle of operation of solar chimneys, Trombe walls and double-skin facades is similar as they are open cavities that exploit solar energy to enhance air movement through, as a result of the stack effect. The Trombe wall – a south-oriented black-painted storage wall covered with a glass pane with an air gap between them – is mainly used for heating the building but with the appropriate operational adjustments, ventilation and cooling can also be achieved. The double-skin facade, i.e. a facade of two glass layers with an air gap between, is used mainly for heating but also for ventilation, daylighting and sound protection.

The solar chimney is usually attached to the building's facade or roof and is connected to internal spaces with a vent (Figure 2.1a). It is important to distinguish between the stack effect ventilation arising purely from the design of the building and that caused by a solar chimney. In the first case, the stack effect is created within the occupied space (Figure 2.1b) and as air temperatures are limited in order to ensure the occupants' comfort, the resulting stack effect is weak and the ventilation rates are relatively low. In the case of a solar chimney, there is, in principle, no limit to the temperature rise within the chimney, since it is isolated from the occupied part of the building. This allows the chimney to be designed in a way that makes the maximum use of solar gain. The air leaving the chimney will be replaced by fresh air entering the building from other low level openings, resulting in ventilation throughout the building.

Solar chimneys are mainly used to promote night ventilation but they can also be used to enhance daytime ventilation. For night ventilation, the chimney is kept closed during the

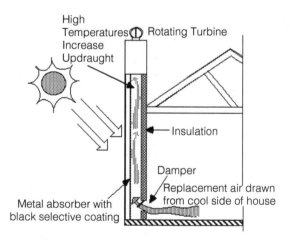

Source: www.greenbuilder.com/sourcebook/PassSolGuide3.html

FIGURE 2.1a Illustration of a solar chimney attached at the wall

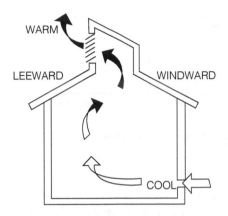

Source: www.azsolarcenter.com/design/pas-3.html
FIGURE 2.1b Illustration of stack effect ventilation

day and the design objective is to maximize solar gain and store it within the fabric of the chimney until ventilation is required. Ventilation usually begins in the late afternoon and continues during the evening, with the duration of the chimney's ventilation operation being dependent on the amount of energy stored in the chimney structure. The hotter the air inside the chimney, the greater the temperature and pressure differences compared with the external environment and thus the greater the air flow. In the mode of daytime ventilation, heat should be immediately transferred to the air in the chimney instead of the fabric of the chimney. This would imply a thermal lightweight construction for daytime operation and a thermal heavyweight structure for night-time operation.

The usual configurations of a solar chimney can be:

● a simple channel, usually rectangular in shape, attached to the wall;
● a solar chimney at the roof level, as part of or at the top of a solar collector.

The external surface of the solar chimney may be:

● thermal heavyweight directly exposed to solar gain;
● heavyweight but externally glazed;
● glazed surface, with the internal wall on the opposite side being the storage surface;
● metal sheet (usually galvanized sheet);
● metal sheet with glazing on its external surface.

The heavyweight external surface with glazing emulates the summer operation of a Trombe wall which also makes use of the solar chimney effect (Figure 2.2). With the Trombe wall, however, we are mainly concerned with increasing winter heat gain and thus a southern orientation is most appropriate.

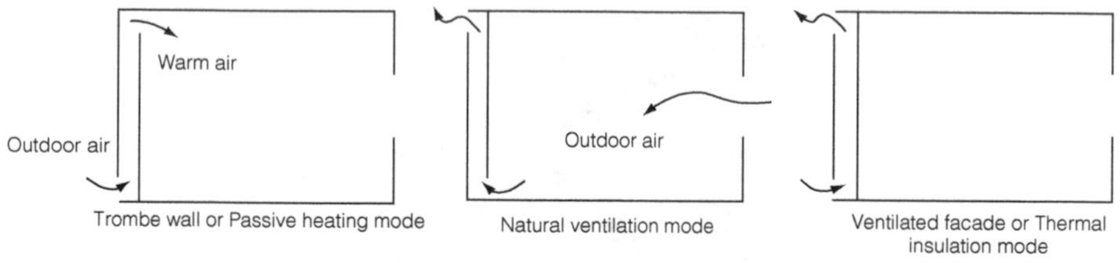

FIGURE 2.2 Different operation modes of a solar chimney

In the case of the solar chimney (Figure 2.2), the orientation must be carefully chosen according to the mode of operation. In general, south-west and west orientations are most useful, with the south-west orientation being the optimum orientation for northern latitudes.

The external surface, exposed to the sun, should have high solar absorptivity and thus a dark colour or a selective surface is appropriate. In the case of a glazed external surface, it is advisable to provide easy access to the glass surface, for maintenance and periodical cleaning, e.g. by using hinged glazing. The inlet opening to the room can be used to regulate the air flow and isolate the chimney from the room. If necessary, the top of the opening should be protected from rain penetration and a mesh can be added to prevent the entry of birds and insects.

Several studies also investigate the operation of a solar chimney in the mode of a ventilated facade (or 'thermal insulation mode') (Figure 2.2). When the cooling load is dominant and the outdoor temperature is higher than the room air temperature, the solar chimney may not be used to enhance the building's natural ventilation because the introduction of outdoor air without precooling causes an increase in the cooling load. Therefore, the air introduced into the chimney is exhausted to the outdoors through the upper opening, resulting in a reduction of the heat gain transmitted to the building.

VENTILATION FLOW IN A CHIMNEY

Ventilation flow rates in buildings are naturally introduced either by wind-induced pressure differences or by buoyancy forces induced by temperature differences between different heights. For a chimney of height H, attached to a room of height H_b (Figure 2.3), the stack pressure ΔPs is given by:

$$\Delta P_s = 273 \cdot \rho_o \cdot g \cdot \left(H \left(\frac{1}{T_a} - \frac{1}{T_{ch.air}} \right) + H_b \left(\frac{1}{T_a} - \frac{1}{T_b} \right) \right)$$ [2.1]

where T_a, $T_{ch.air}$ and T_b are the absolute temperatures of the ambient, chimney and room air.

Considering that the main pressure losses are in the chimney compared with the room, the stack pressure is balanced by the pressure losses through the chimney, ΔP_{loss} which is composed of the inlet and exit pressure losses and the channel friction loss:

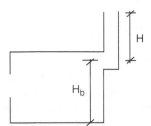

H_b = room height up to chimney inlet
H = inlet to exit chimney height

FIGURE 2.3 Section through a room with a solar chimney

$$\Delta P_{loss} = \Delta P_{inlet} + \Delta P_{exit} + \Delta P_{friction} \qquad [2.2]$$

The pressure losses in fittings are usually expressed as functions of the velocity pressure:

$$\Delta P = k\frac{\rho \cdot v^2}{2} \qquad [2.3]$$

where k is the velocity pressure loss coefficient which depends on the flow pattern and size and shape of the conduit. The friction pressure loss in fully developed flow in a straight channel of length H, with f the friction coefficient, is evaluated using the D'arcy-Weisbach relation:

$$\Delta P_{fr} = f\frac{4 \cdot H}{D_h}\frac{\rho \cdot v^2}{2} \qquad [2.4]$$

Thus the total pressure losses in the chimney are:

$$\Delta P_{loss} = k_{in}\frac{\rho \cdot v_{in}^2}{2} + k_{ex}\frac{\rho \cdot v_{ex}^2}{2} + f\frac{4 \cdot H}{D_h}\frac{\rho \cdot v_{ch}^2}{2} \qquad [2.5]$$

For fully developed laminar viscous internal flows, the coefficient of friction varies inversely with the Reynolds number according to the Poiseuille equation, f = 16/Re. In turbulent flows, the friction loss depends not only on flow conditions as characterized by the Reynolds number but also on the roughness of the surface, increasing with the increase of the surface roughness. Equating the stack pressure with the pressure losses through the chimney, the mass flow may be written as:

$$\overset{\bullet}{m}{}^2 = \frac{C \cdot H + D \cdot H_b}{\dfrac{k_{in}}{A_{in}^2} + \dfrac{k_{ex}}{A_{ex}^2} + f\dfrac{4 \cdot H}{D_h}\dfrac{1}{A_{ch}^2}} \qquad [2.6]$$

where:

$$C = 273 \cdot \rho_o \cdot g\left(\frac{1}{T_a} - \frac{1}{T_{ch,air}}\right) \cdot 2 \cdot \rho \qquad [2.7]$$

$$D = 273 \cdot \rho_0 \cdot g \left(\frac{1}{T_a} - \frac{1}{T_b} \right) \cdot 2 \cdot \rho \qquad [2.8]$$

It is valid to assume that the room effect H_bD is small and thus neglected. If the friction loss inside the chimney is small compared with the pressure losses at exit and inlet, the friction loss can be neglected and the mass flow rate will be:

$$\dot{m} = A_{in} \cdot A_{ex} \sqrt{\frac{C}{k_{in}A_{ex}^2 + k_{ex}A_{in}^2}} \sqrt{H} \qquad [2.9]$$

If the friction loss inside the chimney is high compared with the inlet and exit pressure losses, they may be neglected and the mass flow rate becomes independent of the chimney height:

$$\dot{m} = \frac{A_{ch}}{2} \sqrt{C \cdot D_h} \sqrt{f^{-1}} \qquad [2.10]$$

MODELLING OF SOLAR CHIMNEYS
MODELLING METHODOLOGY

The simulation approaches for the solar chimney's performance reported in the literature can be broadly classified into two categories: ones that use an existing simulation model and ones that are associated with development of a model for prediction of a solar chimney's performance.

In the first approach, the main reported modelling attempts are:

- Application of a thermal and flow model.
  - A step-by-step use of two models, one of which is a thermal model and the other one an air flow model. The thermal simulation model is used to predict the room, air and surface temperatures for a constant pre-specified air flow rate. The air flow model is used to calculate the flow rate for the predicted temperature and vice versa. The calculations are repeated until a convergence between the pre-specified and the predicted flow rate is achieved (Dimoudi, 1987).
  - The TAS thermal model and the FLOW (Melo, 1987) network ventilation model were used in an early study for the evaluation of a rectangular, solar chimney model (Dimoudi, 1987).
- The thermal TRNSYS and the ventilation LOOPDA simulation models were used in a loop to investigate the possibility of applying a double-skin facade, solar chimney system in a restored, old construction factory building (Ballestini et al, 2005).
  - A quasi steady approach was used in conjunction with a simulation model (SERI-RES) to model the ventilation effect of a solar chimney. The minimum constant air change rate for acceptable thermal comfort conditions in the internal space was predicted and it was used to determine the solar chimney temperature. From this temperature, the temperature difference required to drive buoyancy ventilation through the chimney was calculated and, based on this temperature difference, the optimum chimney inlet size was defined (Wilson and Walker, 1993).

● Application of a computational fluid dynamics (CFD) model for prediction of the temperature and air flow distribution inside a solar chimney. The CFD-based models predict in detail the air velocity and temperature distribution inside the solar chimney channel, but they do not usually consider thermal energy storage in the walls. They are based on simplifications for the wall surface boundary conditions: typically, a uniform heat flux or a constant surface temperature and a constant air heat transfer coefficient is considered.

In the second approach, the solar chimney's performance is simulated using a model that simultaneously predicts the air temperature (chimney and/or room) and the associated flow rate. The heat transfer equations are solved for the chimney construction/air channel using different solution schemes. The air flow is usually predicted from the mass flow equations, taking into account the pressure losses inside the channel, at the inlet and exit of the channel.

The heat transfer phenomena developed in a channel between two parallel plates is described in the analytical models. Most models are based on the bulk properties of the air in the channel, while a few consider the detailed behaviour of the boundary layer. Different solution schemes are applied for the solution of the equations, mainly finite difference solution schemes or the thermal resistance approach is used. As the cavity length is much larger than the cavity width, the effect of the side walls is usually neglected and the heat transfer is considered in one-dimensional (horizontal) or two-dimensional (horizontal and vertical) directions.

A uniform air temperature distribution across the same vertical height is assumed in most prediction methods (Chen et al, 2003). But it was argued by Awbi (1994) and Sandberg (1999) that this assumption is applicable to chimneys with a small gap-to-height ratio of less than or close to 1:10. For wider chimneys, the assumption of uniform temperature distribution may no longer be valid. Furthermore, the possible occurrence of reverse flows near the chimney outlets for wide chimneys may also make these theoretical predictions inadequate. Another question raised was about the suitability of the expressions for the pressure loss coefficients at the chimney inlet, outlet and along the chimney channel used in most models, as most of them are based on corresponding data for forced flows (Chen et al, 2003).

A two-part analytical process was also reported to describe the daily operation of a solar chimney (Koronakis, 1992). The daytime operation of the solar chimney when it is closed was simulated with the ESP-r model. For the open operation, an analysis of the boundary layer development inside the chimney channel and associated heat transfer process was used and the corresponding differential equations were solved. The wall temperatures predicted from ESP at the end of the closed mode were the initial conditions for the solution for the differential equations.

REVIEW OF MODELLING APPROACHES
Analytical modelling
The Crank-Nickolson method was used to solve the one-dimensional heat flow finite difference model together with an orifice equation at the chimney inlet and exit for the flow rate estimation (Zriken and Bilgen, 1985).

The flow rate through the chimney was also calculated taking into account the air flow resistance imposed by the solar chimney (Ward and Derradji, 1987).

A simplified model considered one-dimensional heat flow and a single control volume in the chimney channel where the flow rate was associated with the convection coefficient hc = f(u) (Bouchair, 1989). The cooling effect of a solar chimney in terms of the room temperature was also evaluated with a dynamic model based on a finite difference model (Bouchair, 1994).

Bansal et al (1993) developed a steady-state model of a solar chimney consisting of an air heater connected to a conventional chimney. A steady-state, one-dimensional heat flow model, assuming laminar flow inside the channel, was also applied in later work by Bansal et al (2005) for a short, window-sized solar chimney. The equations were solved with the matrix inversion method.

A two-dimensional heat transfer model was developed for the investigation of the heat and flow in a rectangular solar chimney with storage mass in both external and internal surfaces. The differential equations that describe the operation of the system were converted into algebraic ones using a finite difference implicit scheme (Dimoudi, 1997).

A non-steady, one-dimensional heat transfer model solved with a finite difference scheme was applied for the comparison of a solar chimney with a conventional chimney (Afonso and Oliveira, 2000). The model combined the heat transfer equations in the horizontal direction of the solar chimney's fabric with the equations for the flow inside the channel. The buoyancy pressure was initially considered that was equal to the sum of all flow pressure losses between the inlet and outlet: losses across fittings, contractions or expansions and friction losses on the chimney walls. Finally, friction pressure losses were neglected, compared with local losses. A global heat transfer coefficient was considered as a function of forced and free convection coefficients, with the argument that the temperature difference between the air and wall surface can be significant, inducing mixed convection conditions. An iterative procedure was used since the convective heat transfer coefficient was dependent on the temperature and ventilation flow rate at each time instant. Based on the comparison of the simulated results with the experimental results, Afonso and Oliveira concluded that the wind effect should be considered together with the buoyancy pressure in estimating the volumetric flow rate in the chimney.

A thermal network method with steady-state heat transfer equations was set up for a physical model of a solar chimney similar to a Trombe wall but without thermal mass in the internal surface (Ong, 2003). The equations were solved using a matrix inversion solution procedure. The friction flow resistance along the surfaces was assumed negligible compared with the pressure drops at the inlet and outlet of the chimney. The model showed that the air temperature rises along the chimney and the air mass flow rate increases with the increase of the air gap depth. The comparison of the analytical with experimental results (Ong and Chow, 2003) for an air gap of 0.1–0.3 m showed small differences at the lower solar radiation and better agreement for the larger air gap of 0.3 m. This was attributed to the model's assumptions that pressure drops at the inlet and outlet were greater than the pressure loss due to wall friction.

Heat transfer in the horizontal direction was assumed for the analysis of a solar chimney with a glazed external surface and thermal storage mass in the internal wall

(Marti-Herrero and Heras-Celemin, 2007). A uniform temperature was assumed for the glass and the air inside the channel (single node). The air temperature inside the channel was considered as a function of the inlet and exit air temperature as proposed by Hirunlabh et al (1999). A finite difference scheme was used to solve the equations.

CFD modelling

CFD modelling was also applied to investigate in detail the air and temperature distribution inside a solar chimney. As the length of the chimney (or chimney width, as referred to in some studies) is much larger than the cavity width (equal to air space thickness), a two-dimensional flow is usually considered in the analysis. In many numerical studies, pressure losses that occur at the inlet (contraction) and exit (expansion) of the cavity are not considered. For simplicity and minimization of computational efforts, the computational domain is defined by a rectangular cavity with the inlet and exit at the bottom and top, respectively, and thus the inlet and exit have the same dimensions as the cavity. With this approach, the disturbances that occur at the entrance of the channel, in the case of openings at the back wall, are not examined. An effective discharge coefficient is used in some models in order to account for the chimney inlet pressure losses (Awbi and Gan, 1992). Some studies consider the air flow inside the chimney laminar, but in most the flow is considered turbulent. The air turbulence in most studies is represented by the standard k-e model and in several studies by the renormalization group (RNG) k-e turbulence model, which is considered to give good correlations (Harris and Helwig, 2007). The surface heat transfer coefficients are evaluated considering the appropriate correlations of the Nusselt number, available in the literature. A wall surface function and radiation effects between the parallel plates are considered in a few models.

CFD modelling approaches

Two-dimensional laminar air flow and heat transfer with negligible radiation effects were used in a CFD modelling study of a solar chimney positioned at the top of the roof of a space (Barozzi et al, 1992). Reasonable agreement was reported between numerical predictions and experimental measurements of temperature and velocity in a 1:12 model.

A two-dimensional numerical model with a standard k-turbulence model, considering wall function and radiation effects, was applied by Moshfegh and Sandberg (1998). The numerical model was validated with experimental data for natural convection in a tall open cavity – 6.50 m (high) x 0.23 m (wide) x 1.64 m (long) – which was a model of a photovoltaic (PV) integrated facade with back ventilation carried out by Sandberg and Moshfegh (1996).

Rodrigues et al (2000) also applied a two-dimensional model with a standard k-turbulence model and wall function.

A two-dimensional CFD model with a standard k-turbulence model was used by Awbi and Gan (1992). The pressure losses at the inlet were also considered by applying an effective discharge coefficient. The numerical predictions were in good agreement with experimental results obtained by Bouchair (1994). The data obtained from the experimental work carried out by Betts and Bokhari (1996) were used to validate the CFD simulation model in later work by Gan (1998) on a solar chimney as part of the summer operation of a Trombe wall. This study used the RNG k-e model for the description of the turbulent flow inside the chimney. The commercial CFD package FLUENT (Gan, 2006) was also used for a solar chimney represented by a solar-heated open cavity. The air turbulence was represented by the standard k-e turbulence model, and the RNG k-e model was also used for validation. The channel's inlet was assumed at the back wall and thus the flow disturbance at the entrance of the channel was considered. The model was validated with experimental data for natural convection in a tall open cavity – 6.50 m (high) x 0.23 m (wide) x 1.64 m (long) – which was a model of a PV integrated facade with back ventilation carried out by Sandberg and Moshfegh (1996).

The PHOENICS CFD model with the turbulence regime represented by the RNG k-e model was used for an open, rectangular cavity, with inlet and exit at the bottom and top, respectively, representing a glazed, south-facing solar chimney in different inclinations (Harris and Helwig, 2007). The model was validated against the experimental results of Betts and Bokhari (1996) and Bouchair (1994), giving results within 5 per cent of both of these sets of data.

The governing elliptic equations were solved in a two-dimensional domain using a control volume method for a solar chimney with four opaque walls of height 4 m and with a 5 cm air gap thickness (Bacharoudis et al, 2007). For the numerical simulation of the turbulent flow inside the wall solar chimney, six turbulence models were tested:

- the standard k-e model;
- the RNG k-e model;
- the realizable k-e model;
- the Reynolds stress model (RSM); and
- two low-Reynolds (low-Re) models, namely, the Abid and the Lam-Bremhost model.

It was concluded that the use of the k-e models and the use of the Abid low-Re model assures the prediction of realistic velocity and temperature profiles as expected by theory. As the realizable k-e model is likely to provide superior performance for boundary layer flow under strong adverse pressure gradients, the latter has been selected to be used in the simulations. Furthermore, this selection was confirmed from the comparison with the experimental results of a solar chimney with four opaque walls.

CFD and analytical models

In some studies the simultaneous use of a CFD model and a thermal model was applied. The thermal model runs first to predict surface and air temperatures and mean flow rate, and then the results are entered into the CFD model to calculate the new stack flow, giving the new flow rate. The system is of oscillating character. A low flow rate produces high temperatures, with a resulting high flow rate in the next cycle, cooling the surface down again. Therefore, double boundary iteration is successful in giving very accurate results in a few cycles, making it possible to use standard spreadsheet programmes (Harris and Helwig, 2007). Some studies showed good agreement between analytical and CFD modelling in predicting the air flow rate and air temperature.

The performance of a solar chimney integrated into a south facade of a one-storey model office building and the effect on the heating and cooling loads of the building was studied by using a CFD simulation and an analytical model (Miyazaki et al, 2006). The fluid flow inside the chimney was assumed to be turbulent and simulated with the RNG k-e turbulence model, using the FLUENT software package. The CFD model was used to investigate the temperature and flow characteristics inside the chimney as well as the effect of the air gap width. The analytical model was used to simulate the performance of the solar chimney and the heating and cooling loads of the building including the solar chimney. The analytical model was based on the heat balance method and did not take into account the walls' friction loss. The conductive heat fluxes were calculated by the conduction transfer function (CTF) method. The CFD model was validated with results from the literature for a chimney with uniform heat flux. Comparison of both models was performed for the solar chimney, excluding the room, and good agreement was observed for the predicted air flow rate and outlet air temperature in the chimney.

Overall building simulation

The performance of solar chimneys in multistorey apartment buildings was evaluated with the use of two simulation models, the TRNSYS and TAS models (Macias et al, 2006).

The TRNSYS simulation tool was also used to evaluate the performance of an energy upgrade of an existing massive building with the application of solar chimneys among other energy systems (Calderaro and Agnoli, 2007).

The effect of different construction parameters on the thermal performance of a solar chimney attached to a room was investigated with the TRNSYS 15.1 model and a relatively natural ventilation routine. The thermal resistance of the walls, the glazing type and the solar chimney's thickness were examined (Pavlou and Santamouris, 2007).

EXPERIMENTAL WORK

Different configurations of a solar chimney were tested in several studies. The tested configurations should be distinguished into:

- vertical channels;
- chimneys attached at the wall of a building; and
- roof-mounted solar chimneys.

Different construction approaches were examined in each case:

- opaque external surface – opaque internal wall;
- transparent external – opaque internal wall;
- glass external – glass internal surface.

Thermal storage was considered in some cases of the opaque wall.

An evolution of solar chimney configurations was systematically tested by a research team in Thailand. A series of different tests for alternative solar chimney configurations were performed in an *outdoor* single-room test house of 11.5 m² floor area and 25 m³ volume in Thailand: the Trombe wall solar chimney configuration (TW), the roof solar collector (RSC) (Khedari et al, 1997), the modified Trombe wall (MTW) (Khedari et al, 1998), the metallic solar wall (MSW) (Hirunlabh et al, 1999), the bio-climatic roof (BCR) (Waewsak et al, 2003) and the glazed solar chimney wall (GSCW) (Chantawong et al, 2006). The last one was tested in small test house of 2.8 m³.

The experimental work as found in the literature is presented below.

VERTICAL CHANNEL
Transparent external – opaque internal wall
A solar chimney configuration represented by an air solar collector system was investigated in controlled conditions (Burek and Habeb, 2007). The test rig comprised a vertical channel (1.025 m high x 0.925 m wide) open at the top and bottom, with one side covered by a transparent material (Perspex) and the other one comprising an aluminium 'absorber plate', painted matt black. The channel depth varied from 0.02 to 0.11 m. The solar intensity was supplied by an electric heating mat (*uniform heat flux*) behind the absorber plate, with variable heat input from 200 to 1000 W/m². It was shown that the mass flow rate depends on the heat input and the channel depth. Dimensionless expressions were also derived as described in a following section.

Opaque external and internal walls
The flow and temperature conditions inside a rectangular chimney (2 m high x 1 m long), constructed from mild steel plates, with the inlet opening at the bottom of the back surface and the exit at the top of the channel were investigated inside a laboratory (Dimoudi, 1997). The side walls were made of rigid polystyrene board. The effect of the depth of the channel (0.1 m up to 0.5 m) and the height of the inlet opening (0.1 to 0.2 m) on the chimney's performance was studied. The external surface of the chimney, painted black, was irradiated by a simple solar simulator constructed for the needs of the experiment. Three, 1000 W halogen heat lamps, each one hosted in an almost parabolic reflector and dimmed independently, were used in a position away from the irradiated surface chosen to give an approximately uniform irradiance on the surface. The results were used for validation of a model and derivation of expressions with dimensionless correlations. It was found that the mass flow rate per unit area of the chimney is highest for the narrower channel (d = 0.10 m) and decreases with increases of the channel depth.

Experiments on a solar chimney consisting of a rectangular channel with *uniform heat flux* in one wall only, a variable chimney gap (10 to 60 cm), different heat flux (200 to 600 W/m²) and inclination angles (15° to 60° relative to the vertical position) were reported (Chen et al, 2003). The channel with internal dimensions of 1.5 m high, 0.62 m long and a variable air gap width was constructed with insulated Plexiglass for the three walls and with stainless steel shims for the heated surface. The air temperature was measured at different points along the height and width inside the channel. The air velocity was measured along the width at 1.1 m from the inlet while the air flow patterns inside the channel were visualized by a Drager smoke tube. The results showed:

- An increase of the air flow with the increase of the chimney gap and *no optimum gap width* was found.
- *Reverse flow* was observed at the chimney outlet for chimney gaps *starting from 30 cm*, but limited near the chimney outlet. With a further increase in the chimney gap, the reverse flow was further enhanced down into the chimney channel.
- A maximum air flow rate was achieved at an *inclination angle around 45°* for a 20 cm gap and 1.5 m high chimney, which is about 45 per cent higher than that for a vertical chimney.
- In vertical chimneys with large gaps both the *air temperature* and *air velocity* inside the chimney are highly *non-uniform* and the velocity profile has a maximum very close to the heated wall. It was thus argued that most prediction methods that assume uniform air temperature at a certain height can substantially over-predict the air flow rate for a chimney configuration similar to the tested one, especially for vertical chimneys with large gaps. This may be attributed to the underestimation of the pressure losses at the chimney outlet by using loss coefficients obtained for normal forced flows.
- The *air temperature along the chimney generally increased with height* but the temperatures on the heated surface and the opposite one do not increase linearly and even drop above the middle of the chimney height.

A solar chimney with four opaque walls made of plasterboard with a 5 cm air gap was constructed and put at each wall and orientation of a model room (4 x 6 x 4 m) located *outdoors* at the campus of the Technological Education Institution of Chalkida, Greece (Bacharoudis et al, 2007). The presented results concern the operation of a vertical channel, as the inlet and outlet in the channel are connected with the external environment and connection with the interior of the room will be studied in a later stage. The research focused on the study of the thermo-fluid phenomena occurring inside the wall solar chimneys, and a numerical investigation of the buoyancy-driven flow field and heat transfer inside the wall solar chimneys was performed.

SOLAR CHIMNEY ATTACHED AT BUILDING'S WALL
Glass external – opaque internal wall chimney
A modified wall solar chimney, the metallic solar wall (MSW), was tested (Hirunlabh et al, 1999), consisting of a glass cover, air gap, a black metallic plate insulated at the back with micro-fibre and plywood; it was tested outdoors in a single-room test house of 11.5 m^2 floor area and 25 m^3 volume in Thailand. It was found that for the 2 m high by 1 m wide MSW the 14.5 cm air gap produced the highest air mass flow rate, of about 0.01–0.02 kg/s, and that the room temperature during the tests was close to that of the ambient air.

A solar chimney with a south-facing glazed surface and thermal storage in the other walls was experimentally compared with a conventional brick wall chimney *outdoors* in Portugal (Afonso and Oliveira, 2000). The two chimneys of the same size – 0.2 m wide x 1.0 m long x 2.0 m high – were attached on the roof in each room of a test cell, of total test cell floor area of 12 m^2 (4 x 3 m). The opaque walls of the chimneys were constructed from brick (10 cm thick) and 5 cm external insulation was added at the solar chimney; the test cell structure was made of concrete, externally insulated. The air flow rates in both compartments of the test cell were measured with a tracer gas technique. The experimental results were used for validation of a simulation model.

Experiments on a solar chimney with a glazed external surface and a painted matt black opaque internal wall, *without thermal storage*, were carried out *outdoors* in Malaysia (Ong and Chow, 2003). The chimney, 2 m high x 0.45 m long with *variable air gaps* of 0.1, 0.2 and 0.3 depths, was attached at a rectangular box, 2.00 m high x 0.48 m wide x 1.02 m deep, constructed of 22 mm thick rigid polyurethane sheets laminated on both sides with 1 mm thick steel sheet cladding. A 50 mm thick polystyrene sheet was attached at the back of the heat-absorbing wall to provide additional heat insulation. The height of the inlet to the room was 0.1 m. The air velocities increased with the air gap, varying between 0.25 and 0.39 ms^{-1} at air gaps between 0.1 and 0.3 m, respectively, at a radiation intensity of 650 Wm^{-2}. This showed that a solar chimney with a 0.3 m air gap was able to provide 56 per cent more ventilation than one with a 0.1 m air gap. No reverse air flow circulation was observed up to a 0.3 m air gap. The results were used for verification of a simulation model.

A solar chimney, consisting of air heaters, was tested, among other passive cooling techniques, in *outdoor* conditions in a metal structure in Egypt (Amer, 2006). Two air heaters (without thermal storage) used in a series to form a solar chimney, with a total aperture 2.0 m (high) x 0.5 m (long), were attached at the south wall of a test room (1 x 1 x 1 m). Longitudinal fins were provided in the air gap to improve the heat transfer. The results showed that without passive cooling measures, the average indoor temperature was about 16°C higher than outdoors, whereas the use of the solar chimney or application of evaporative cooling led to a reduction of the indoor temperature to within ±1°C of the outdoor ambient temperature, and white painting or insulation of the roof reduced the air temperature to 6 and 3°C above the outdoor temperature, respectively.

Opaque wall systems

The modified Trombe wall (MTW), consisting of a masonry wall as an outer layer, an air gap and a gypsum board as an inner layer, was investigated under *real weather conditions* in Thailand (Khedari et al, 1998). It was concluded that the MTW was effective in reducing the heat transmitted to the room but the induced ventilation rate was less than the required ventilation for thermal comfort.

A small window-sized solar chimney, with absorber length less than 1 m embedded in a regular window, without major structural changes, was investigated in *outdoor* conditions in India (Mathur et al, 2006). The set-up, which resembled a typical glass window with a black shutter behind the glazing, was installed on a cubical wooden chamber measuring 1 x 1 x 1 m. As windows are usually kept closed during the sunshine hours in hot climatic conditions to avoid direct heat gain in the building envelope, the ventilation potential of this approach was investigated with these tests. An aluminium sheet was attached on the exposed side of the shutter to act as a solar radiation absorber. Nine different combinations of absorber height and air gap were investigated. There is the potential of inducing a flow rate of 5.6 ACH (air changes per hour) in a room of 27 m^3, at solar radiation of 700 W/m² on a vertical surface with the stack height to air gap ratio of 2.83 for a 1 m high chimney.

Glass – glass wall

The glazed solar chimney walls (GSCW), consisting of two glass panes (0.74 m high x 0.50 m wide) with a 0.1 m intermediate air gap and openings at the bottom of the room side glass pane and at the top of the external glass pane, were placed into the southern wall of a small test room, 1.4 x 1.4 and 2.0 m high (Chantawong et al, 2006). The size of the inlet/outlet openings was 0.05 x 0.5 m. It was concluded that the induced air flow rate was about 0.13–0.28 m³/s, the temperature difference between the room and the exterior was less than that with a single-layer clear glass window and that the reduction of daylight due to the double glass layer was negligible. The experimental results were used for validation of a simulation model developed for the GSCW.

ROOF SYSTEMS

In roof systems, the solar chimney is attached on the roof (Figure 2.4) and different configurations and construction materials were tested.

The roof solar collector (RSC) system, consisting of Monier concrete tiles in the external surface, an air gap and gypsum or plywood board in the internal surface, was tested *outdoors* in Thailand. Tests of the roof solar collector systems (Khedari et al, 1997) showed that the application of gypsum board is better than plywood regarding resistance to heat loss. Monitoring of the system during the period September to November showed that the temperature, velocity and flow rate inside the air gap ranged from 30–36°C, 0.5–1.3 m/s and 0.08–0.15 m³/s/m² of solar chimney, respectively. The optimum roof tilt of 25–30° and an air duct length of less than 1 m were recommended.

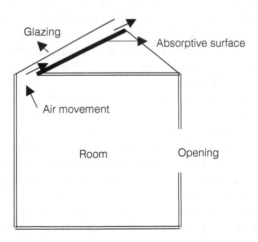

FIGURE 2.4 Illustration of a roof-mounted solar chimney system

The effect of the size of the air gap on the air flow rate was then studied in four different configurations of the roof solar collector system (Khedari et al, 2000a). It was shown that the 14 cm air gap gives a 15 per cent higher volumetric air flow rate compared with the 8 cm air gap. An equal size of inlet and exit openings induces the highest air flow rate. It was also shown that with only a roof solar collector (RSC) system, there is little potential of inducing sufficient natural ventilation to satisfy residents' comfort and that additional devices such as a Trombe wall should be used together with the roof solar collector system in a real building, in order to improve cooling 'ventilation' efficiency.

A modified solar collector system was also tested outdoors in Thailand, the so-called BSCR–BCR (bio-climatic roof) composed of a combination of Monier concrete and transparent acrylic tiles on the outer side, an air gap of 0.14 m and a combination of gypsum with an aluminium foil board and translucent sheets on the room side (Waewsak et al, 2003). The transparent acrylic tile area was about 15 per cent of the 1.5 m^2 (1.0 x 1.5 m) surface area of the BCR unit, while the ratio of translucent sheet to gypsum with an aluminium foil board was approximately 70:30. The experimental results showed that the BSRC–BCR can reduce roof heat gain significantly, provide sufficient natural lighting and induce high air change. The measured air changes per hour were about 13–14 and 5–7 in summer and winter, respectively, two times higher than those induced by the conventional roof solar collector (RSC) system.

A modification of a roof solar chimney was investigated *outdoors* in Israel (Drori and Ziskind, 2004). A configuration proposed in earlier work by Ziskind et al (2002) was experimentally tested in a one-storey prefabricated structure. The structure, with internal dimensions of 5.95 m (long) x 2.35 m (wide) x 2.35 m (high), was constructed from metallic walls, well insulated and covered on the inside and partially on the outside by wood. The chimney part is a horizontal metal sheet mounted above the roof of the building forming a horizontal duct connected to the inner space of the building. Air is heated inside the duct while fresh air is sucked into the home through a vertical duct at the northern wall, with an air gap of 20 cm and openings at the top and the floor level. The results showed that effective ventilation was achieved and the air temperature inside the structure typically followed the ambient temperature. By comparison, in the conventional structure with no alterations, the indoor temperature was considerably higher than the ambient one, especially during the afternoon hours. The results were verified with the FLUENT CFD simulation software.

In another institute in Thailand, an experimental study of a solar chimney attached to the 45° tilted roof (the roof solar collector) of a test cell resembling an actual room size (3.8 x 2.8 x 2.4 m) was conducted under *real weather* conditions (Chungloo and Limmeechokai, 2007). The south roof was constructed from terracotta roof tiles on the outer side, with a 0.15 m air gap and gypsum board on the inner side. The effect of the solar chimney alone and additional water roof spraying was investigated during the tests. Emphasis was given to removing the wind effect and investigating the ventilation due to the solar chimney alone. It was shown that without the wind effect significant ventilation was provided in the test cell, in the order of 1.13–2.26 ACH, but this is still considered low. It was thus argued that by using only a solar chimney, the ventilation is insufficient and the wind ventilation effect must be included in an overall natural ventilation approach. However, the experimental results recommended further studies on stack effect, especially on natural ventilation with low Reynolds number. A temperature reduction of 1.0–3.5°C compared with the outdoor ambient air temperature of 32–40°C was achieved with the solar chimney operating alone, while additional spraying on the roof can reduce indoor temperatures by 2.0–6.2°C.

COMBINATIONS OF DIFFERENT SOLAR CHIMNEY SYSTEMS

Tests on alternative solar chimney configurations in an outdoor test room in Thailand showed that all the configurations minimized the amount of solar flux absorbed by the building's fabric, therefore acting as good insulation, and also induced natural ventilation which improved indoor thermal comfort. However, the resulting number of air changes was rather low, varying, depending on climate conditions, between 3 and 5 ACH, which is not sufficient to completely satisfy room occupants' comfort, considering that a higher number is required in naturally cooled houses (above 20 ACH) (Khedari et al, 2000a). Thus, a combination of three different wall solar chimney configurations of 2 m² each and two roof solar collectors (RSCs) of 1.5 m² each were also experimentally investigated in the same building (Khedari et al, 2000b). With a solar chimney surface area of 6–9 m², corresponding to a ratio of 0.24–0.36 m²/m³ of solar chimney area to house volume, the average ACH, between 12 am and 2 pm, was about 15 and the room temperature was about 2–3°C above the ambient one. It was argued that regarding thermal comfort, the induced air motion of about 0.04 m/s at the living level cannot satisfy occupants, as a higher air velocity is needed when the indoor air temperature is about 35–37°C. It was suggested that a larger chimney area could provide greater indoor air velocity.

The performance of a wall and roof solar chimney system in an air-conditioned building was tested under real weather conditions in Thailand (Khedari et al, 2003). Two configurations of solar chimney were applied – the roof solar collector composed of Monier concrete tile, a 14 cm air gap and gypsum board, and the modified Trombe wall composed of a masonry wall, 14 cm air gap and gypsum board. The inlet size was a variable that was used to control the induced air flow rate. The experiments were carried out for a period of six months (March to September) and the results for a common house and the solar chimney house (SCH) were compared for relatively similar ambient conditions. It was demonstrated that a SCH could reduce the average daily electrical consumption of the air-conditioning by 10–20 per cent. The appropriate size of inlet opening of a solar chimney was 5 x 5 cm, which induced about 3–8 m³/h/solar chimney length unit.

APPLICATIONS IN EXISTING BUILDINGS

The use of solar chimneys for the natural ventilation of buildings is found in some historical buildings, the so-called 'Scirocco rooms' in Italy, dating back to the 16th century. The solar chimneys were combined with underground corridors and water features to enhance cooling and ventilation of these buildings (Di Cristofalo et al, 1989).

There are very few reports in the literature on monitoring the performance of solar chimneys in existing buildings. Measurements of system performance in a university building in Japan (Shinada et al, 2003, as stated in Miyazaki et al, 2006) and the energy savings from the use of natural ventilation and the performance of the system were presented in a report.

A solar chimney in the mode of a double-skin facade was proposed to be constructed on the south-west facade in an existing industrial listed building in Cavaso del Tomba (60 km north of Venice), Italy (Ballestini et al, 2005). A glass facade (40 m wide by 10 m high), attached to the outside of the existing building envelope, was separated by an air gap of 0.65 m. A glazed facade will be provided in both the upper and lower parts with automatic louvres in order to allow and regulate winter and summer ventilation. During summer, the top and bottom louvres should be kept open to guarantee the necessary air changes to the double skin so as to avoid risk of overheating. Also, the negative pressure created by the stack effect in the double-skin gap will allow cross-ventilation, permitting exhaustion of part of the air from the inner rooms during almost all hours in summer.

A solar chimney system was constructed at a new multi-floor project of social housing in Spain (Macias et al, 2006). Solar chimneys constructed from concrete, with west orientation, were connected to each floor. A swinging flap at the top of the chimney was kept closed during the day and open during the night to allow ventilation of the apartments. The chimneys in combination with the high thermal mass of the building construction were used to enhance the effectiveness of night ventilation.

Solar chimneys were proposed to be constructed at an existing library building in Rome (Calderaro and Agnoli, 2007). The solar chimneys were constructed from brick and at the top were externally covered by metal plates and glass. The solar chimneys, combined with the high thermal storage structure of the building and supportive indirect evaporative systems, were expected to improve indoor conditions in the building during the summer.

DIMENSIONLESS CORRELATIONS

Some studies used the data from experiments on solar chimneys to derive dimensionless correlations as easy calculation tools for designers to predict the flow rate through a chimney without the need to use complex modelling tools. The correlations attempt to predict the flow rate as a function of the outside ambient conditions (solar heat input or ambient temperature), the ventilated space room air and the geometry of the chimney.

For a rectangular opaque solar chimney connected to a room, correlations were derived relating the ventilation rate caused by buoyancy-driven flow through a solar chimney to the geometry of the chimney and characteristic temperatures (Dimoudi, 1997). The dimensional analysis was undertaken for three dependent variables:

- the volumetric flow rate, V';
- the mass flow rate per unit area of chimney, M'; and
- the exit air temperature, T_e.

Considering the physical phenomena occurring in the chimney, and following the Buckingham method of analysis, the following functional equation was obtained:

$$\Pi 1 = f1\,(\Pi 2,\, \Pi 3,\, \Pi 4,\, \Pi 5,\, \Pi 5,\, \Pi 6) \qquad [2.11]$$

The dimensionless groups are described in Table 2.1, where H, d, h represent the chimney height, the air gap depth and the inlet height, β is the coefficient of volumetric thermal expansion of the air, g is the acceleration due to gravity, ρ, μ, c_p and λ represent the density, viscosity, specific heat and thermal conductivity of the air, and T_i and T_e are the absolute temperature of the inlet and exit air in the chimney channel.

The experimental data from tests for the configuration described in Dimoudi (1997) were used to validate the analysis. The experimental data were therefore correlated using a power law relationship and the following correlations were derived:

$$N_v = 2.0 \cdot 10^{-11} Ra_H^{0.4} \left(\frac{h}{H}\right)^{-0.8} \left(\frac{d}{H}\right)^{0.20} \tag{2.12}$$

$$N_m = 8.51 \cdot 10^{-10} Ra_H N_T^{0.60} \left(\frac{h}{H}\right)^{-0.4} \left(\frac{d}{H}\right)^{-0.80} \tag{2.13}$$

$$N_{Te} = 2.70 \cdot 10^3 Ra_H^{-0.2} N_{Ti}^{0.8} \tag{2.14}$$

where the Rayleigh number (Ra_H) is a function of the Grashof number (Gr_H) and the Prandtl number (Pr), $Ra_H = Gr_H Pr$ (Table 2.1).

A simple tool based on dimensional analysis was proposed for the evaluation of thermal performance of a solar chimney as part of a ventilated facade: external opaque panel, air channel with outdoor–outdoor buoyancy flow, insulation and internal wall (Balocco, 2004). Fourteen non-dimensional numbers were defined in a power series:

$$M1 = aM2^b M3^c M4^d M5^e M6^f M7^g M8^h M9^i \times M10^j M11^m M12^n M13^p M14^q \tag{2.15}$$

The constants were determined and correlation validity was tested using experimental data from a PV-covered ventilated facade (Sandberg and Moshfegh, 1996). The correlation was also compared with results from a simple simulation tool developed by Balocco.

Experimental work carried out in a solar chimney represented by a solar collector thermosyphoning system (Burek and Habeb, 2007) showed that the mass flow rate through the channel (represented by the Reynolds number Re(s) = us/v) was a function

TABLE 2.1 Dimensionless groups for a solar chimney

$$\Pi_1 = \frac{\dot{V}}{g\beta\Delta TH^5} = N_v \qquad \Pi_4 = \frac{c_p \mu}{\lambda} = Pr \qquad N_m = \frac{\dot{M} \cdot H}{\mu}$$

$$\Pi_2 = \frac{\rho^2 g\beta\Delta TH^3}{\mu^2} = Gr_H \qquad \Pi_5 = \frac{h}{H} \qquad N_{Te} = \frac{c_p T_e}{g\beta\Delta THPr}$$

$$\Pi_3 = \frac{c_p T_i}{g\beta\Delta THPr} = N_{Ti} \qquad \Pi_6 = \frac{d}{H}$$

of both the heat input (Q) (represented by the modified Rayleigh number, $Ra^* = Pr$ $(gbq_{in}H^4/v^2k))$ and the channel depth(s):

$$m \infty Q^{0.572} \quad \text{and} \quad m \infty s^{0.712} \qquad [2.16]$$

as the results of a multivariant regression analysis of the experimental data gave:

$$Re(s) = 0.00116(Ra^*)^{0.572}(s/H)^{0.712} \qquad [2.17]$$

The thermal efficiency of the system (as a solar collector) was independent of the channel depth but was a function of the heat input:

$$n \infty Q^{0.298} \qquad [2.18]$$

DESIGN RECOMMENDATIONS

Average flow rates for a solar chimney are always higher than flow rates for a conventional chimney and the solar contribution is higher in warmer months (Afonso and Oliveira, 2000).

In another study, the advantage of the solar chimney as a passive heating device rather than a cooling device was shown by simulation results for an office building in Japan (Miyazaki et al, 2006).

Outdoor tests of a glazed chimney without thermal mass showed that air velocities increased with the air gap and varied between 0.25 and 0.39 ms^{-1}, for air gaps between 0.1 and 0.3 m, respectively, at radiation intensity up to 650 Wm^{-2} (Ong and Chow, 2003). Thus, a solar chimney with a 0.3 m air gap was able to provide 56 per cent more ventilation than one with a 0.1 m air gap. No reverse air flow circulation was observed up to a 0.3 m air gap.

The opposite performance, meaning that the air velocity generally decreased with the increase of the gap width, was shown by Hocevar and Casperson (1979). They found that in a 2.2 m high solar chimney similar to a Trombe wall the average velocities were about 0.4 ms^{-1} with an air gap of 0.05 m and 0.21 ms^{-1} with an air gap of 0.15 m. It was indicated that most of the flow regimes during the tests were laminar in nature.

For an air gap greater than 0.2 m, the mass flow rate induced by buoyancy was almost independent of the air gap width of the solar chimney. The inlet and outlet pressure losses are dominant rather than the wall friction losses (Miyazaki et al, 2006).

Bouchair (1989, 1994) found that in the 1.95 m high solar chimney, without thermal mass, made of metallic plates electrically heated to attain uniform temperature, the highest air flow rate was found at a gap of 0.2 m and reduced for a gap greater than 0.3 m. The inlet height did not affect the flow rate in the 0.1 m chimney gap but the flow increased with inlet height for air gaps of 0.3–0.5 m. Bouchair also observed that the air flow inside the chimney was turbulent and downward flow occurred in the case of a chimney with a 0.5 m air gap.

Smoke visualization studies by Chen et al (2003) observed air flow disturbances near the chimney inlet and also reverse flow at the chimney outlet especially for chimney gaps larger than 0.4 m. They showed that air flow rates increased continuously with chimney length/gap ratio up to 2.5, but they could not determine any optimum air gap for maximum air flow rate.

Two parameters were identified as important for satisfying the required average flow rate – chimney section and chimney height. The average flow rate changes linearly with chimney section. For a given solar collection area, an increase in chimney width is more effective than a height increase (Afonso and Oliveira, 2000).

A large chimney area was required to reduce the cooling load of an office building in Japan as shown in simulation studies. A chimney length of 4 m was required to reduce the cooling load of the building compared with a conventional design, between 10 am and 12 pm on an average day in May (Miyazaki et al, 2006).

Due to the variable, almost random, nature of wind, the a solar chimney can be designed without considering the wind effect, which will underestimate the real ventilation rates (Afonso and Oliveira, 2000).

The contribution of thermal storage is significant and thus external insulation of the brick wall was suggested, otherwise the solar efficiency, compared with a conventional chimney, is reduced by more than 60 per cent; an insulation thickness of 5 cm was considered as sufficient (Afonso and Oliveira, 2000).

Increase of the wall storage thickness does not significantly affect the average flow rate, with the maximum rate obtained for 10 cm. The optimum storage thickness depends on the building use pattern. For day ventilation, a small thickness is suggested, as a greater thickness allows higher thermal energy storage in the wall, decreasing the flow rate during sunshine hours. Whereas for night operation, a greater thickness should be adopted (Afonso and Oliveira, 2000).

It was shown, based on CFD modelling of a solar chimney as part of the summer operation of a Trombe wall, that in order to maximize the ventilation rate, the interior surface of the opaque wall should be insulated for summer cooling. This would also prevent undesirable overheating of room air due to convection and radiation heat transfer from the wall (Gan, 1998).

Increased insulation of the inner wall of the solar chimney resulted in increased mass flow rate, but caused a rise in the cooling load and reduction of the heating load of the building due to the increased thermal resistance (Miyazaki et al, 2006).

CONCLUSIONS

Ventilation of buildings is an effective means of reducing the cooling load and providing thermal comfort for the occupants. Solar chimneys are configurations used to enhance natural ventilation. They are constructed as part of the building, either attached at the wall or on the roof.

A lot of research has been undertaken on testing and modelling the performance of a solar chimney. Some of the research focused on the study of the fluid and thermal field inside the channel of the chimney, whereas other studies investigated the effect of the chimney on the thermal and flow conditions inside the attached building.

The main advantages of solar chimneys are (Dimoudi, 1997):

- They do not generally prescribe the architectural form of the building as they are relatively small and can be easily integrated into the structure.
- They are simple devices without special architectural and construction needs and use common construction materials.

- As the preferred orientation for evening operation is west or south-west, they have the advantage of leaving the south face free for other passive devices; they may be combined with other passive solar systems on the south facade, e.g. Trombe wall or ventilated facade.
- They are suitable for any type of building.

The main disadvantages of solar chimneys are:

- A solar chimney, as a passive device, cannot generally respond to changes in the internal environment and thus it is unsuitable if control of the indoor environment is essential.
- They may not sufficiently cover the cooling needs during the whole cooling period and thus alternative techniques and backup systems may be required.
- Solar chimneys are more suited to open layouts or for individual rooms as the pressure loss associated with air flows between rooms will affect the total flow.

AUTHOR CONTACT DETAILS

Argiro Dimoudi: Department of Environmental Engineering, School of Engineering, Democritus University of Thrace, Vass. Sofias 12, 67 100 Xanthi, Greece; adimoudi@env.duth.gr

REFERENCES

Afonso, C. and Oliveira, A. (2000) 'Solar chimneys: Simulation and experiment', *Energy and Buildings*, vol 32, no 1, pp71–79

Amer, E. (2006) 'Passive options for solar cooling of buildings in arid areas', *Energy*, vol 31, no 8–9, pp1332–1344

Awbi, H. B. and Gan, G. (1992) 'Simulation of solar-induced ventilation', in *Proceedings of 2nd World Renewable Energy Congress (WREC 92) – Renewable Energy Technology and the Environment*, Reading (UK), 13–18 September, vol 4, pp2016–2030

Awbi, H. B. (1994) 'Design consideration for natural ventilated buildings', *Renewable Energy*, vol 5, no 5–8, pp1081–1090

Bacharoudis, E., Vrachopoulos, M., Koukou, M., Margaris, D., Filios, A. and Mavrommatis, St. (2007) 'Study of the natural convection phenomena inside a wall solar chimney with one wall adiabatic and one wall under a heat flux', *Applied Thermal Engineering*, vol 27, no 13, pp2266–2275

Ballestini, G., De-Carli, M., Masiero, N. and Tombola, G. (2005) 'Possibilities and limitations of natural ventilation in restored industrial archaeology buildings with a double-skin facade in Mediterranean climates', *Building and Environment*, vol 40, no 7, pp983–995

Balocco, C. (2004) 'A non-dimensional analysis of a ventilated double façade energy performance', *Energy and Buildings*, vol 36, no 1, pp35–40

Bansal, N. K., Mathur, R. and Bhandari, M. S. (1993) 'Solar chimney for enhanced stack ventilation', *Building and Environment*, vol 28, no 3, pp373–377

Bansal, N. K., Mathur, J., Mathur, S. and Meenakshi, J. (2005) 'Modeling of window-sized solar chimneys for ventilation', *Building and Environment*, vol 40, no 10, pp1302–1308

Barozzi, G. S., Imbabi, M. S. E., Nobile, E. and Sousa, A. C. M. (1992) 'Physical and numerical modeling of a solar chimney based ventilation system for buildings', *Building and Environment*, vol 27, no 4, pp433–445

Betts, P. I. and Bokhari, I. H. (1996) 'Experiments on natural convection of air in a tall cavity', in, *IAHR Workshop on Flow modeling* Paris, vol V, pp25–26

Bouchair, A. (1989) *Solar Induced Ventilation in the Algerian and Similar Climates*, PhD thesis, Civil Engineering Department, University of Leeds, UK

Bouchair, A. (1994) 'Solar chimney for promoting cooling ventilation in Southern Algeria', *Building Services Engineers Research Technology*, vol 15, no 2, pp81–93

Burek, S. A. M. and Habeb, A. (2007) 'Air flow and thermal efficiency characteristics in solar chimneys and Trombe walls', *Energy and Buildings*, vol 39, no 2, pp128–135

Calderaro, V. and Agnoli, St. (2007) 'Passive heating and cooling strategies in approaches of retrofit in Rome', *Energy and Buildings*, vol 39, no 8, pp875–885

Chantawong, P., Hirunlabh, J., Zeghmati, B., Khedari, J., Teekasap, S. and Win, M. M. (2006) 'Investigation on thermal performance of glazed solar chimney walls', *Solar Energy*, vol 80, no 3, pp288–297

Chen, Z. D., Bandopadhayay, P., Halldorsson, J., Byrjalsen, C., Heiselberg, P. and Li, Y. (2003) 'An experimental investigation of a solar chimney model with uniform wall heat flux', *Building and Environment*, vol 38, no 7, pp893–906

Chungloo, S. and Limmeechockai, B. (2007) 'Application of passive cooling systems in the hot and humid climate: The case study of solar chimney and wetted roof in Thailand', *Building and Environment*, vol 42, no 9, pp3341–3351

Di Cristofalo, S., Orioli, S., Silvestrini, G. and Alessandro, S. (1989) 'Thermal behavior of "Scirocco rooms" in ancient Sicilian villas', *Tunneling and Underground Space Technology*, vol 4, no 4, pp471–473

Dimoudi, A. (1987) *An Evaluation of the Use of Solar Chimneys to Promote the Natural Ventilation of Buildings*, MSc thesis, School of Mechanical Engineering, Cranfield Institute of Technology, UK

Dimoudi, A. (1997) *Investigation of the Flow and Heat Transfer in a Solar Chimney*, PhD thesis, School of Civil and Structural Engineering, Bath University, UK

Drori, U. and Ziskind, G. (2004) 'Induced ventilation of a one-storey real-size building', *Energy and Buildings*, vol 36, no 9, pp881–890

Gan, G. (1998) 'A parametric study of Trombe walls for passive cooling of buildings', *Energy and Buildings*, vol 27, no 1, pp37–43

Gan, G. (2006) 'Simulation of buoyancy-induced flow in open cavities for natural ventilation', *Energy and Buildings*, vol 38, no 5, pp410–420

Harris, D. J. and Helwig, N. (2007) 'Solar chimney and building ventilation', *Applied Energy*, vol 84, no 2, pp135–146

Hirunlabh, J., Kongduang, W., Namprakai, P. and Khedari, J. (1999) 'Study of natural ventilation of houses by a metallic solar wall under tropical climate', *Renewable Energy*, vol 18, no 1, pp109–119

Hocevar, C. J. and Casperson, R. L. (1979) 'Thermo-circulation data and instantaneous efficiencies for Trombe walls', in *Proceedings of 4th National Passive Solar Conference*, Kansas City, Missouri, US, pp163–167

IEA (International Energy Agency) (2003) *Energy Information Administration (EIA)*, IEA Annual Report, www.eia.doe.gov/iea/

JRAIA (Japan Refrigeration and Air Conditioning Industry Association) (2008) *Estimates of World Demand for Air Conditioners, 2000–2008*, www.jraia.or.jp

Khedari, J., Hirunlabh, J. and Bunnag, T. (1997) 'Experimental study of a roof solar collector towards the natural ventilation of new houses', *Energy and Buildings*, vol 26, no 2, pp159–164

Khedari, J., Lertsatitthanakorn, C., Pratinthong, N. and Hirunlabh, J. (1998) 'The modified Trombe wall: A simple ventilation means and an efficient insulating material', *International Journal of Ambient Energy*, vol 19, no 2, pp104–110

Khedari, J., Mansirisub, W., Chaima, S., Pratinthong, N. and Hirunlabh, J. (2000a) 'Field measurements of performance of roof solar collector', *Energy and Buildings*, vol 31, no 3, pp171–178

Khedari, J., Boonsri, B. and Hirunlabh, J. (2000b) 'Ventilation impact of a solar chimney on indoor temperature fluctuation and air change in a school building', *Energy and Buildings*, vol 32, no 1, pp89–93

Khedari, J., Rachapradit, N. and Hirunlabh, J. (2003) 'Field study of performance of solar chimney with air-conditioned building', *Energy*, vol 28, no 11, pp1099–1114

Koronakis, P. (1992) 'Solar chimney dynamic performance under typical Mediterranean summer conditions', *International Journal of Sustainable Energy*, vol 13, no 2, pp73–84

Macias, M., Mateo, A., Schuler, M. and Mitre, E. M. (2006) 'Application of night cooling concept to social housing design in dry hot climate', *Energy and Buildings*, vol 38, no 9, pp1104–1110

Marti-Herrero, J. and Heras-Celemin, M. R. (2007) 'Dynamic physical model for a solar chimney', *Solar Energy*, vol 81, no 5, pp614–622

Mathur, J., Bansal, N. K., Mathur, S., Jain, M. and Anupma (2006) 'Experimental investigations on solar chimney for room ventilation', *Solar Energy*, vol 80, no 8, pp927–935

Melo, C. (1987) 'FLOW – An algorithm for calculating air infiltration into buildings', in *Proceedings of the 3rd International Congress on Building Energy Management, ICBEM '87, Building Energy Management*, Lausanne (Switzerland), 28 September–2 October, pp5–12

Miyazaki, T., Akisawa, A. and Kashiwagi, T. (2006) 'The effects of solar chimneys on thermal load mitigation of office buildings under the Japanese climate', *Renewable Energy*, vol 31, no 7, pp987–1010

Moshfegh, B. and Sandberg, M. (1998) 'Flow and heat transfer in the air gap behind photovoltaic panels', *Renewable and Sustainable Energy Reviews*, vol 2, no 3, pp287–301

Ong, K. S. (2003) 'A mathematical model of solar chimney', *Renewable Energy*, vol 28, no 7, pp1047–1060

Ong, K. S. and Chow, C. C. (2003) 'Performance of a solar chimney', *Solar Energy*, vol 74, no 1, pp1–17

Pavlou, K. and Santamouris, M. (2007) 'Study on the thermal performance of a solar chimney', in *Proceedings of 2nd PALENC/28th AIVC Conference on Building Low Energy Cooling and Advanced Ventilation Technologies in the 21st Century*, Crete (Greece), 27–29 September, pp774–776

Rodrigues, A. M., Canha da Piedade, A., Lahellec, A. and Grandpeix, J. Y. (2000) 'Modeling natural convection in a heated vertical channel for room ventilation', *Building and Environment*, vol 35, no 5, pp455–469

Sandberg M. (1999) 'Cooling of building integrated photovoltaics by ventilation air', in *Proceedings of HybVent Forum '99*, The University of Sydney, Darlington, New South Wales (Australia), 28 September, pp10–18

Sandberg, M. and Moshfegh, B. (1996) 'Investigation of fluid flow and heat transfer in a vertical channel heated from one side by PV elements: Part II – Experimental study', *Renewable Energy*, vol 8, no 1–4, pp254–258

Santamouris, M. (ed) (2007) *Advances in Passive Cooling*, Earthscan, London

Santamouris, M. and Asimakopoulos, D. (eds), Argiriou, A., Balaras, C., Dascalaki, E., Dimoudi, A., Mantas, D. and Tselepidaki, I. (1996) *Passive Cooling of Buildings*, James & James, London

Shinada, Y., Kimura, K., Song, S. K., Katsuragi, H. and Enomoto, J. (2003) 'Field study on natural ventilation system using natural energy in a university school building and its measured results in the second year after school opening', in *Proceedings of JSES/JWEA Joint Conference* (in Japanese), 6–7 November, pp553–556

Waewsak, J., Hirunlabh, J., Khedari, J. and Shin, U. C. (2003) 'Performance evaluation of the BSRC multi-purpose bio-climatic roof', *Building and Environment*, vol 38, no 11, pp1297–1302

Ward, I. and Derradji, M. (1987) 'Air movement within buildings cooled by passive solar means', in *Proceedings of European Conference on Architecture*, CEC, Munich (Germany), 6–10 April, pp59–64

Wilson, M. and Walker, J. (1993) 'Optimization of solar chimney design', in *Proceedings of 3rd European Conference on Architecture*, Florence (Italy), 17–21 May, pp435–439

Ziskind, G., Dubovsky, V. and Letan, R. (2002) 'Ventilation by natural convection of a one-storey building', *Energy and Buildings*, vol 34, no 1, pp91–102

Zriken, Z. and Bilgen, E. (1985) 'Thermal performance of solar chimney in Northeastern African climate', *Alternative Energy Sources VII*, vol 2, Miami (FL), December, pp225–232

earthscan

publishing for a sustainable future

3

Optimization and Economics of Solar Cooling Systems

Ursula Eicker and Dirk Pietruschka

Abstract
The two main market-available thermal cooling technologies with regeneration temperatures below 100°C are evaluated in this chapter. For closed cycle absorption chillers and open desiccant cooling systems, efficiencies, costs and optimization potentials are analysed. Measurements and simulation studies from realized demonstration projects are presented. If properly designed, both technologies offer significant primary energy savings. However, as coefficients of performance (COPs) are generally lower than for electrically driven compressor chillers, care has to be taken to reduce auxiliary energy demand. While measured average thermal COPs are between 0.6 and 0.7 for absorption chillers, desiccant units can reach higher values, as they often operate with evaporative cooling only. The electrical COPs can be as high as 11 for absorption systems with efficient cold distribution and recooling units and is about 7–8 for desiccant systems with an air-based distribution system. The total costs of both desiccant and absorption cooling systems are dominated by capital costs so that high full-load hours are crucial for an economic performance.

■ *Keywords* – solar cooling; desiccant cooling; absorption cooling; ACM; control optimization

INTRODUCTION
Solar- or waste heat-driven thermal cooling plants can provide summer comfort conditions in buildings at low primary energy consumption. Today, the dominant cooling systems are electrically driven compression chillers, which have a world market share of about 90 per cent. The average coefficient of performance (COP) of installed systems is about 3.0 or lower and only the best available equipment can reach an annual average COP above 5.0. To reduce the primary energy consumption of chillers, thermal cooling systems offer interesting alternatives, especially if primary energy-neutral heat from solar thermal collectors or waste heat from cogeneration units can be used. The main technologies for thermal cooling are closed cycle absorption and adsorption machines, which use either liquids or solids for the sorption process of the refrigerant. The useful cold is in both cases produced through the evaporation of the refrigerant in exact analogy to the electric chillers.

doi:10.3763/aber.2009.0303 ■ © 2009 Earthscan ■ ISSN 1751-2549 (Print), 1756-2201 (Online) ■ www.earthscanjournals.com

For air-based cooling systems, desiccant cooling cycles are useful, as they directly condition the inlet air to the building.

Thermal cooling systems are mainly powered by waste heat or fossil fuel sources. In 2005, solar cooling systems in Europe had a total capacity of 6 MW only (Nick-Leptin, 2005).

The type of solar thermal collector required for the sorption material regeneration depends on the heating temperature level, which in closed systems is a function of both cold water and cooling water temperatures. In open sorption systems, the regeneration heating temperature depends on the required dehumidification rate, which is a function of ambient air conditions.

Commercial adsorption systems, either open air-based systems or closed adsorption, are designed for heating temperature ranges around 70°C. Single-effect (SE) absorption chillers start at operation temperatures of about 70°C. Commercial machines are often designed for average heating temperatures of 85–90°C.

For the often used single-effect machines, the ratio of cold production to input heat (COP) is in the range of 0.5–0.8. Solar fractions therefore need to be higher than about 50 per cent to start saving primary energy (Mendes et al, 1998). The exact value of the minimum solar fraction required for energy saving depends not only on the performance of the thermal chiller, but also on other components such as the cooling tower: a thermal cooling system with an energy efficient cooling tower performs better than a compression chiller at a solar fraction of 40 per cent, a low-efficiency cooling tower increases the required solar fraction to 63 per cent. These values were calculated for a thermal chiller COP of 0.7, a compressor COP of 2.5 and an electricity consumption of the cooling tower between 0.02 and 0.08 kWh electricity per kWh of cold (Henning, 2004).

Double-effect absorption cycles have considerably higher COPs around 1.1–1.4, but require significantly higher driving temperatures between 120 and 170°C (Wardono and Nelson, 1996), so that the energetic and economic performance of the solar thermal cooling system is not necessarily better (Grossmann, 2002).

In recent years, the consideration of auxiliary energy consumption of thermal chiller systems has increased in importance. Due to the low COPs, the amount of rejected heat is significantly higher than for compression chillers, which requires additional electrical energy for pumps and fans as well as increased water consumption. Complete primary energy balances have to be established to compare the efficiency of thermal cooling technologies with those of purely electrical systems.

As the focus of this work is on solar driven thermal chillers, the optimum use of the fluctuating solar energy source has to be analysed. This issue mainly concerns control strategies, which are responsible for auxiliary energy system setpoints, shortening of start-up times in the morning with rising irradiance and storage management in general. These aspects will be discussed in the chapter.

Based on measurements and simulation results from solar cooling systems, the energetic and economic performance of solar-powered absorption and desiccant cooling systems will be presented. The goals are to analyse the solar contribution to the total energy demand of the thermal chiller system and to specify the associated costs.

TECHNOLOGY OVERVIEW OF ABSORPTION COOLING

Absorption cooling is a mature technology with the first machine developed in 1859 by Ferdinand Carré. For the closed cycle process, a binary working fluid that consists of the refrigerant and an absorbent is necessary. Carré used the working fluid ammonia/water (NH_3/H_2O). Today the working pair lithium bromide as absorbent and water as the refrigerant is most commonly used for building climatization ($H_2O/LiBr$). In contrast to the ammonia/water system with its pressure levels above ambient pressure, the water/lithium bromide absorption cooling machine (ACM) works under vacuum because of the low vapour pressure of the refrigerant water.

In 1945, the company Carrier Corp, US, developed and introduced the first large commercial single-effect ACM using water/lithium bromide with a cooling power of 523 kW. In 1964, the company Kawasaki Heavy Industry Co, Japan produced the first double-effect (DE) water/lithium bromide ACM (Hartmann, 1992). The DE ACM is equipped with a second generator and condenser to increase the overall COP by reusing the high temperature input heat also for the lower temperature generator. Absorption chillers today are available in a range of 10 to 20,000 kW. In the past few years, some new developments were made in the small and medium-scale cooling range of 5 to 50 kW for water/lithium bromide and ammonia/water absorption chillers (Albers and Ziegler, 2003; Jakob et al, 2003; Storkenmaier et al, 2003; Safarik and Weidner, 2004; Jakob, 2005).

While absorption cooling has been common for decades, heat pump applications have only become relevant in recent years, due to the improvement in the performance figures; small gas-driven absorption heat pumps achieve COPs of approximately 1.5, i.e. using 1 kWh of the primary energy of gas, 1.5 kWh of heat can be produced using environmental energy, which is better than the condensing boilers presently available on the market with maximum COPs of about 1.0. Different manufacturers are producing absorption heat pumps with 10–40 kW output which achieve COPs of about 1.3 at heating temperatures of 50°C. However, today, absorption chillers are mainly used as cooling machines rather than as heat pumps.

Absorption cooling machines are categorized either by the number of effects or by the number of lifts: effects refer to the number of times high temperature input heat is used by the absorption machine. In general, increasing the number of effects is meant to increase the COP using higher driving temperature levels. Lifts refer to the number of generator/absorber pairs to successively increase the refrigerant concentration in the solution and to thus reduce the required heat input temperature level.

The most important restrictions of single-effect absorption cooling machines are the limitation of the temperature lifting through the solution field, the fixed coupling of the driving temperature with the temperature of heat source and heat sink and the COP not being larger than 1.0 independent of temperature lifting (Ziegler, 1998; Schweigler et al, 1999). The aim of multistage processes is to overcome these restrictions.

Several types of absorption cooling machines are available on the market: single-effect and double-effect ACMs with the working pair water/lithium bromide ($H_2O/LiBr$), as well as single-effect and double-lift (DL) ACMs with the working pair ammonia/water (NH_3/H_2O).

The single-effect/double-lift (SE/DL) cycle with the working pair water/lithium bromide is a novel technique. Further cycle designs such as triple-effect (TE) ACM, other multistage cycles and the use of the described cycles for solar cooling have been investigated by a number of researchers (Kimura, 1992; Lamp and Ziegler, 1997; Ziegler, 1999).

Typical performance characteristics for the closed ACM cycles described above are stated in Table 3.1 for water/lithium bromide and in Table 3.2 for ammonia/water.

HISTORY OF SOLAR COOLING WITH ABSORPTION CHILLERS

In the 1970s, the company Arkla Industries Inc, US (now owned by Robur SpA, Italy) developed the first commercial, indirectly driven, single-effect H_2O/LiBr ACM for solar cooling with two different nominal cooling capacities. The driving heat temperatures were in the range of 90°C and the cooling water temperature was 29°C for 7°C cold water temperature. The machine was installed in demonstration projects more than 100 times in the US (Loewer, 1978; Lamp and Ziegler, 1997; Grossmann, 2002). Arkla and also Carrier Corp, US then developed a small-size single-effect H_2O/LiBr ACM that could work with air cooling. There was no market success mainly due to the high investment costs for solar cooling. Carrier Corp further decreased the driving temperature of a water-cooled single-effect H_2O/LiBr ACM by using a falling film generator with a large surface area. The driving heat temperature was 82°C and the cooling water temperature was 28°C for 7°C cold water temperature (Lamp and Ziegler, 1997). The production of these ACMs was stopped and the technology's licence was given to the Japanese company Yazaki. Up to the beginning of the 1990s, the company Yazaki offered H_2O/LiBr ACMs with 5–10 kW cooling power (such as the WFC-600 with 7 kW) which were used for solar cooling projects. However, also due to low demand, the production was stopped.

At the beginning of the 1980s, Arkla developed a double-effect H_2O/LiBr ACM in which the lower temperature generator was supplied with solar energy, while in fossil mode the double-effect generator was fired using the higher COP. Due to the lack of demand on the market for solar cooling, the production of this cooling machine was

TABLE 3.1 Water lithium bromide absorption chiller characteristics

CYCLE TYPE	SE	DE	SE/DL
cold temperature / °C	6–20	6–20	6–20
heating temperature / °C	70–110	130–160	65–100
cooling water temperature / °C	30–35	30–35	30–35
COP / –	0.5–0.7	1.1–1.3	0.4–0.7

TABLE 3.2 Performance of NH_3/H_2O absorption chillers and diffusion–absorption machines with auxiliary gas

CYCLE TYPE	SE	DL	SE
auxiliary gas	–	–	hydrogen/helium
cold temperature / °C	−30–20	−50–20	−20–20
heating temperature / °C	90–160	55–65	100–140
cooling water temperature/ °C	30–50	30–35	30–50
COP / –	0.4–0.6	0.4	0.2–0.5

stopped and the technology was also licensed to the Japanese company Yazaki. They sold the machines for several years, but they are no longer available today.

In the medium-sized performance range, the most project experiences for solar cooling exist for the single-effect water/lithium bromide ACM WFC-10 from Yazaki, Japan, with a cooling power range of 35–46 kW. The operation is generally rated as dependable and unproblematic.

After the market failure of low-power systems some decades ago, there has been an increased interest in low-power absorption chillers during the past decade. A range of manufacturers – including many from Europe – now offer single-effect thermal chillers with cooling power below 10 kW. The Swedish company ClimateWell alone installed several hundred units in Spain over the past two years, while in 2007 the total number of solar cooling systems was estimated at only 200 (Preisler, 2008: case study Rococo, Arsenal Research).

There are no general rules yet for the dimensioning of solar cooling systems and planners often do not have adequate tools to determine the energy yield and solar fraction. The ratios between solar collector surface area and cooling power or storage volume and collector surface in the various demonstration projects vary strongly (Figure 3.1). Under comparable climatic conditions – Austria and Germany, for example – less than one and more than five square metres of collectors have been installed per kilowatt of cooling power.

Also the ratio of storage volume (in litres of water) to installed collector surface varies by more than a factor 20 in the different demonstration projects. While warm water solar thermal systems or heating support systems have typical storage volumes of between 50

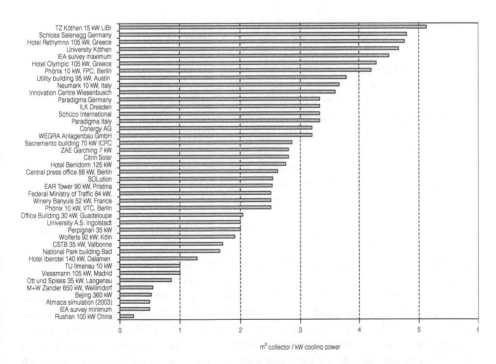

FIGURE 3.1 Collector surface per kW of cooling power in various demonstration projects

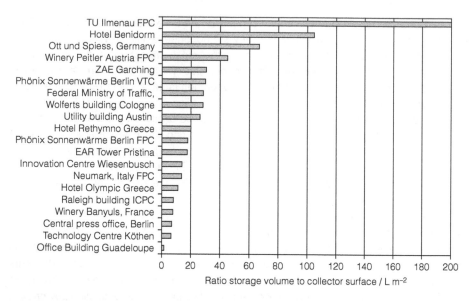

FIGURE 3.2 Ratio of hot water storage volume in litres per square metre of collector surface for different solar cooling demonstration projects

and 100 litres per square metre of collector surface, in the solar cooling projects, storage volumes are often much lower with less than 30 litres per square metre of collector. However, there are also some projects with significant storage volumes over 100 litres per square metre (Figure 3.2).

The ratio of installed collector surface to the cooled building surface spreads over a wide range of values from less than 10 per cent up to 30 per cent of the building surface area (Figure 3.3). It is clear that the solar contribution to the total cooling demand must vary significantly. Unfortunately very few published results of measured solar fractions are available today.

For some of the solar cooling systems described above, total investment costs are available. Depending on system size and technology chosen, the total investment costs vary between €1900 and €6000 per kilowatt cooling power installed (Figure 3.4).

DESIGN, PERFORMANCE AND OPTIMIZATION POTENTIALS OF SOLAR-DRIVEN ABSORPTION COOLING SYSTEMS

Dynamic simulation showed that the design and dimensioning of a solar thermal system for cooling applications depends not just on the maximum cooling load, but on the characteristics of the cooling load distribution over the year (Assilzadeh et al, 2005; Mittal et al, 2005; Eicker and Pietruschka, 2008; Bujedo et al, 2008). On the plant side, the overall performance of the ACM is significantly influenced by the control of the cooling temperature at the absorber and condenser inlet, the required cold water temperature and the heating temperature on the generator side. The solar fraction of the required heating

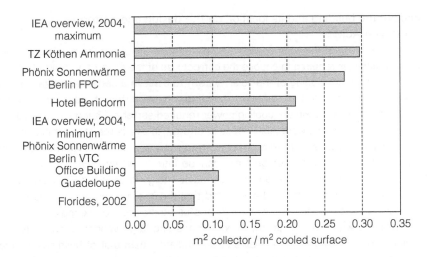

FIGURE 3.3 Ratio of installed collector surface area to building surface area in various demonstration projects

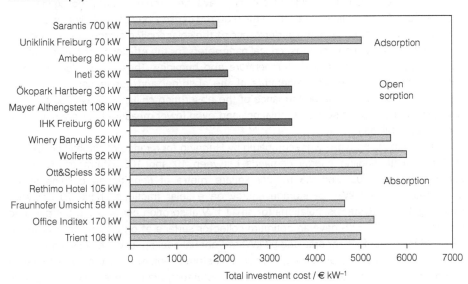

FIGURE 3.4 Total investment costs for different solar cooling systems implemented over the past decades in Europe

energy depends not just on the collector and hot storage size, but also on the specific mass flow rate of the collector field, the control of the collector pump and the charge and discharge control of the hot storage. These influencing factors on the performance of solar-driven absorption chillers are discussed separately below, starting with the influence of the building and its cooling load characteristics and location, and then dealing with the influence of the control of the solar system.

Influence of the building and its location on the design and performance of solar-driven ACM

For the dimensioning of the solar cooling system, information about the maximum cooling load of the building is required to choose the correct size of the absorption chiller, but is not sufficient for the dimensioning of the solar system to achieve a certain solar fraction. The main reason for this is that different buildings have different daily cooling load characteristics with the maximum cooling power required earlier or later during the day and the problem of a fixed timely availability of solar energy. To overcome these shifts between maximum cooling load and maximum solar irradiation or cloudy days with low solar irradiation, hot or cold water storage tanks can be introduced into the system. Another problem results from the fact that the available solar energy depends not only on the time of the day, but also on the time of the year, and that the cooling load of some buildings is not that closely related to the ambient conditions as others are. For example, the cooling load of buildings with medium to low internal loads and large window areas is typically much more closely linked to the available solar irradiation than that of buildings with low window areas or good shading protection but much higher internal loads. To reach the same solar fraction, the second type of building needs much larger collector fields and hot water storage tanks although both building types may have the same maximum cooling load.

The location has a clear influence on the required cooling load of the buildings but also has a distinctive influence on the performance of the solar cooling system. Apart from the available solar irradiation, the most critical parameters for the efficiency of the absorption cooling system are the ambient temperature and humidity. These two values strongly influence the cooling water temperature of the absorber and condenser which has a significant influence on the performance of the absorption chiller. However, so far, very few results of complete system simulations and even fewer measured performance data have been published.

Different building types were compared by Henning for a range of climatic conditions in Europe with cooling energy demands between 10 and 100 $kWhm^{-2}a^{-1}$. Collector surfaces between 0.2 and 0.3 m^2 per square metre of conditioned building space combined with 1–2 kWh of storage energy gave solar fractions above 70 per cent (Henning, 2004). System simulations for an 11 kW absorption chiller using the dynamic simulation tool TRNSYS gave an optimum collector surface of only 15 m^2 for a building with 196 m^2 of useful floor area and 90 $kWhm^{-2}$ annual cooling load, i.e. less than 0.1 m^2 per square metre of useful building floor area. A storage volume of 0.6 m^3 was found to be optimum (40 L/m^2 collector), which at 20 K useful temperature level only corresponds to 14 kWh or 0.07 kWh per square metre of useful building floor area (Florides et al, 2002). Another system simulation study (Atmaca and Yigit, 2003) considered a constant cooling load of 10.5 kW and a collector field of 50 m^2: 75 L storage volume per square metre of collector surface was found to be optimum. Larger storage volumes were detrimental to performance. Attempts have also been made to relate the installed collector surface to the installed nominal cooling power of the chillers in real project installations. The surface areas varied between 0.5 and 5 m^2kW^{-1} of cooling power with an average of 2.5 m^2kW^{-1}.

A recent simulation study discussed the influence of building loads on solar thermal system dimensioning (Eicker and Pietruschka, 2008). Here the orientation, location and internal loads of the building were varied. The peak values of the daily cooling loads are highest for an office with a large western window front and lowest for an office with the main windows orientated to the east (Figure 3.5). The phase shift between the maxima of the curves is about 2 hours with the maximum of the east- and south-orientated buildings at around 3 pm and of the west-orientated building at around 5 pm, which demonstrates the typical phase shift between the maximum solar radiation (1 pm in summertime) and the cooling load of the buildings. A wide range of specific cooling energies is covered, ranging from about 10 kWhm^{-2} for an office with low internal loads in a moderate climate up to 70 kWhm^{-2} for the same building in Madrid and high internal loads.

If a given cooling machine designed to cover the maximum load is used for different building cooling load profiles, the influence of the specific load distribution and annual cooling energy demand can be clearly seen (Figure 3.6). An office building with low internal loads and the main windows facing south requires about 2 m^{2}kW^{-1} solar thermal collector area for a solar fraction of 80 per cent at the location Madrid. If the building orientation is changed, the peak cooling power also changes. To adjust for these differences, the building surface areas were adapted to give the same maximum cooling load. The different orientations then change the required collector area by about 10–15 per cent.

More significant is the impact of the building internal loads, which strongly increases the full-load hours for a given cooling peak power. Here the required collector surface area is nearly double with 3.6 m^{2}kW^{-1}, although the required maximum power is still the same. Due to the longer operating hours of the solar thermal cooling system, the specific collector yield is 20 per cent higher and a mean solar thermal system efficiency of 45 per cent can be obtained for solar cooling operation alone. If the office building with low internal loads is placed in Stuttgart/Germany with a more moderate climate, the collector yield drops by 60 per cent to about 300 kWhm^{-2}a^{-1}.

The storage volumes are comparable to typical solar thermal systems for warm water production and heating support (between 40 and 100 L/m^2 of collector aperture area,

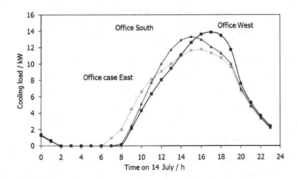

FIGURE 3.5 Hourly cooling loads for an office building with low internal loads with its main front window facing in different directions

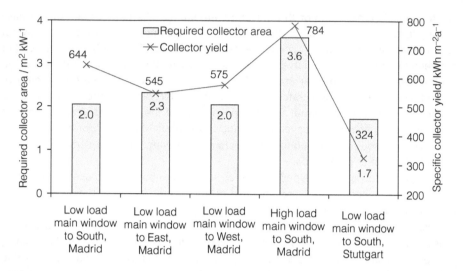

FIGURE 3.6 Required collector area per maximum cooling load and resulting collector yield for 80 per cent solar fraction

depending on the cooling load file). They increase with cooling energy demand for a given location. In moderate climates with only occasional cooling energy demand, the storage volumes are generally higher.

For the location Madrid, the collector surface area required to cover 1 MWh of cooling energy demand is between 1.8 and 2.7 m² per MWh, depending on building orientation chosen. The lower the cooling energy demand, the higher the required collector aperture area per MWh. This is very clear for the building in Stuttgart with a low total energy demand of about 10 kWhm⁻²a⁻¹, where 5.5 m² solar thermal collector apperture area per MWh is necessary to cover the energy demand (Figure 3.7).

The work shows that dynamic system simulations of the building and solar cooling system so far are the only reliable possible way to determine the correct solar thermal system size to reach a certain solar fraction of the total energy requirement.

Influence of system configuration and control strategy

Apart from the cooling load characteristic of the buildings, the design and performance of solar-driven absorption cooling systems depend on the chosen system configuration and control including the chillers, the cooling tower, the installed cooling distribution system and the solar heating system (Sumath, 2003; Henning, 2004; Kohlenbach, 2007). A cold distribution with high cold water temperatures, i.e. in chilled suspended ceilings or thermally activated concrete structures, allows the absorption chiller to work on a much higher COP. Less collector area is therefore required to achieve the same solar fraction than in the case of air-based distribution systems with low cold water temperatures.

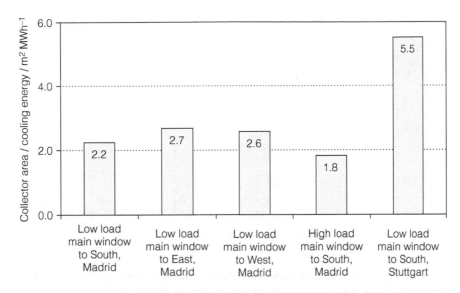

FIGURE 3.7 Required collector area for 80 per cent solar fraction related to the annual cooling energy demand

The decision for a wet or dry cooling tower/recooler has a similar but even more significant effect on the required collector area since the cooling water temperature strongly influences the thermal COP of the absorption chiller. Some of the installed systems operate at a constant high generator temperature which results in a slightly higher thermal COP but lower solar fractions compared with systems where the generator inlet temperature is allowed to vary according to the temperature level available in the hot storage tank. Different control options are summarized in Table 3.3 and the results of the analysed cases are shown in Figures 3.8 and 3.9. If a wet cooling tower is used, the required collector surface area can be reduced by about 20 per cent to 1.8 m² per kW, if the cold water temperature level is high and a variable generator inlet temperature is used

TABLE 3.3 Analysed system configurations and control options for a low internal load building

ANALYSED CASES	CONTROL OPTIONS						
	COOLING TOWER		COLD DISTRIBUTION TEMPERATURE		GENERATOR INLET TEMPERATURE	
	Type	Setpoint cooling water	6°C/ 12°C	15°C/ 21°C	85°C constant	70–95°C variable
Case 1	wet	27	x		x	
Case 2	wet	27	x			x
Case 3	wet	27		x		x
Case 4	dry	27		x		x

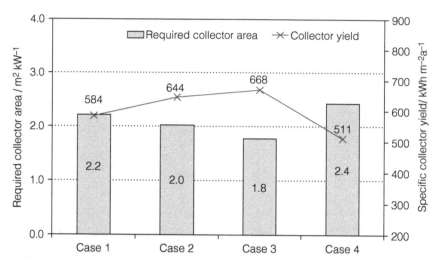

FIGURE 3.8 Required collector area related to maximum cooling load and resulting collector yield for 80 per cent solar fraction (office building in Madrid, Spain with low internal loads)

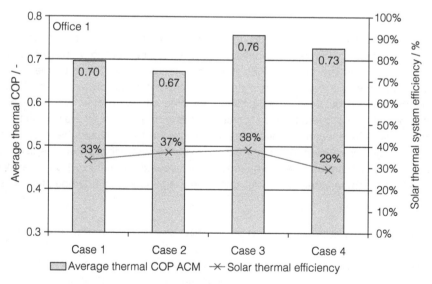

FIGURE 3.9 Average annual COP and solar thermal system efficiency for different system configurations and control strategies

(case 3). If dry recooling is used at the same cold water and generator temperature levels, this leads to 33 per cent more collector surface area (2.4 m² per kW cooling power). In addition, the solar thermal system efficiency is highest at nearly 40 per cent for high cold water temperature levels and variable generator inlet temperatures.

Auxiliary energy consumption

High electricity consumption caused by inefficient pumps, bad hydraulic design and suboptimal control in combination with low solar fractions through insufficient system design are critical for the environmental and economic performance of installed absorption cooling systems, especially if they are compared with highly efficient electrically driven compression chillers (Kohlenbach, 2007). Most of the recent publications on the performance of solar cooling systems focus on the thermal COP and the solar fraction reached. The electricity consumption of the absorption chiller, the cooling tower and all pumps are not discussed in detail. However, since this technology is now more and more leaving the prototype and demonstration status towards a proven technology which needs to compete with highly efficient electric compression chillers, the awareness of the importance of this issue is increasing significantly. Within TASK 38, therefore, standardized monitoring requirements are currently developed to overcome the problem of missing electricity and heat meters in demonstration installations. (The IEA SHC Task 38 'Solar Air-Conditioning and Refrigeration' is a four-year task initiated by the International Energy Agency Solar Heating and Cooling Programme.) In parallel, these monitoring guidelines are applied to a large number of demonstration installations to get reliable monitoring data of existing systems, which are not yet available.

To demonstrate the effect of different control strategies on the overall performance of solar-driven absorption chillers, the results of a recent study (Pietruschka et al, 2007) on the performance of a solar-driven absorption cooling system installed in an office building in Rimsting, Germany are discussed. This system includes a market-available 15 kW LiBr absorption chiller, two 1 m³ hot water storage tanks, one 1 m³ cold storage tank, 37 m² flat plate collectors and 34 m² solar vacuum tube collectors, all facing south with an inclination of 30°, a 35 kW wet recooling tower and an additional dry recooling system (Figure 3.10). For the distribution of the cooling energy, chilled ceilings and fan coils are used with 16°C supply

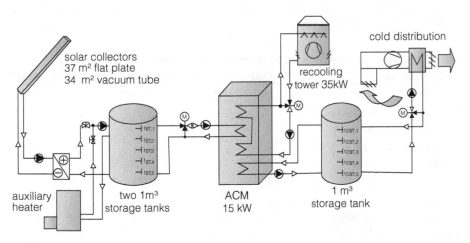

FIGURE 3.10 Schematic drawing of a solar-driven absorption cooling system used for control strategy optimization

and 18°C return temperature and an automated supply temperature increase for dew point protection. The office building with 566 m² of conditioned space has a maximum cooling load of 18 kW and a total annual cooling energy demand of 8.7 MWh/a (16 kWh/m²a).

The installed absorption chiller is able to provide the required cooling power of 18 kW if the setpoint of the recooling temperature is reduced by 3 K from 30°C design conditions to 27°C. For the analysis of different control strategies on the overall performance of the installed system, a detailed dynamic simulation model developed in INSEL (Schumacher, 1991; INSEL, 2008) was used, which also considers the electricity consumption of all installed components (fans, pumps, etc.). (INSEL is a modular simulation environment for all kinds of renewable energy systems.) Five main cases with different control options have been analysed (see Table 3.4). Three of them operate with wet cooling towers, cases 4 and 5 with dry recooling.

The setpoint of the recooling water temperature is either controlled by a three-way valve or by fan speed control of the cooling tower. Values below 27°C (30°C for a dry cooling tower) are only provided as long as reachable at the given ambient conditions with the chosen recooling technology. The generator inlet temperature is either constant or variable according to the temperatures in the hot and cold storage tanks.

An additional case 6 is defined and analysed as a reference system for a compression chiller using manufacturer's data with an electrical COP of 4.0 at 27°C recooling temperature. The recooling is done with a dry recooler with constant fan speed control, which tries to keep the 27°C if possible.

For a detailed analysis of the system performance, annual simulations were carried out for the regarded cases with different control options as defined in Table 3.4 and for the reference compression chiller system of case 6. The main results are summarized in three graphs shown in Figures 3.11, 3.12 and 3.13. To compare the efficiency of the system with varied control strategies, three different COPs are used:

1 the standard thermal COP_{th} which is defined as the cooling energy produced divided by the heating energy used;

TABLE 3.4 Analysed control options of the absorption cooling system

ANALYSED CASES		CASE 1	CASE 2	CASE 3	CASE 3.1	CASE 3.2	CASE 4	CASE 5
Cooling tower	type	wet	wet	wet	wet	wet	dry	dry
Cooling water temp. control	3-way-valve	x					x	
	fan speed		x	x	x	x		x
Cooling water setpoint	27°C	x	x	x			x	x
	24°C				x			
	21°C		x			x		
Generator temperature setpoint	90°C	x	x					
	70–90°C			x	x	x		
	70–95°C						x	x
Control dist. pump	yes		x	x	x	x	x	x
ΔT-control	no	x						

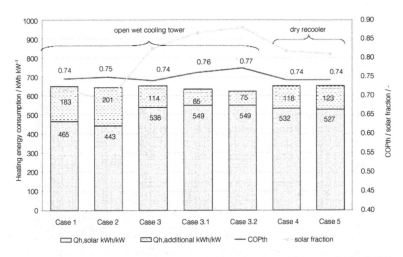

FIGURE 3.11 Comparison of heating energy consumption related to maximum cooling load, COP$_{th}$ and solar fraction

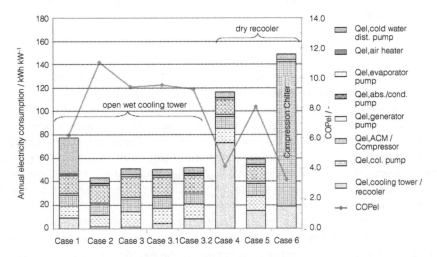

FIGURE 3.12 Comparison of electricity consumption related to maximum cooling load and electrical COP for different control strategies and system temperatures

2 the total electrical COP$_{el}$ which is simply the cooling energy produced divided by the total electricity consumption of the solar cooling system;

3 the total primary energy (PE)-related COP which is defined as the provided cooling power divided by the sum of consumed electricity and auxiliary thermal energy multiplied by the PE factors of 3.0 for electricity and 1.1 for a gas boiler:

$$COP_{PE,total} = Q_{cool}/(3Q_{el} + 1.1Q_{h,aux}) \qquad [3.1]$$

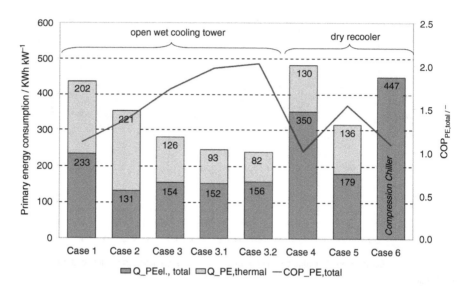

FIGURE 3.13 Comparison of primary energy consumption related to maximum cooling power and primary energy-based COP

The average thermal COP is in most cases around 0.74 and can be slightly increased to 0.77 by reducing the recooling temperature setpoint (Figure 3.11). Therefore the required heating energy is nearly constant for all analysed cases at a value of slightly below 650 kWh per kW maximum cooling load. A more significant influence of the different control strategies is visible for the solar fraction reached. Cases with constant generator inlet temperature reach the lowest solar fractions of around 70 per cent. If the generator inlet temperature is allowed to vary according to the temperature in the hot storage tank and the required cooling load, this value can be increased up to 88 per cent for the case with the lowest recooling temperature setpoint.

The electricity consumption strongly depends on whether the cold water distribution pump of the building and the ventilator of the cooling tower are controlled according to the load or not. For the absorption cooling system, the electricity consumption related to the maximum cooling load varies between 42 and 120 kWhkW⁻¹ with the highest value for systems with dry recoolers without fan speed control (Figure 3.12). The electricity consumption of these systems is quite close to the electricity consumption of a system with a compression chiller which consumes 150 kWhkW⁻¹. Highly efficient systems with a controlled distribution pump and wet cooling towers with fan speed control consume only 50 kWhkW⁻¹ which is one third of that consumed by a system with a compression chiller. The electrical COPs vary between 6.5 and 11 for the cases with a wet cooling tower and between 4.2 and 8 for the cases with a dry cooling tower. The compression chiller reaches an overall electrical COP of 3.6 which is only slightly below the worst absorption cooling system.

If the primary energy-related COP and the primary energy consumption are analysed (Figure 3.13), it becomes obvious that badly controlled absorption cooling systems do not save any primary energy compared with compression chillers. Only systems with highly efficient pumps, cooling towers with fan speed control and ΔT-mass flow control of the cold distribution pump are able to significantly save primary energy. The primary energy-related COP of these systems is between 1.5 and 2 which means that in the best case nearly half of the primary energy consumption of a system with a compression chiller can be saved with well-designed and well-controlled solar cooling systems.

To avoid additional heating energy consumption and to increase the primary energy savings, some systems are designed for purely solar-driven operation. However, these systems are either designed with very large collector areas and huge hot storage volumes or an auxiliary electrical cooling system has to be used. To start the operation of absorption chillers, a certain temperature level needs to be available either in the hot storage tank or in the collector circuit in the case of direct solar-driven systems without a hot storage tank. Large hot water storage tanks require time to be heated up in the morning, therefore such systems suffer from a late start-up of the absorption chiller in the morning. Direct solar-driven systems without hot water storage only need to heat up the collector circuit and therefore can start the absorption chiller operation much earlier. On the other hand, these systems often have the problem of unstable operation of the absorption chiller on cloudy days caused by system shutdowns and start-ups due to insufficient heat supply. To combine the advantages of both systems, the hot storage tank can either be partitioned in a smaller upper part and a bigger lower part or bypassed completely. Both possibilities require the implementation of intelligent storage charge and discharge control strategies. In a recent study (Pietruschka et al, 2008) the solar cooling system described above was considered to be purely solar driven and analysed for different storage charge and discharge strategies. The results show that through the implementation of a storage bypass, the start-up time can be significantly reduced by 1 hour 40 minutes in the worst case and nearly 2 hours in the best case. For the partitioned storage case (300 litre top and 1700 litre bottom volume), the start-up time is between 12 and 20 minutes later. On a cloudless summer day, the start-up of the solar cooling system is at 9.23 am (bypass) or 9.43 am (partitioned storage), instead of 11.04 am (full storage). The overall best performance is reached for a system with combined bypass and partitioned storage control with system start-up at 9.23 am and stable system operation on cloudy days. Compared with a system that always uses the full storage volume, the combined system with partitioned storage and storage bypass can produce between 19 per cent (cloudless days) and 33 per cent (cloudy days) more cooling energy per day.

TECHNOLOGY OVERVIEW: DESICCANT COOLING

Desiccant cooling systems are an interesting technology for sustainable building climatization, as the main required energy is low temperature heat, which can be supplied by solar thermal energy or waste heat. Desiccant processes in ventilation mode use fresh air only, which is dried, precooled and humidified to provide inlet air at temperature levels between 16–19°C. The complete process is shown in Figure 3.14 with the fresh air side and the exhaust air side. Outside air (1) is dried in the sorption wheel (2), precooled in the heat

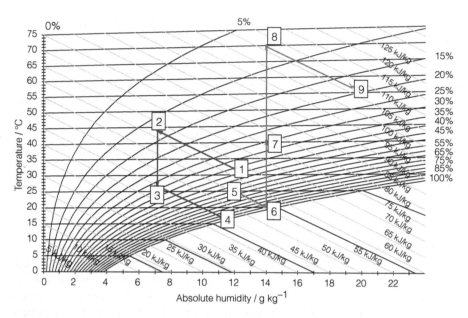

FIGURE 3.14 Process steps of a desiccant cooling system in an hx-diagram

recovery device with the additionally humidified cool space exhaust air (3) and afterwards brought to the desired supply air status by evaporative cooling (4). The space exhaust air (5) is maximally humidified by evaporative cooling (6) and warmed in the heat recovery device by the dry supply air (7). In the regeneration air heater, the exhaust air is brought to the necessary regeneration temperature (8), takes up the water adsorbed on the supply air side in the sorption wheel, and is expelled as warm, humid exhaust air (9). If room air is recirculated, the desiccant wheel is used to dry the room exhaust air, which is then precooled using the rotating heat exchanger and humidified to provide the cooling effect. Regeneration of the desiccant wheel and precooling of the dried recirculation air is done by ambient air, which is first humidified, then passes the rotating heat exchanger, is heated to the necessary regeneration temperature and finally used to regenerate the desiccant wheel.

The concept of desiccant cooling was developed in the 1930s and early attempts to commercialize the system were carried out unsuccessfully. Pennington patented the first desiccant cooling cycle (Pennington, 1955), which was then improved by Carl Munters in the 1960s (Munters, 1960). Good technology overviews are given by Mei et al (1992) or Davanagere et al (1999). The most widely used desiccants are silica gel, lithium chloride or molecular sieves, for example zeolites. Solid desiccants such as silica gel adsorb water in its highly porous structure. Lithium chloride solution is used to impregnate, for example, a cellulose matrix or simpler cloth-based constructions and can then be used to absorb water vapour from the air stream (Hamed et al, 2005).

The thermal COP is defined by the cold produced divided by the regeneration heat required. For the hygienically needed fresh air supply, the enthalpy difference between ambient air and room supply air can be considered as useful cooling energy. If the building has higher cooling loads than can be covered by the required fresh air supply, then the useful cooling energy has to be calculated from the enthalpy difference between room exhaust and supply air, which is mostly lower. The thermal COP is obtained from the ratio of enthalpy differences (state points are given in brackets):

$$COP_{thermal} = \frac{q_{cool}}{q_{heat}} = \frac{h_{amb(1)} - h_{supply(4)}}{h_{waste(9)} - h_{reg(8)}}$$ [3.2]

Related to ambient air, COPs can be near to 1.0 if regeneration temperatures are kept low, and reduces to 0.5 if the ambient air has to be significantly dehumidified. COPs obtained from room exhaust to supply air are lower, between 0.35 and 0.55 (Eicker, 2003). The maximum COP of any heat-driven cooling cycle is given for a process in which the heat is transferred to a Carnot engine and the work output from the Carnot engine is supplied to a Carnot refrigerator. For driving temperatures of 70°C, ambient air temperatures of 32°C and room temperatures of 26°C, the Carnot COP is 5.5.

$$COP_{Carnot} = \left(1 - \frac{T_{ambient}}{T_{heat}}\right)\left(\frac{T_{room}}{T_{amb} - T_{room}}\right)$$ [3.3]

However, as the desiccant cycle is an open cycle with mass transfer of air and water, several authors suggested to rather use a reversible COP as the upper limit of the desiccant cycle, which is calculated from Carnot temperatures given by the enthalpy differences divided by entropy differences (Lavan et al, 1982). This allows the calculation of equivalent temperatures for the heat source (before (7) and after the regenerator heater (8)), evaporator (room exhaust (6) minus supply (4)) and condenser (waste air (9) minus ambient (1)):

$$T_{equivalent} = \frac{\sum m_{in}h_{in} - \sum m_{out}h_{out}}{\sum m_{in}s_{in} - \sum m_{out}s_{out}}$$ [3.4]

Under conditions of 35°C ambient air temperature and 40 per cent relative humidity, defined by the US Air-Conditioning and Refrigeration Institute (so called ARI conditions), reversible COPs of 2.6 and 3.0 were calculated for ventilation and recirculation mode (Kanoğlu et al, 2007). The real process, however, has highly irreversible features such as adiabatic humidification. Furthermore, the specific heat capacity of the desiccant rotor increases the heat input required.

An effective heat exchange is crucial for the process between the dried fresh air (state 2) and the humidified exhaust air (state 6), as the outside air is dried at best in an isenthalpic process and is warmed up by the heat of adsorption. For a rather high heat exchanger efficiency of 85 per cent, high humidification efficiencies of 95 per cent and a dehumidification efficiency of 80 per cent, the inlet air can be cooled from design condition of 32°C and 40 per cent relative humidity to below 16°C (Figure 3.15).

Simple models have been used to estimate the working range of desiccant cooling systems, for example to provide room conditions not just for one setpoint, but for a range of acceptable comfort conditions (Panaras et al, 2007). The performance of the desiccant

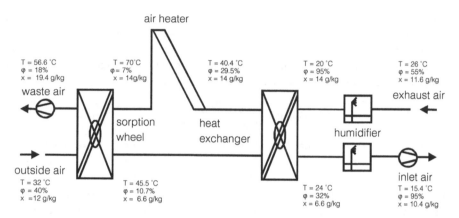

FIGURE 3.15 Temperature and humidity levels in a high-performance desiccant cooling process

rotor itself can be evaluated by complex heat and mass transfer models based on Navier-Stokes equations (Maclaine-Cross, 1988; Gao et al, 2005). This allows the evaluation of the influence of flow channel geometry, sorption material thickness, heat capacity, rotational speed, fluid velocity, etc. They are mostly too time consuming to be used in full system simulations, including solar thermal collectors, where mostly simpler models are available based either on empirical fits to measured data (Beccali et al, 2003) or on models of dehumidification efficiency.

OPERATING EXPERIENCES DESICCANT COOLING

There are only a few desiccant plants implemented that are powered by solar energy, so operational experience from such plants is very scarce. A common conclusion from three systems driven by solar air collectors in Germany and Spain (IHK building in Freiburg/Germany, Factory Maier in Althengstett/Germany and Public Library in Mataró/Spain) was that the solar collector yield during the summer period was below 100 kWhm^{-2}a^{-1} and thus rather low. This is not due to the efficiency of the solar air collectors, which is quite high, around 50 per cent for temperature levels of 70°C, but rather to the low number of hours during which regeneration is really needed: if the ambient air is not very humid and temperature levels moderate, the pure evaporative cooling effect is sufficient to cool the ambient air. A detailed analysis of the Maier factory building in Germany showed that out of 421 operation hours of the system in the month of July, only 83 hours needed full regeneration mode, while during 90 hours pure evaporative cooling was sufficient. The dominant operation mode even in July was free cooling with 178 operating hours (Figure 3.16).

From the 83 hours of required regeneration, the collector provided energy during 53 hours, i.e. during 64 per cent of the time and produced a total of 11 kWhm^{-2}month^{-1}, which is about the same amount as the waste heat delivered from the factory machinery. The mean collector efficiency during the operation hours is 48 per cent, which is a good performance value (Figure 3.17).

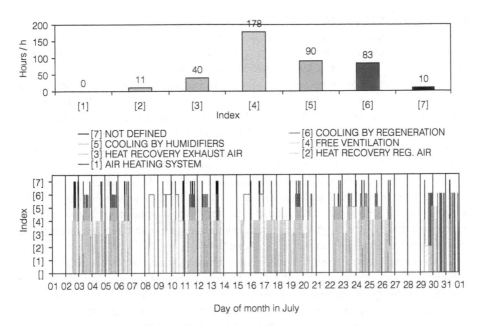

FIGURE 3.16 Operation hours of different DEC modes for all days in July for the system of the Maier factory in Althengstett, Germany

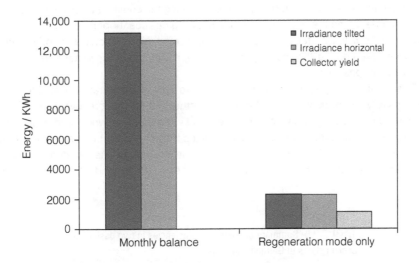

FIGURE 3.17 Measured irradiance on collector surface and horizontal plane during July and energy yield during regeneration hours only in the Maier factory building

COMPONENT AND CONTROL STRATEGY ANALYSIS

To reach the design supply air temperatures in desiccant cooling plants, the component performance must be excellent. The measurements in two demonstration plants in Althengstett/Germany and Mataró/Spain showed that heat recovery efficiencies of rotating heat exchangers were only between 62 and 68 per cent at rotation rates of 600 turns per hour. At a measured mass flow ratio between supply and exhaust air of 1.16, the manufacturer value given was 73 per cent. As this component is crucial for the precooling of dried process air, better heat recovery efficiencies should be achieved.

The simple contact evaporators reach 85–86 per cent humidification efficiency, compared with 92 per cent given by the manufacturer. The dehumidification efficiency of sorption rotors, which is defined as the measured absolute dehumidification compared with the maximum possible dehumidification down to the relative humidity of the regeneration air, is 80 per cent at a ratio of regeneration to process air of 66 per cent. The component parameters from the experimental analysis of the two desiccant cooling units in Spain and Germany are summarized in Table 3.5.

Different control strategies have been compared by Ginestet et al (2003) to study the influence of air volume flow and regeneration temperature. As the increase of regeneration temperature does not linearly lower the supply air temperature, the study concluded that increased air flow rates are preferable to increased thermal input, if the cooling demand is high. Mean calculated COPs for the climatic conditions of Nice were between 0.3 and 0.4. Henning et al (1999) also remarked that increasing the air flow is useful in desiccant cooling mode, but that the minimum acceptable flow rate should be used in adiabatic cooling or free ventilation mode to reduce electricity consumption. When thermal collectors with liquid heat carriers are used in combination with a buffer storage, Bourdoukan et al (2007) suggested operating the cooling system only in adiabatic cooling mode during the morning and then allowing desiccant operation in the afternoon, using heat from the buffer storage. However, in many applications, dehumidification is already required during the morning hours. Also, if cheaper air collectors are used, heat storage is not possible.

Desiccant cooling system controllers operate with a temperature cascade system, where usually the exhaust air humidification together with the heat recovery wheel are first switched on, then the supply air humidifier, finally the sorption wheel and if needed the auxiliary energy system. The heat recovery losses from the exhaust air side are avoided if the supply air humidifier is switched on first (for example in the Althengstett project). However, as the room air humidity cannot exceed set values (around 10–12 g/kg), the direct

TABLE 3.5 Summary of measured efficiencies in two desiccant cooling plants in Germany and Spain

	MATARÓ	ALTHENGSTETT
Dehumidification efficiency	80	80
Humidification efficiency	86	85
Heat recovery efficiency	68	62

humidification of supply air has limits. The regeneration air temperature has to be limited for certain technologies to avoid damage to the sorption rotor. For example, for a cellulose rotor with LiCl solution, the maximum allowed regeneration temperature is 72°C.

Each component has specific time delays between the control signal (release) and showing effect in operation. Simple contact humidifiers need about 5–10 minutes after release to reduce the air temperature at the outlet significantly and the cooling effect does not stop until 45 minutes after the release is switched off (Figure 3.18).

This dynamic and partially slow response of the components is often not adequately considered in the control strategy of desiccant cooling plants. For example, supply air temperatures in one of the German plants was shown to fluctuate by about 6 K, when supply air humidifiers and heat recovery strategies work against each other. Full humidification of the three-stage supply air humidifier leads to a temperature decrease of the supply air below the minimum allowed temperature of 17°C in less than 30 minutes (Figure 3.19).

The humidifiers are consequently switched off and the rotational speed of the sorption rotor is increased to raise the temperature level in heat recovery mode. Furthermore, the rotating heat exchanger is switched on too early (about 10 minutes before the exhaust air humidifier), so that supply air temperatures increase well above 20°C and then drop again below 15°C. These control problems can be avoided if the response time of the humidifiers is taken into account and if the sorption wheel is not used in heat recovery mode for compensation of too low temperatures. Furthermore, the rotating heat exchanger should only switch on together with or 5 minutes after the exhaust air humidifier.

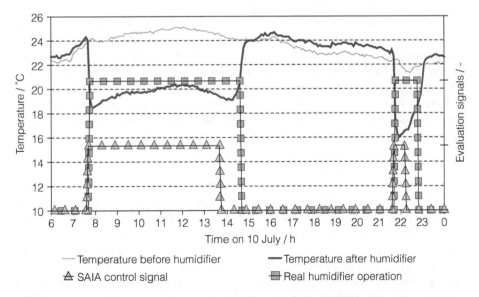

FIGURE 3.18 Time delay between the control signal and the reaction of the humidifier together with the signal representing the real operation of the humidifier, when the temperature difference is at least 3 K

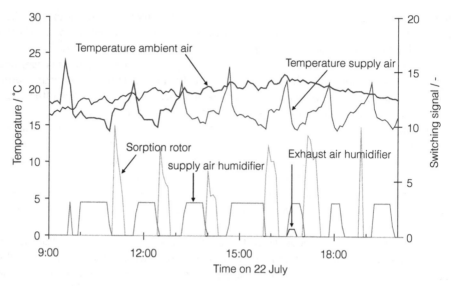

FIGURE 3.19 Fluctuations of supply air temperature due to control strategy

Fixed high temperature levels for regeneration lead to a high auxiliary energy demand. If temperature setpoints are fixed, auxiliary energy for heating will always be used during the mornings, when solar thermal energy is not yet available at sufficient temperature levels. With air-based systems without thermal storage this is difficult to avoid. It is therefore advisable to operate the desiccant mode with low purely solar-based regeneration temperatures and admit lower dehumidifaction rates. Temporary increases of room humidity could then be reduced when solar energy is available. To use the solar cooling energy also for long operation hours until midnight would only be possible if the storage mass of the building itself could be used. In conclusion it can be stated that in order to achieve higher solar fractions, the control strategy of the desiccant system must be adapted more to the solar air collector system, allowing full regeneration operation whenever solar irradiance is available, and using the building's heat and humidity storage capacity to cool and dry it down more than required by static setpoints.

ENERGY ANALYSIS AND COPs

The COP of the desiccant cooling process strongly depends on the ambient boundary conditions: if the dehumidification required is low, the regeneration temperatures can also be low and the COP is high. In cases where just evaporative cooling is sufficient, the thermal COP is infinite. Typical average COPs over a summer period are near 1.0. If the heating energy is related only to the hours with full desiccant operation, the COP is on average around 0.5. Full desiccant cooling operation is mainly required in the summer months, when the humidification process alone is not sufficient for exterior air cooling. These results were obtained from one year of detailed monitoring results of the Maier factory building (Figure 3.20).

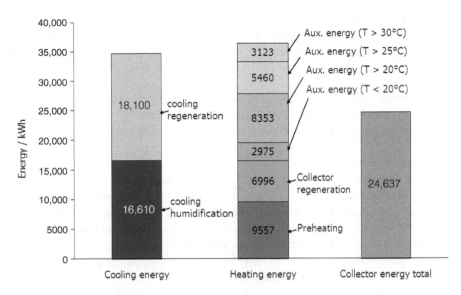

FIGURE 3.20 Summer cooling energy produced in the Maier factory building and heating energy for different ambient air temperature ranges. The collector energy is also used for the heating of old factory buildings

Under German climatic conditions, about half of the cooling energy is usually obtained with full desiccant operation, the rest of the operating time only humidification is sufficient. An interesting fact is that auxiliary energy is often necessary when outside temperatures are rather low (below 25°C). Although the temperature reduction between outside air and inlet air is small for these conditions, the dehumidification process requires a high amount of energy. Auxiliary cooling would be preferable under such conditions. If that is not possible, the controller must ensure that the regeneration air temperature is kept at its minimum and not set to fixed values.

The thermal auxiliary energy demand obviously depends on the dimensions of the solar collector field and the climatic boundary conditions. As regeneration air heating is very energy intensive with sometimes little cooling effect, it is recommended to use auxiliary cooling systems. If the electrical COPs are analysed, the additional pressure drops across the humidifiers, sorption wheel and air collector fields have to be considered. Compared with a conventional air-conditioning system, this means for the Althengstett case about 20 per cent higher electricity consumption.

Related to the produced cold, the additional electricity demand corresponds to an average electrical COP of 7.5. If the total electrical energy consumption for an air-based distribution system is considered, the ratio of produced cold to total electricity demand is much lower and typically below 2.0. This is a consequence of an air-based cooling distribution system. The situation can only be improved if variable volume flows are possible at partial load conditions.

COST EVALUATION

To plan and design energy systems such as solar cooling systems, economic considerations form the basis for decision making. The costs in energy economics can be divided in three categories:

1 capital costs, which contain the initial investment including installation;
2 operating costs for maintenance and system operation; and
3 the costs for energy and other material inputs into the system.

The analysis presented here is based on the annuity method, where all cash flows connected with the solar cooling installation are converted into a series of annual payments of equal amounts. The annuity, a, is obtained by first calculating the net present value of all costs occurring at different times during the project, i.e. by discounting all costs to the time t = 0, when the investment takes place. The initial investment costs, P (t = 0), as well as further investments for component exchange in further years P(t) result in a capital value, CV, of the investment, which is calculated using the inflation rate, f, and the discount or basic interest rate, d:

$$CV = \sum P(t) \frac{(1 + f)^t}{(1+d)^t} \qquad [3.5]$$

Annual expenses for maintenance and plant operation, EX, which occur regularly during the lifetime, N, of the plant, are discounted to the present value by multiplication of the expenses with the present value factor, PVF . The system lifetime was set to 15 years.

$$PVF(N,f,d) = \frac{1+f}{d-f}\left[1-\left(\frac{1+f}{d-f}\right)^N\right] \qquad [3.6]$$

In the case of solar cooling plants, no annual income is generated, so that the net present value, NPV, is simply obtained from the sum of discounted investment costs, CV, and the discounted annual expenses. It is here defined with a positive sign to obtain positive annuity values:

$$NPV = CV + EX \times PVF(N,f,d) \qquad [3.7]$$

Annual expenses include the maintenance costs and the operating energy and water costs. For maintenance costs, some standards, like VDI 2067, use 2 per cent of the investment costs. Some chiller manufacturers calculate maintenance contracts with 1 per cent of the investment costs. For large absorption chillers, some companies offer constant cost maintenance and repair contracts: the costs vary between 0.5 per cent for large machines (up to 700 kW) up to 3 per cent for smaller power. Repair contracts are even more expensive with 2 per cent for larger machines up to 12 per cent for a 100 kW machine. In the calculations shown here, 2 per cent maintenance costs are used.

To obtain the annuity, a, as the annually occurring costs, the NPV is multiplied by a recovery factor, r_f, which is calculated from a given discount rate, d, and the lifetime of the plant, N. The cost per kWh of cold is the ratio of the annuity divided by the annual cooling energy produced.

$$a = NPV \times r_f(N, d) = NPV \times \frac{d(1+d)^N}{(1+d)^N - 1}$$ [3.8]

COST AND ECONOMICS OF ABSORPTION CHILLERS

An own market study gives an overview of the investment costs for the chillers (Figure 3.21). In addition to the chiller investment costs, the annuity of the solar thermal system depends on the surface area of the collector field, the storage costs and the costs for system integration and mounting. Cost information for the solar thermal collectors and storage volumes were obtained from a German database for small collector systems, from the German funding programme Solarthermie 2000 for flat plate collector surface areas above 100 m² and for vacuum tube collectors from different German distributors (Figures 3.22 and 3.23).

Maintenance costs and the operating costs for electric pumps were set to 2 per cent. Major unknowns are the system integration and installation costs, which depend a lot on the building situation, the connection to the auxiliary heating or cooling system, the type of cooling distribution system and so on. Due to the small number of installations, it is difficult to obtain reliable information about installation and system integration costs. Two prices were calculated: the first with very low installation costs of 5 per cent of total investment plus 12 per cent for system integration. A second price is based on 25 per cent installation and 20 per cent system integration costs.

The total costs per MWh of cold produced, C_{total}, are obtained by summing the chiller cost, $C_{chiller}$, the solar costs, C_{solar}, the auxiliary heating costs, C_{aux}, and the costs for the

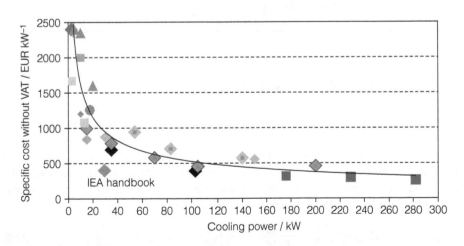

Note: The symbols represent data from different manufacturers. In addition, the IEA handbook quotes average investment costs of absorption chillers (Henning, 2004).

FIGURE 3.21 Investment costs of low-power thermal absorption chillers

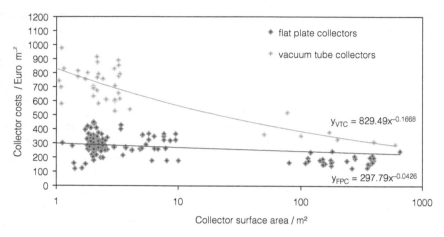

FIGURE 3.22 Specific collector costs without VAT as a function of size of the installation

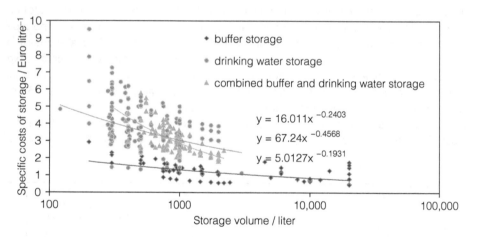

FIGURE 3.23 Costs for different storage tank sizes and systems

cooling water production, $C_{cooling}$. The costs for heating have to be divided by the average COP of the system to relate the cost per MWh heat to the cold production and multiplied by the solar fraction, s_f, for the respective contributions of solar and auxiliary heating. For the cooling water, the costs per MWh of cooling water were taken from literature (Gassel, 2004) and referred to the MWh of cold by multiplication with (1 + 1/COP) for removing the evaporator heat (factor 1) and the generator heat with a factor 1/COP.

$$C_{total} = C_{chiller} + \frac{s_f C_{solar}}{COP} + \frac{(1-s_f)C_{aux}}{COP} + C_{cooling}\left(1 + \frac{1}{COP}\right) \qquad [3.9]$$

A value for $C_{cooling}$ of €9 per MWh cooling water was used and the auxiliary heating costs $C_{heating}$ were set to €50 per MWh of auxiliary heat.

The chosen system technology (dry or wet chiller, low or high temperature distribution system, control strategy) influences the costs only slightly (7 per cent difference between the options), if the operating hours are low (Figure 3.24). If the operating hours increase, the advantage of improving the control strategy or increasing the temperature levels of the cooling distribution system become more pronounced (16 per cent difference between the different cases, see Figure 3.25).

For very low operating hours such as an office building in the Stuttgart climate with low internal loads (only about 300 full-load hours), the costs are between €790 and €860 per MWh, 60 per cent of which are due to the chiller investment costs only (Figure 3.26).

The calculated solar thermal energy prices are between €85 and €258 per MWh for solar cooling applications, depending on the operating hours and the location. They go down as far as €76 per MWh for an office in Madrid with high internal loads and a high temperature cooling distribution system. These costs are getting close to economic operation compared with fossil fuel heating supply.

Depending on the control strategy and building type, the total costs for a solar thermal absorption cooling system in a Southern European location such as Madrid/Spain are between €170–320 per MWh. If the system integration and mounting costs are assumed to be 45 per cent of total investment costs instead of 17 per cent, the costs per MWh of cold are in the range of €300–390 per MWh for an office in Madrid with low internal loads and €200 per MWh for the best case of an office with longer operating hours (Figure 3.27).

In the case of an absorption chiller connected to a district heating system, heat prices can be lowered while the costs associated to the system components (mostly chiller and

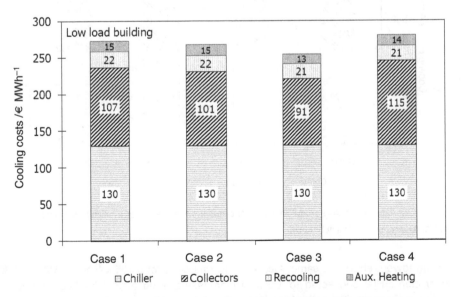

FIGURE 3.24 Cooling costs per MWh of cold for different system technology options and control strategies. Low load building in Madrid

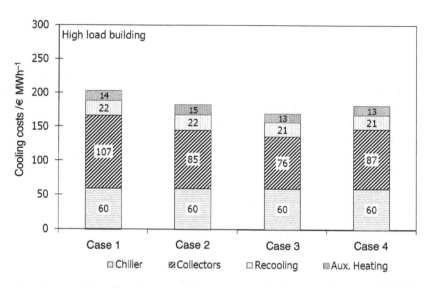

FIGURE 3.25 Cooling costs for different operation strategies and cooling distribution systems for the office with high internal loads. High load building in Madrid

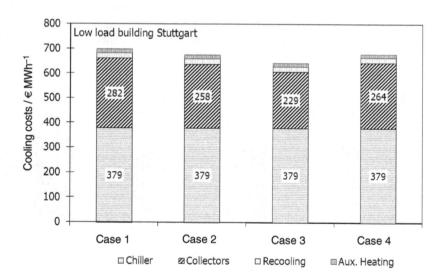

FIGURE 3.26 Cooling costs per MWh of cold for different system technology options and control strategies. Low load building in Stuttgart

cooling tower) remain the same. A case study was done for a 105 kW Yazaki WFC-SC 30 absorption chiller connected to a 6 MW biomass-powered district heating network in Ostfildern/Germany. The cost of the absorption chiller itself is only 25 per cent of the total capital costs (Figure 3.28). With a heat price of €20 per MWh, total cooling costs for 2000

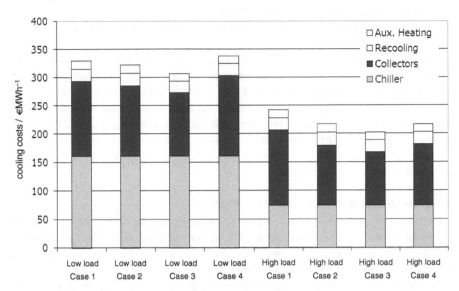

FIGURE 3.27 Cost distribution for a solar thermal absorption chiller system with mounting and integration costs of 45 per cent of total investment cost

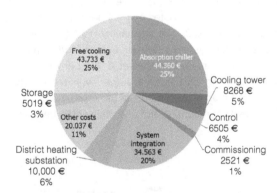

FIGURE 3.28 Distribution of capital costs of an absorption cooling system connected to a district heating network including a free cooling system

full-load hours are €169 per MWh. 62 per cent of the cooling costs are capital costs, 10 per cent are operating costs and 28 per cent are energy costs. At electricity costs of €120 per MWh, the specific costs for a corresponding vapour-compression chiller with a COP of 4 would be approximately €115 per MWh.

By comparison, Schölkopf and Kuckelkorn (2004) calculated the cost of conventional cooling systems for an energy efficient office building in Germany to be €180 per MWh. 17 per cent of the costs were for the electricity consumption of the chiller. The total annual cooling energy demand for the 1094 m² building was 31 kWhm⁻²a⁻¹. Our own comparative calculations for a 100 kW thermal cooling project showed that the compression chiller system costs without cold distribution in the building were between €110 and 140 per MWh.

Henning (2004) also investigated the costs of solar cooling systems compared with conventional technology. The additional costs for the solar cooling system per MWh of saved primary energy were between €44 per MWh in Madrid and €77 per MWh in Freiburg, Germany for large hotels. It is clear that solar cooling systems can become economically viable only if both the solar thermal and the absorption chiller costs decrease. This can be partly achieved by increasing the operation hours of the solar thermal system and thus the solar thermal efficiency by also using the collectors for warm water production or heating support.

COSTS OF DESICCANT COOLING SYSTEMS

The costs for desiccant cooling units are often full system costs, as the desiccant unit already includes parts of the conventional distribution system such as fans, fresh air channels, exhaust air channels, filters, etc. Furthermore, the machine itself contains humidifiers and heat exchangers, which are also available when a conventional air-handling unit is installed.

As an example the total capital, consumption and operation related costs were analysed for the Althengstett demonstration system with a volume flow of 18 000 m³h⁻¹. The investment costs are dominated by the desiccant unit itself and its system integration, which together make up 72 per cent of the total investment costs. The solar air collector field including tubing and mounting contributes only 14 per cent to the total investment costs at €300/m² collector area (Figure 3.29). The total investment costs per m³h⁻¹ are €12.4.

For another well analysed system of the IHK in Freiburg/Germany with a smaller volume flow of 10 000 m³h⁻¹, the costs were higher at €16.6 per m³h⁻¹. Also in this system, the desiccant evaporative cooling (DEC) unit together with the air channels, mounting and control was responsible for about two thirds of the total investment costs (46 per cent for the unit itself, 17 per cent for mounting and channels and 6 per cent for control). The solar

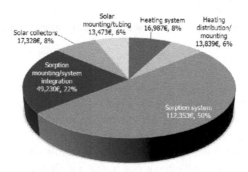

FIGURE 3.29 Total investment costs for hardware, tubing and system integration for the solar-powered desiccant cooling plant in Althengstett with 18 m³/h volume flow

air collector field was responsible for 10 per cent of the total costs (Hindenburg, 2002). A third system with 105 m² of solar air collectors connected in series to a ventilated photovoltaic facade and shed roof was constructed in Mataró, Spain with a total volume flow of 12,000 m³h⁻¹. As a part of a European Union demonstration project, the unit was constructed in Germany and then mounted and connected to the existing building management system by the Spanish project partners. Here the DEC unit accounted for 33 per cent of the total costs of €179,300, and a high percentage – 23 per cent – was used for the unit control and its connection to the existing building management system. The collectors were responsible for 12 per cent of the total investment costs, but mounting and system integration was expensive at 15 per cent. In total, the price per cubic metre and hour of air flow was similar to the German system at €15 per m³h⁻¹.

From the total investment costs in the Maier factory project, an annuity of €26,070 capital related costs results. If the funding for the DEC investment costs of €100,000 is taken into account, the annuity reduces to €14,122. Costs for heat, electricity and water consumption occur together with the demand charge to provide a given electrical power. In total, the annual consumption costs for the desiccant cooling system were calculated to be €3147, about 40 per cent less than for a conventional air-conditioning system (Figure 3.30). In the Spanish Mataró project, the savings calculated from electrical peak power cost reduction were €4200 per year.

In addition, operation-related costs for maintenance and repair occur. Repair costs are usually between 1 and 3 per cent of investment costs; for the calculations, 2 per cent was chosen. Maintenance costs are in a similar range – 76 per cent of the total annual costs are capital costs, the rest is for operation and consumption costs (Figure 3.31).

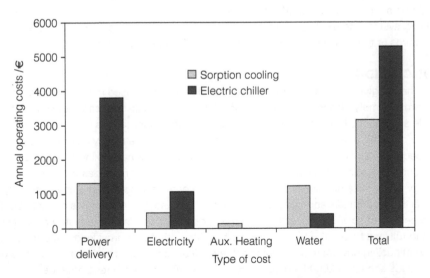

FIGURE 3.30 Annual consumption costs for the installed desiccant cooling system compared with an electric chiller

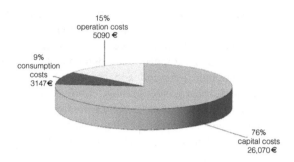

FIGURE 3.31 Distribution of total annual costs for the installed system including subsidy on the capital costs

The annual heating energy saving through the solar thermal collector field of about €1500 can be subtracted from the total annual costs. The remaining cooling costs for the investigated year with 34,710 kWh of cold production result in a specific cold price of €0.94 per kWh without funding and €0.6 per kWh including the investment funding.

By comparison, the costs for a conventional air-handling unit with humidification and an electric compression chiller were calculated in this project at €0.65 per kWh. The high price per kilowatt hour is largely due to the low total cooling demand in the building. At a nominal power of the system of approximately 100 kW, the cooling energy corresponds to only about 350 full-load hours. In climates with higher cooling demand and 1000 load hours, the price could then go down to €0.3 per kWh. An Austrian research team compared the costs of a district heating-powered small DEC system (6000 m³h⁻¹) with an air-handling unit with an electric compression chiller. For 960 full-load hours, they obtained cooling costs for the DEC unit of about €0.55 per kWh compared with €0.51 to 0.56 per kWh for the electrical cooling system (depending on the tariff structure). Also here the capital costs are about two-thirds of the total annual costs of the system (Simader and Rakos, 2005).

CONCLUSIONS

The chapter gives an overview of the two main technologies used in solar thermal cooling systems, namely closed cycle absorption chillers and open desiccant cooling systems. While absorption chillers can be used in a wide range of applications with cold distribution based on water or air systems, desiccant cooling systems are mainly recommended if the needed fresh air flow in a building is high. For all solar thermal cooling technologies, the reduction of auxiliary electrical energy consumption is a major goal, as otherwise primary energy savings are not significant. This means an optimized control strategy especially for partial load operation and good design of the heat source and heat sink circuits. The surface area of solar thermal collectors for an absorption cooling system strongly depends on the full-load hours of the system. For a building with low internal loads, about 2 m² per kilowatt cooling power are sufficient to achieve 80 per cent solar fraction; for buildings with high loads and up to 2000 full-load hours, about 4 m² per kW are recommended.

The total system costs for commercially available solar cooling systems in a Southern European location are between €180–320 per MWh, depending on the cooling load file and the chosen control strategy. The total costs are dominated by the costs for the solar thermal system and the chiller itself. For a more moderate climate such as Germany with a low cooling energy demand, the costs rise to €680 per MWh.

In desiccant cooling systems, capital costs clearly dominate the total costs. About two-thirds of the investment costs are due to the desiccant air-conditioning and distribution system, while the solar collectors are just 10–15 per cent of the investment. The investment costs are between €12–17 per m^3h^{-1} volume flow. Depending on the full-load hours, this results in a cooling price of between €300 and 900 per MWh. In moderate climates, the number of regeneration hours with solar thermal operation is limited, as often evaporative cooling is sufficient. Here the specific collector summer yield is rather low and new control strategies need to be found to optimize the solar thermal contribution.

ACKNOWLEDGEMENTS

We acknowledge the financial support of the European Union within the project POLYCITY, contract TREN/05FP6EN/S07.43964/51381.

The authors would like to thank the following colleagues from the research centre zafh.net at the University of Applied Sciences Stuttgart: Martin Huber, who provided the building load files using TRNSYS and Juergen Schumacher, who supported the solar cooling modelling work in INSEL. The description of the history of solar cooling systems is taken from the PhD work of Dr Uli Jakob. The analysis of the desiccant cooling system in Althengstett was carried out in the PhD work of Dr Uwe Schürger. The analysis of the district heating absorption chillers was done by Eric Duminil.

AUTHOR CONTACT DETAILS

Ursula Eicker and Dirk Pietruschka: University of Applied Sciences Stuttgart, Schellingstrasse 24, D-70174 Stuttgart, Germany; ursula.eicker@hft-stuttgart.de

REFERENCES

Albers, J. and Ziegler, F. (2003) 'Analysis of the part load behaviour of sorption chillers with thermally driven solution pumps', in *Proceedings of the 21st IIR International Congress of Refrigeration, International Institute of Refrigeration (IIR)*, Washington DC, 17–22 August

Assilzadeh, F., Kalogirou, S. A., Ali, Y. and Sopian, K. (2005) 'Simulation and optimisation of a LiBr solar absorption cooling system with evacuated tube collectors', *Renewable Energy*, vol 30, no 8, pp1143–1159

Atmaca, I. and Yigit, A. (2003) 'Simulation of solar-powered absorption cooling system', *Renewable Energy*, vol 28, no 8, pp1277–1293

Beccali, M., Butera, F., Guanella, R. and Adhikari, R. S. (2003) 'Simplified models for the performance evaluation of desiccant wheel dehumidification', *International Journal of Energy Research*, vol 27, no 1, pp17–29

Bourdoukan, P., Wurtz, E., Sperandio, M. and Joubert, P. (2007) 'Global efficiency of direct flow vacuum collectors in autonomous solar desiccant cooling: Simulation and experimental results', in *Proceedings of the Building Simulation Confererence IBPSA*, Bejing

Bujedo, L., Rodriguez, J., Marticnez, P. J., Rodriguez, L. R. and Vicente, J. (2008) 'Comparing different control strategies and configurations for solar cooling', *Proceedings of EUROSUN 2008*, Lisbon, Portugal, 7–8 October

Davanagere, B. S., Sherif, S. A. and Goswami, D. Y. (1999) 'A feasibility study of solar desiccant air conditioning system – Part I: Psychrometrics and analysis of the conditioned zone', *International Journal of Energy Research*, vol 23, no 1, pp7–21

Eicker, U. (2003) *Solar Technologies for Buildings*, John Wiley and Sons

Eicker, U. and Pietruschka, D. (2008) 'Design and performance of solar powered absorption cooling systems in office buildings', *Energy and Buildings*, vol 41, no 1, pp81–91, doi:10.1016/j.enbuild.2008.07.015

Florides, G. A., Kalogirou, S. A., Tassou, S. A. and Wrobel, L. C. (2002) 'Modelling, simulation and warming impact assessment of a domestic-size absorption solar cooling system', *Applied Thermal Engineering*, vol 22, no 12, pp1313–1325

Gao, Z., Viung, C. M. and Tomlinson, J. J. (2005) 'Theoretical analysis of dehumidification process in a desiccant wheel', *Heat Mass Transfer*, vol 41, no 11, pp1033–1042

Gassel, A. (2004) *Kraft-Wärme-Kälte-Kopplung und solare Klimatisierung*, Habilitationsschrift TU Dresden

Ginestet, S., Stabat, P. and Marchio, D. (2003) 'Control design of open-cycle desiccant cooling systems using a graphical environment tool', *Building Service Engineering Research and Technology*, vol 24, no 4, pp257–269

Grossmann, G. (2002) 'Solar powered systems for cooling, dehumidification and air conditioning', *Solar Energy*, vol 72, no 1, pp53–62

Hamed, A. M., Khalil, A., Kabeel, A. E., Bassuoni, M. M. and Elzahaby, A. M. (2005) 'Performance analysis of dehumidification rotating wheel using liquid desiccant', *Renewable Energy*, vol 30, pp1689–1712

Hartmann, K. (1992) 'Kälteerzeugung in Absorptionsanlagen (Cold production of absorption plants)', *Die Kälte und Klimatechnik*, vol 45, no 9

Henning, H.-M. (2004) *Solar-Assisted Air-Conditioning in Buildings – A Handbook for Planners*, Springer-Verlag, Wien

Henning, H.-M., Hindenburg, C., Erpenbeck, T. and Santamaria, I. S. (1999) 'The potential of solar energy use in desiccant cooling cycles', in *Proceedings of the International Sorption Heat Pump Conference (ISHPC)*, München

Hindenburg, C. (2002) 'Anlagenplanung und Betrieb einer offenen sorptionsgestützten Klimaanlage', in *Proceedings of 2nd Symposium Solares Kühlen in der Praxis*, HFT Stuttgart

INSEL (2008) 7.0 Block reference manual, www.insel.eu

Jakob, U. (2005) *Investigations into Solar Powered Diffusion-Absorption Cooling Machines*, PhD thesis, De Montfort University, Leicester

Jakob, U., Eicker, U., Taki, A. H. and Cook, M. J. (2003) 'Development of an optimised solar driven diffusion-absorption cooling machine', in *Proceedings of the ISES Solar World Congress 2003*, International Solar Energy Society (ISES), Gothenburg, Sweden, 16–19 June

Kanoğlu, M., Bolattürk, A. and Altuntop, N. (2007) 'Effect of ambient conditions on the first and second law performance of an open desiccant cooling process', *Renewable Energy*, vol 32, no 6, pp931–946

Kimura, K. (1992) 'Solar absorption cooling', in A. A. M. Sayigh and J. C. McVeigh (eds), *Solar Air Conditioning and Refrigeration*, 1st edn, Oxford, Pergamon Press, pp13–65

Kohlenbach, P. (2007) *Solar Cooling with Absorption Chillers: Control Strategies and Transient Chiller Performance*, dissertation, Technische Universität Berlin

Lamp, P. and Ziegler, F. (1997) 'Solar cooling with closed sorption systems', in *Proceedings of the Workshop Solar Sorptive Cooling, Forschungsverbund Sonnenenergie (FVS)*, Hardthausen, Germany, 16–17 October, pp79–92

Lavan, Z., Monnier, J.-B. and Worek, W. M. (1982) 'Second law analysis of desiccant cooling systems', *Journal of Solar Energy Engineering*, vol 104, pp229–236

Loewer, H. (1978) 'Solar-Kühlung in der Klimatechnik (Solar refrigeration for air conditioning installations)', *Ki Klima-Kälte-Ingenieur*, vol 6, no 4, pp155–162

Maclaine-Cross, I. L. (1988) 'Proposal for a desiccant air conditioning system', *ASHRAE Transactions*, vol 94, no 2, pp1997–2009

Mei, V. C., Chen, F. C., Lavan, Z., Collier, R. K. and Meckler, G. (1992) *An Assessment of Desiccant Cooling and Dehumidification Technology*, Report prepared by the Oak Ridge National Laboratory, Contract No DE-AC05-840R21400

Mendes, L.F, Collares-Pereira, M. and Ziegler, F. (1998) 'Supply of cooling and heating with solar assisted heat pumps: An energetic approach', *International Journal of Refrigeration*, vol 21, no 2, pp116–125

Mittal, V., Kasana, K. S. and Thakur, N. S. (2005) 'Performance evaluation of solar absorption cooling system of Bahal (Haryana)', *Journal of the Indian Institute of Science*, September–October, vol 85, no 1, pp295–305

Munters, C. G. (1960) *Air Conditioning System*, US Patent 2,926,502

Nick-Leptin, J. (2005) 'Political framework for research and development in the field of renewable energies', in *Proceedings of the International Conference Solar Air Conditioning*, Staffelstein

Panaras, G., Mathioulakisa, E. and Belessiotisa, V. (2007) 'Achievable working range for solid all-desiccant air-conditioning systems under specific space comfort requirements', *Energy and Buildings*, vol 39, no 9, pp1055–1060

Pennington, N. A. (1955) *Humidity Changer for Air Conditioning*, US Patent 2,700,537

Pietruschka, D., Jakob, U., Eicker, U. and Hanby, V. (2007) 'Simulation based optimisation and experimental investigation of a solar cooling and heating system', in *Proceedings of the Solar Air Conditioning 2nd International Conference*, Tarragona, Spain

Pietruschka, D., Jakob, U., Eicker, U. and Hanby, V. (2008) 'Simulation based optimisation of a newly developed system controller for solar cooling and heating systems', in *Proceedings EUROSUN 2008*, Lisbon, Portugal, 7–8 October

Preisler, A. (2008) *ROCOCO – Final Report*, Project No TREN/05/FP6EN/SO7.54855/020094

Safarik, M. and Weidner, G. (2004) 'Neue 15kW H₂O-LiBr Absorptionskälteanlage im Feldtest für thermische Anwendungen (New 15kW H_2O-LiBr absorption cooling machine in filed test for thermal applications)', in *Proceedings of the 3rd Symposium Solares Kühlen in der Praxis*, Fachhochschule Stuttgart – Hochschule für Technik, Germany, vol 65, pp159–171

Schoelkopf, W., Kuckelkorn, J. (2004). 'Verwaltungs- und Bürogebäude–Nutzverhalten und interne Wärmequellen' in *Proceedings of the OTTI Kolleg Klimatisierung von Büro-und Verwaltungsgebäuden Regensburg*

Schumacher, J. (1991) *Digitale Simulation regenerativer elektrischer Energieversorgungssysteme*, dissertation, Universität Oldenburg, www.insel.eu

Schweigler, C., Storkenmaier, F. and Ziegler, F. (1999) 'Die charakteristische Gleichung von Sorptionskälteanlagen', in *Proceedings of the 26th Deutsche Klima-Kälte-Tagung*, Berlin

Simader, G. and Rakos, C. (2005) *Klimatisierung, Kühlung und Klimaschutz: Technologien, Wirtschaftlichkeit und CO_2 Reduktionspotentiale*, Austrian Energy Agency, Vienna

Storkenmaier, F., Harm, M., Schweigler, C., Ziegler, F., Alebers, J., Kohlenbach, P. and Sengewald, T. (2003) 'Small-capacity water/LiBr absorption chiller for solar cooling and waste-heat driven cooling', in *Proceedings of the 21st IIR International Congress of Refrigeration*, International Institute of Refrigeration (IIR), Washington DC, 17–22 August

Sumath, K. (2003) 'Study on a solar absorption air-conditioning system', in *Proceedings of the International Congress of Refrigeration*, Washington, DC

Wardono, B. and Nelson, R. M. (1996) 'Simulation of a double effect LiBr/H₂O absorption cooling system', *ASHRAE Journal*, October, pp32–38

Ziegler, F. (1998) *Sorptionswärmepumpen, Forschungsberichte des Deutschen Kälte-und Klimatechnischen Vereins Nr 57*, Stuttgart

Ziegler, F. (1999) 'Recent developments and future prospects of sorption heat pump systems', *International Journal of Thermal Science*, vol 38, pp191–208

4

Artificial Neural Networks and Genetic Algorithms in Energy Applications in Buildings

Soteris A. Kalogirou

Abstract

The major objective of this chapter is to illustrate how artificial neural networks (ANNs) and genetic algorithms (GAs) may play an important role in modelling and prediction of the performance of various energy systems in buildings. The chapter initially presents artificial neural networks and genetic algorithms and outlines an understanding of how they operate by way of presenting a number of problems in the different disciplines of energy applications in buildings including environmental parameters, renewable energy systems, naturally ventilated buildings, energy consumption and conservation, and HVAC systems. The various applications are presented in a thematic rather than a chronological or any other order. Results presented in this chapter are testimony of the potential of artificial neural networks and genetic algorithms as design tools in many areas of energy applications in buildings.

■ *Keywords* – artificial neural networks; genetic algorithms; environmental parameters; renewable energy systems; naturally ventilated buildings; energy consumption and conservation; HVAC systems

INTRODUCTION

Analytic computer codes are often used for the estimation of the flow of energy and the performance of energy systems in buildings. The algorithms employed usually require the solution of complex differential equations. Thus, a lot of computer power and a considerable amount of time are required to give accurate predictions. As data from building energy systems are inherently 'noisy', they make good candidate problems to be handled with artificial neural networks (ANNs) and genetic algorithms (GAs).

When dealing with research and design associated with energy in buildings, there are often difficulties encountered in handling situations where many variables are involved. To be able to adequately model and predict the behaviour of building energy systems, it is often necessary to consider non-linear multivariate interrelationships, in a 'noisy' environment. In addition, the performance of building energy systems depends on the

doi:10.3763/aber.2009.0304 ■ © 2009 Earthscan ■ ISSN 1751-2549 (Print), 1756-2201 (Online) ■ www.earthscanjournals.com

environment – the parameters involved are ambient temperature, solar radiation, and wind speed and direction, the strength and duration of which are highly variable.

Analytical techniques have been very successful in studying the behaviour of thermal systems. However, although the analytical models have been valuable in understanding principles and useful where less than optimal designs were acceptable, with the advent of digital computers, numerical methods became much more attractive than analytical solutions as they could handle more complex and realistic situations.

However, numerical methods also have limitations. They cannot easily account for practical problems and they tend to perform well in analysing a situation, but not so well as a designer's tool for quickly looking at options. Moreover, the number of variables that can be considered is still limited and numerical solutions cannot usually be obtained directly. It is often the case that complex systems for which there is no exact model of behaviour or complex systems whose expected performances are completely unknown, need to be designed. The complexity is usually due to the multi-parameter and multi-criteria aspects of a system's design which are not easily handled using rules of thumb, analytical methods, physical models or numerical methods.

Many of the energy problems experienced in buildings are exactly the types of problems and issues for which the artificial intelligence approach appears to be most suited. In these computation models, attempts are made to simulate the powerful cognitive and sensory functions of the human brain and to use this capability to represent and manipulate knowledge in the form of patterns. Based on these patterns, ANNs, for example, model input–output functional relationships and can make predictions about other combinations of unseen inputs.

ANNs are collections of small individually interconnected processing units. Information is passed between these units along interconnections. An incoming connection has two values associated with it, an input value and a weight. As will be explained in more detail later, the output of the unit is a function of the summed value. It should be pointed out that ANNs are implemented on computers that are not programmed to perform specific tasks. Instead, they are trained with representative data sets until they learn the patterns used as inputs. Once trained, new patterns may be presented to an ANN for prediction or classification. An ANN can automatically learn to recognize patterns in data from real systems or from physical models, computer simulations or other sources. It can handle many inputs and produce answers that are in a form suitable for designers. Neural networks have the potential for making better, quicker and more practical predictions than any of the traditional methods. Neural network analysis is based on the past history data of a system and is therefore likely to be better understood and appreciated by designers than other theoretical and empirical methods.

Genetic algorithms are inspired by the way living organisms adapt to the harsh realities of life in a hostile world and apply the principles of evolution and inheritance. The algorithm imitates the process of evolution of a population by selecting only fit individuals for reproduction. Therefore, a GA is an optimum search technique based on the concepts of natural selection and survival of the fittest. It works with a fixed-size population of possible solutions to a problem, called individuals, which are evolving in time. A GA utilizes three principal genetic operators – selection, crossover and mutation.

ANNs and GAs may be used to provide innovative ways of solving design issues and will allow designers to get an almost instantaneous expert opinion on the effect of a proposed change to a design. Kalogirou has presented a review of ANNs in energy systems (Kalogirou, 2000a) and in renewable energy systems (Kalogirou, 2001). He also presented a review of artificial intelligence for the modelling and control of combustion processes (Kalogirou, 2003).

The objective of this chapter is to briefly introduce artificial neural networks and genetic algorithms and to present various applications in energy systems in buildings. The applications are presented in a thematic rather than a chronological or any other order. The majority of the applications presented are related with ANNs. This will show the capability of these tools in the prediction and modelling of building energy systems. In fact, artificial intelligence also comprises fuzzy logic systems, but a review of the applications of these systems in buildings was presented by Kolokotsa (2007) and will not be repeated again here.

ARTIFICIAL NEURAL NETWORKS

The concept of neural network analysis was discovered nearly 55 years ago, but it is only in the past 25 years that applications software has been developed to handle practical problems. The history and theory of neural networks have been described in a large number of published works and will not be covered in this chapter except for a very brief overview of how neural networks work.

ANNs are good for some tasks while lacking in some others. Specifically, they are good for tasks involving incomplete data sets, fuzzy or incomplete information, and for highly complex and ill-defined problems, where humans usually decide on an intuitional basis. ANNs can learn from examples and are able to deal with non-linear problems. Furthermore, they exhibit robustness and fault tolerance. The tasks that ANNs cannot handle effectively are those requiring high accuracy and precision as in logic and arithmetic.

ANNs have been applied successfully in various fields of mathematics, engineering, medicine, economics, meteorology, psychology, neurology and many others. Some of the most important ones are: in pattern, sound and speech recognition; in the analysis of electromyographs and other medical signatures; in the identification of military targets; and in the identification of explosives in passenger suitcases. They have also been used in weather and market trends forecasting, in the prediction of mineral exploration sites, in electrical and thermal load prediction, in adaptive and robotic control and many others. Neural networks are used for process control because they can build predictive models of the process from multidimensional data routinely collected from sensors.

A neural network is a massively parallel distributed processor that has a natural propensity for storing experiential knowledge and making it available for use (Haykin, 1994). It resembles the human brain in two respects:

1 the knowledge is acquired by the network through a learning process; and
2 inter-neuron connection strengths known as synaptic weights are used to store the knowledge.

Artificial neural network models may be used as an alternative method in engineering analysis and predictions. ANNs mimic somewhat the learning process of a human brain. They operate like a 'black box' model, requiring no detailed information about the system. Instead, they learn the relationship between the input parameters and the controlled and uncontrolled variables by studying previously recorded data. ANNs can also be compared to multiple regression analysis except that with ANNs no assumptions need to be made about the system to be modelled. Neural networks usually perform successfully where other methods do not, and have been applied in solving a wide variety of problems, including non-linear problems that are not well suited to classical methods of analysis, such as pattern recognition. Another advantage of using ANNs is their ability to handle large and complex systems with many interrelated parameters. They seem to simply ignore excess data that are of minimal significance and concentrate instead on the more important inputs. Instead of complex rules and mathematical routines, artificial neural networks are able to learn the key information patterns within a multidimensional information domain. In addition, neural networks are fault tolerant, robust and noise immune (Rumelhart et al, 1986).

Artificial neural networks try to imitate the human brain. In fact, the human brain is the most complex and powerful structure known today. Artificial neural networks are composed of a number of elements operating in parallel. These elements are inspired by biological nervous systems. A schematic diagram of typical multilayer feed-forward neural network architecture is shown in Figure 4.1. Although two hidden layers are shown, their number can be one or more than two, depending on the problem examined. The network usually consists of an input layer, some hidden layers and an output layer. In its simple form, each single neuron is connected to other neurons of a previous layer through adaptable synaptic weights. The number of input and output parameters and the number of cases influence the geometry of the network. The network consists of an 'input' layer of neurons, where each neuron corresponds to each input parameter, a 'hidden' layer or layers of neurons, and an output layer where each neuron corresponds to each output parameter. A neuron, also called processing element, is the basic unit of a neural network and is shown in Figure 4.2 – it performs summation and activation functions to determine

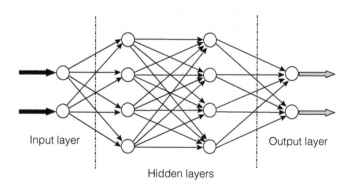

FIGURE 4.1 Schematic diagram of a fully connected multilayer feed-forward neural network

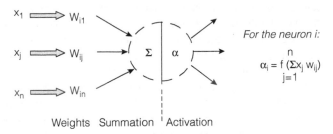

For the neuron i:

$$\alpha_i = f\left(\sum_{j=1}^{n} x_j\, w_{ij}\right)$$

Weights   Summation   Activation

FIGURE 4.2 Information processing in a neural network unit

the output of the neuron. Usually the biggest challenge faced when designing a neural network is to find the right number of neurons in the hidden layer. This depends on the number of the inputs and outputs and also on the number of training cases. Too many hidden layer neurons can result in 'overtraining' (or lack of generalization) and lead to large 'verification' errors. Too few neurons can result in large 'training' and 'verification' errors. Knowledge is usually stored as a set of connection weights (presumably corresponding to synapse efficacy in biological neural systems).

Training is the process of modifying the connection weights in some orderly fashion using a suitable learning method. The network uses a learning scheme, in which an input is presented to the network along with the desired output and the weights are adjusted so that the network attempts to produce the desired output. After training, the weights contain meaningful information whereas before training they are random and have no meaning.

The way information is processed through a single node is shown in Figure 4.2. The node receives weighted activation of other nodes through its incoming connections. First, these are added up and the result is passed through an activation function; the outcome is the activation of the node. For each of the outgoing connections, this activation value is multiplied with the specific weight and transferred to the next node.

A training set is a group of matched input and output patterns used for training the network, usually by suitable adaptation of the synaptic weights. The outputs are the dependent variables that the network produces for the corresponding input. All the information the network needs to learn is supplied to the network as a data set. Starting from an initially randomized weighted network system, input data is propagated through the network to provide an estimate of the output value. When each pattern is read, the network uses the input data to produce an output, which is then compared with the training pattern which represents the correct or desired output. If there is a difference, the connection weights (usually but not always) are altered in such a direction that the error is decreased. After the network has run through all the input patterns, if the error is still greater than the maximum desired tolerance, the ANN runs again through all the input patterns repeatedly until all the errors are within the required tolerance. When the training reaches a satisfactory level, the network holds the weights constant and uses the trained network to make decisions, identify patterns, or define associations in new input data sets not used to train it.

There are several algorithms that can be used to achieve the minimum error in the shortest time. Additionally, there are many alternative forms of neural networking systems and many different ways in which they may be applied to a given problem. The suitability of an appropriate strategy for application is very much dependent on the type of problem to be solved and the variability of the data in the training data set.

The most popular learning algorithm is back-propagation (BP) and its variants (Werbos, 1974; Rumelhart et al, 1986), which is also one of the most powerful learning algorithms in neural networks. Back-propagation training is a gradient descent algorithm. It tries to improve the performance of the neural network by reducing the total error by changing the weights along its gradient. The training of all patterns of a training data set is called an epoch. It should be noted that the training set has to be a representative collection of input–output examples. A flow chart of the BP learning algorithm is shown in Figure 4.3.

With reference to Figure 4.3 for the training of the neural network with BP, the following steps are followed after the network structure is decided:

1 Initialize weights and threshold values by assigning small random values to each one.
2 Present a continuous input vector X_1, X_2, X_3,...X_n and the desired outputs O_1, O_2, O_3,...O_n after these are normalized to a suitable range, either [-1,1] or [0,1].
3 Compute the output of each node in the hidden layer using:

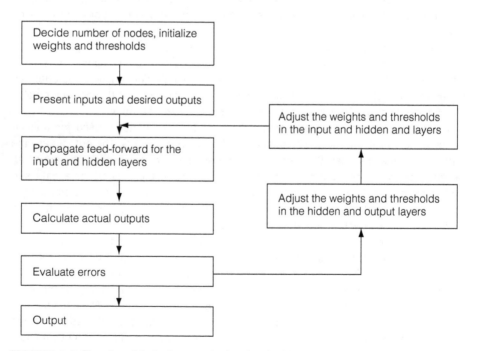

FIGURE 4.3 Flow chart of the back-propagation learning algorithm

$$h_j = f\left(\sum_{i-1}^{n} W_{ij}X_i - \theta_j\right)$$ [4.1]

where h_j is the vector of hidden-layer neurons, i is the input-layer neurons, W_{ij} are the weights between the input and hidden layer and θ_j is the threshold between the input and hidden layers.

4 Compute the outputs of each node in the output layer by using:

$$O_k = f\left(\sum_{i=1}^{m} W_{kj}X_j - \theta_k\right)$$ [4.2]

where k represents the output layer, W_{kj} the weights connecting the hidden and output layers, θ_k the threshold connecting the hidden and output layers and f(x) is the activation function. A number of such functions can be used, but usually the logistic function is used given by:

$$f(x) = \frac{1}{1 + e^{-x}}$$ [4.3]

5 Estimate the output layer error between the target and the observed outputs:

$$\delta_k = O_k(1 - O_k)(Y - O_k)$$ [4.4]

where δ_k is the vector of errors for each output neuron and Y is the target activation of the output layer.

6 Estimate the hidden layer error by using:

$$\delta_j = h_j(1 - h_j)\sum_{k=1}^{m} \delta_k W_{kj}$$ [4.5]

where δ_j is the vector of errors for each hidden layer neuron.

7 Adjust the weights and thresholds in the output layer:

$$W_{kj}(t+1) = W_{kj(t)} + \alpha\delta_k h_j + n\left(W_{kj(t)} - W_{kj(t-1)}\right)$$ [4.6]

and

$$\theta_{k(t+1)} = \theta_{k(t)} + \alpha\delta_k$$ [4.7]

where α is the learning rate and n is the momentum factor, specified from the beginning, and used to allow the previous weight change to influence the weight change in this time period, t.

8 Adjust the weights and thresholds in the hidden layer:

$$W_{ji(t+1)} = W_{ji(t)} + \alpha\delta_j h_i + n\left(W_{ji(t)} - W_{ji(t-1)}\right)$$ [4.8]

and

$$\theta_{j(t=1)} + \theta_{j(t)} + \alpha\delta_j$$ [4.9]

9 Repeat steps 2 to 8 for all patterns in the data set until the output layer error is within a specified tolerance.

Neural networks obviate the need to use complex mathematically explicit formulae, computer models, and impractical or costly physical models. Some of the characteristics that support the success of artificial neural networks and distinguish them from conventional computational techniques are as follows (Nannariello and Fricke, 2001):

● The direct manner in which artificial neural networks acquire information and knowledge about a given problem domain (learning interesting and possibly non-linear relationships) through the 'training' phase.
● Neural networks can work with numerical or analogue data that would be difficult to deal with by other means because of the form of the data or because there are so many variables.
● Neural network analysis can be conceived of as a 'black box' approach and the user does not require sophisticated mathematical knowledge.
● The compact form in which the acquired information and knowledge is stored within the trained network and the ease with which it can be accessed and used.
● Neural network solutions can be robust even in the presence of 'noise' in the input data.
● The high degree of accuracy reported when artificial neural networks are used to generalize over a set of previously unseen data (not used in the 'training' process) from the problem domain.

While neural networks can be used to solve complex problems they do suffer from a number of shortcomings. The most important of them are:

● The data used to train neural networks should contain information, which ideally, is spread evenly throughout the entire range of the system.
● There is limited theory to assist in the design of neural networks.
● There is no guarantee of finding an acceptable solution to a problem.
● There are limited opportunities to rationalize the solutions provided.

NETWORK PARAMETERS SELECTION
When building the neural network model, the process has to be identified with respect to the input and output variables that characterize the process. The inputs include measurements of the physical dimensions, measurements of the variables specific to the environment, equipment and controlled variables modified by the operator. Variables that do not have any effect on the variation of the measured output should not be used. These can be estimated by the contribution factors of the various input parameters. These factors indicate the contribution of each input parameter to the learning of the neural network and depending on the software employed are estimated by the network, after training.

To apply the ANN method, the first step is to collect the required data and prepare them in a spreadsheet format with various columns representing the input and output parameters. Three types of data files are required – a training data file, a test data file and a validation data file. The former and the latter should contain representative samples of all the cases the network is required to handle, whereas the test file may contain about 10 per cent of the cases contained in the training file. During training, the network is tested against the test file to determine the accuracy and training is stopped when the mean average error remains unchanged for a number of epochs. This is done in order to avoid overtraining, which is the case where the network learns perfectly the training patterns but is unable to make predictions when an unknown training set is presented to it.

The basic operation that has to be followed to successfully handle a problem with ANNs, is to select the appropriate architecture and the suitable learning rate, momentum, number of neurons in each hidden layer and the activation function. This is a laborious and time-consuming method but as experience is gathered some parameters can be predicted easily thus shortening tremendously the time required.

GENETIC ALGORITHMS

The genetic algorithm is a model of machine learning, which derives its behaviour from a representation of the processes of evolution in nature. This is done by the creation within a machine/computer of a population of individuals represented by chromosomes. Essentially these are a set of character strings that are analogous to the chromosomes that we see in the DNA of human beings. The individuals in the population then go through a process of evolution.

GAs are mostly used for multidimensional optimization problems in which the character string of the chromosome can be used to encode the values for the different parameters to be optimized.

In practice, this genetic model of computation can be implemented by having arrays of bits or characters to represent the chromosomes. Simple bit manipulation operations allow the implementation of the principal genetic operators, i.e. selection, crossover, mutation and other operations.

When the GA is executed, the fitness of all the individuals in the population is evaluated. Then a new population is created by performing operations such as crossover, fitness-proportionate reproduction and mutation on the individuals whose fitness has just been measured. Subsequently, the old population is discarded and a new iteration is done using the new population. One complete iteration is referred to as a generation. The logic of the standard genetic algorithm is shown in Figure 4.4 (Zalzala and Fleming, 1997).

With reference to Figure 4.4, in each generation, individuals are selected for reproduction according to their performance with respect to the fitness function. Essentially, selection gives a higher chance of survival to better individuals. Subsequently, genetic operations are applied in order to form new and possibly better offspring. The algorithm is terminated either after the fitness function value remains the same for a user-specified number of generations or when the optimal solution is reached. More details on genetic algorithms can be found in Goldberg (1989), Davis (1991) and Michalewicz (1996).

```
                        Standard Genetic Algorithm
   Begin (1)
       t = 0 [start with an initial time]
       Initialize Population P(t) [initialize a usually random population of individuals]
       Evaluate fitness of Population P(t) [evaluate fitness of all individuals in
                                    population]
       While (Generations < Total Number) do begin (2)
           t = t + 1[increase the time counter]
           Select Population P(t) out of Population P(t-1) [select sub-population for
                                                    offspring production]

           Apply Crossover on Population P(t)
           Apply Mutation on Population P(t)
           Evaluate fitness of Population P(t) [evaluate new fitness of population]
       end (2)
   end (1)
```

FIGURE 4.4 The structure of the standard genetic algorithm

The first generation (generation 0) of this process operates on a population of randomly generated individuals. From there on, the genetic operations, with the aid of the fitness measure, operate to improve the population, i.e. during each step of the reproduction process, the individuals in the current generation are evaluated by a fitness function value, which is a measure of how well the individual solves the problem. Then each individual is reproduced in proportion to its fitness; the higher the fitness, the higher its chance to participate in mating (crossover) and to produce offspring. A small number of newborn offspring undergo the action of the mutation operator. After many generations, only those individuals who have the best genetics, from the point of view of the fitness function, survive. The individuals that emerge from this 'survival of the fittest' process are the ones that represent the optimal solution to the problem specified by the fitness function and the constraints. Therefore, genetic algorithms are suitable for finding the optimum solution in problems where a fitness function is present. As described, genetic algorithms use a 'fitness' measure to determine which of the individuals in the population survive and reproduce. Thus, survival of the fittest causes good solutions to progress.

A genetic algorithm works by selective breeding of a population of 'individuals', each of which could be a potential solution to the problem. The genetic algorithm is seeking to breed an individual, which either maximizes, minimizes or is focused on a particular solution of a problem.

The larger the breeding pool size or population, the greater its potential to produce a better individual. However, as the fitness value produced by every individual is compared with all other fitness values of all other individuals on every reproductive cycle, larger breeding pools take a longer time. After testing all of the individuals in the population, a new 'generation' of individuals is produced for testing.

During the setting up of the GA, the user has to specify the adjustable chromosomes, i.e. the parameters that would be modified during evolution to obtain the maximum value of the fitness function. Additionally, the user has to specify the ranges of these values called constraints.

A genetic algorithm is not gradient based and uses an implicitly parallel sampling of the solutions space. The population approach and multiple sampling means that it is less possible to become trapped to local minima than when using traditional direct approaches and it is possible to navigate a large solution space with a highly efficient number of samples. Although not guaranteed to provide a globally optimum solution, GAs have been shown to be highly efficient at reaching a very near optimum solution in a computationally efficient manner (Kalogirou, 2003).

The genetic algorithm parameters to be specified by the user are as follows.

Population size
This is the size of the genetic breeding pool, i.e. the number of individuals contained in the pool. If this parameter is set to a low value, there would not be enough different kinds of individuals to solve the problem satisfactorily. If there are too many, a good solution will take longer to be found because the fitness function must be calculated and compared with the fitness function for every individual in every generation.

Crossover rate
This determines the probability that the crossover operator will be applied to a particular chromosome during a generation. This parameter is usually near 90 per cent.

Mutation rate
This determines the probability that the mutation operator will be applied to a particular chromosome during a generation. This parameter should be very small, usually near 1 per cent.

Generation gap
This determines the fraction of those individuals that do not pass into the next generation. It is sometimes desirable that individuals in the population be allowed to pass into the next generation, which is especially important if individuals selected are the most fit ones in the population. This parameter is usually near 95 per cent.

Chromosome type
Populations are composed of individuals, and individuals are composed of chromosomes, which are equivalent to variables. Chromosomes are composed of smaller units called genes. There are two types of chromosomes, continuous and enumerated.

Continuous chromosomes are implemented in the computer as binary bits. The two distinct values of a gene, 0 and 1, are called alleles. Multiple chromosomes make up the individual. Each partition is one chromosome, each binary bit is a gene, and the value of each bit (1, 0, 0, 1, 1, 0) is an allele. The genes in a chromosome can take a wide range of values between the minimum and maximum values of the associated variables. One variation of continuous chromosomes is the 'integer chromosome' which isused in problems where it is required to take only integer values of chromosomes and genes.

Enumerated chromosomes consist of genes that can have more allele values than just 0 and 1, and these values are usually visible to the user. These are suitable for a category of problems, usually called combinatorial problems, the most famous example of which is the travelling salesman problem (TSP). In the TSP, a salesman must visit some number of cities, say eight, and he only wants to visit each city once. In this problem, the objective is to find the shortest route through all cities in order to minimize his travelling expenses. The values of chromosomes range from a minimum to a maximum value, and the order of their genes in the chromosome is important. In the TSP example with eight cities, there will be eight genes, each with values ranging from 1 to 8. The entire chromosome represents the path through the cities, therefore each individual has only one chromosome, comprised of the entire list of cities. For example, suppose a particular chromosome in the population is 3, 4, 5, 7, 8, 2, 1, 6. This would represent the solution where the salesman will go first to city 3, then to city 4 and so on. Each partition (city) is a gene and each value (3, 4, 5, 7, 8, 2, 1, 6) is an allele.

It should be noted that the chromosomes in the TSP problem should not contain duplicate genes, i.e. genes with the same allele values, otherwise the salesman will travel to the same city more than once. Usually two different types of enumerated chromosomes exist – 'repeating genes' and 'unique genes'. For the TSP, unique genes have to be used but some other problems may require repeating genes where chromosomes can be 2, 3, 2, 4, 5, 2, 3, 5 or even 3, 3, 3, 3, 3, 3, 3, 3.

COMBINATION OF GAs WITH ANNs

Genetic algorithms can also be combined with neural networks. They can be used to find the size of weights that could minimize the error function of a neural network or they can be used for a particular type of network called the general regression neural network (GRNN), which is known for the ability to train quickly on sparse data sets. GRNNs can have multidimensional input and could fit multidimensional surfaces through data. GRNNs work by measuring how far (in terms of the Euclidean distance) a given sample pattern is from patterns in the training set in N-dimensional space, where N is the number of inputs in the problem.

A GRNN is a four-layer feed-forward neural network, which is based on the non-linear regression theory consisting of the input layer, the pattern layer, the summation layer and the output layer (see Figure 4.5). There are no training parameters such as learning rate and momentum as there are in back-propagation networks, but there is a smoothing factor that is applied after the network is trained. The smoothing factor determines how tightly the network matches its predicted output to the training data. While the neurons in the first three layers are fully connected, each output neuron is connected only to some processing units in the summation layer. The summation layer has two different types of processing units, the summation units and a single division unit. The number of the summation units is always the same as the number of the GRNN output units. The division unit only sums the weighted activations of the pattern units of the hidden layer, without using any activation function. Each of the GRNN output units is connected only to its corresponding summation unit and to the division unit (there are no weights in these connections). The function of the output units is obtained by a simple division of the signal coming from the summation unit by the signal coming from the division unit. The

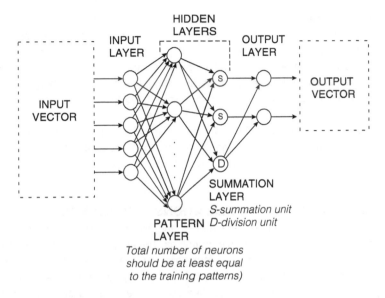

FIGURE 4.5 General regression neural network architecture

summation and output layers together basically perform a normalization of the output vector, thus making GRNN much less sensitive to the proper choice of the number of patterns. For the GRNN, an activation function is only required in the input layer slab. More details on GRNN have been reported by Tsoukalas and Uhrig (1977) and Ripley (1996).

As the hidden layer consists of one neuron for each pattern in the training set, for GRNN networks, the number of neurons in the hidden pattern layer is usually equal to the number of patterns in the training set. This number can be made larger if one wants to add more patterns later, but it cannot be made smaller. As in all other networks, the number of neurons in the input layer is equal to the number of input parameters and the number of neurons in the output layer corresponds to the number of outputs.

The training of the GRNN is quite different from the training used in other neural networks. It is completed after presentation of each input–output vector pair from the training data set to the GRNN input layer only once.

The GRNN may be trained using a genetic algorithm that is used to find the appropriate individual smoothing factors for each input as well as an overall smoothing factor, i.e. the genetic algorithm is seeking to breed an individual that minimizes the mean squared error of the test set, which can be calculated by:

$$E = \frac{1}{p}\sum_{p}\left(t_p - o_p\right)^2 \qquad [4.10]$$

where E is the mean squared error, t the network output (target) and o the desired output vectors over all patterns (p) of the test set.

After testing all the individuals in the pool, a new 'generation' of individuals is produced for testing. Unlike the back-propagation algorithm, which propagates the error through the network many times seeking a lower mean squared error between the network's output and the actual output or answer, GRNN training patterns are only presented to the network once.

At the end of training, the individual smoothing factors may be used as a sensitivity analysis tool; the larger the factor for a given input, the more important that input is to the model, at least as far as the test set is concerned. Inputs with low smoothing factors are candidates for removal for a later trial. Individual smoothing factors, however, are unique to each network. The numbers are relative to each other within a given network and they cannot be used to compare inputs from different networks. Therefore, if the number of input, output or hidden neurons is changed, the network must be retrained. This may occur when more training patterns are added because GRNN networks require one hidden neuron for each training pattern.

ANNs AND GAs IN ENERGY APPLICATIONS IN BUILDINGS

ANNs have been used by various researchers and by the author for modelling and prediction in the field of energy systems in buildings. The applications include models for predicting solar radiation, ambient temperature and wind, which are the environmental parameters affecting the performance of a building's energy systems; renewable energy systems that can be applied in buildings, like solar water heaters and photovoltaics; naturally ventilated buildings; energy consumption and conservation; and various aspects of heating, ventilating and air-conditioning (HVAC) systems, like modelling, control, operation and fault diagnosis. The applications of GAs in buildings include optimization of buildings, energy estimation and the determination of the heat transfer coefficient. This chapter presents various such applications in a thematic rather than a chronological or any other order. Table 4.1 presents the numbers of applications presented in each category.

APPLICATIONS OF ARTIFICIAL NEURAL NETWORKS IN BUILDINGS
Environmental parameters prediction
In this section, applications of ANNs for the modelling and prediction of solar radiation, ambient temperature and wind speed are presented. A summary of the ANN applications in environmental parameters prediction is shown in Table 4.2.

Prediction of global radiation and irradiance
Alawi and Hinai (1998) have used ANNs to predict solar radiation in areas not covered by direct measurement instrumentation. The input data to the network are the location,

TABLE 4.1 Number of applications presented

TECHNOLOGY	AREA	APPLICATIONS PRESENTED
ANNs	Environmental parameters	20
	Renewable energy systems	8
	Naturally ventilated buildings	1
	Energy consumption and conservation	8
	HVAC systems	7
GAs	All areas	3

TABLE 4.2 Summary of applications of ANNs in environmental parameters prediction

PARAMETER	REMARKS	REFERENCES
Solar radiation	Solar irradiance, global radiation, daily insolation, solar potential, direct and diffuse radiation prediction, mapping solar potential.	Alawi and Hinai (1998); Mohandes et al (1998a); Kemmoku et al (1999); Cao and Cao (2005); Lopez et al (2005); Argiriou et al (2000); Reddy and Ranjan (2003); Sozen et al (2004a, 2004b, 2004c, 2005); Soares et al (2004).
Ambient temperature	Prediction of minimum temperature and ambient temperature.	Schizas et al (1991, 1994); Tasadduq et al (2002); Arca et al (1998); Argiriou et al (2000).
Wind speed	Wind speed prediction and wind classification.	Kalogirou et al (1999a); Mohandes et al (1998b); Kretzschmar et al (2004).

month, mean pressure, mean temperature, mean vapour pressure, mean relative humidity, mean wind speed and mean duration of sunshine. The ANN model predicts solar radiation with an accuracy of 93 per cent and mean absolute error of 7.3 per cent.

Mohandes et al (1998a) used data from 41 collection stations in Saudi Arabia. From these, data for 31 stations were used to train a neural network and the data for the other 10 for testing the network. The input values to the network were latitude, longitude, altitude and sunshine duration. The results for the testing stations obtained were within 16.4 per cent and indicate the viability of this approach for spatial modelling of solar radiation.

Kemmoku et al (1999) used a multistage ANN to predict the insolation of the next day. The input data to the network were the average atmospheric pressure, predicted by another ANN, and various weather data of the previous day. The results obtained show a prediction accuracy of 20 per cent.

Cao and Cao (2005) used artificial neural network combined with wavelet analysis for the forecast of solar irradiance. This method is characteristic of the preprocessing of sample data using wavelet transformation for the forecast, i.e. the data sequence of solar irradiance is first mapped into several time-frequency domains, and then a recurrent BP network is established for each domain. The forecasted solar irradiance is exactly the algebraic sum of all the forecasted components obtained by the respective networks, which correspond respectively to the time-frequency domains. Discount coefficients were applied to take account the different effects of different time-steps on the accuracy of the ultimate forecast when updating the weights and biases of the networks in network training. As proved by the authors, the combination of recurrent BP networks and wavelet analysis gives more accurate forecasts of solar irradiance. An example of the forecast of day-by-day solar irradiance is presented, in which the historical day-by-day records of solar irradiance for Shanghai were used. The results show that the accuracy of the method was more satisfactory than that of other methods.

A very important factor in the assessment of solar energy resources is the availability of direct irradiance data of high quality. However, this component of solar radiation is seldom measured and thus must be estimated from data of global solar irradiance, which is registered in most radiometric stations. Lopez et al (2005) employed the Bayesian framework for ANN, known as the automatic relevance determination (ARD) method, to obtain the relative relevance of a large set of atmospheric and radiometric variables used for estimating hourly direct solar irradiance. In addition, the authors analysed the viability of this novel technique applied to select the optimum input parameters to the neural network. A multilayer feed-forward perceptron is trained on these data. The results reflect the relative importance of the inputs selected. Clearness index and relative air mass were found to be the most relevant input variables to the neural network, as was expected, proving the reliability of the ARD method. Moreover, the authors show that this novel methodology can be used in unfavourable conditions, in terms of a limited amount of available data, giving successful results.

Argiriou et al (2000) developed a predictive tool for solar irradiance by applying the feed-forward neural network model and supervised training with the method of back-propagation with momentum term. The inputs to this model were the measured solar irradiance at regular time intervals of 15 minutes. The ANN developed was trained using part of real meteorological data for 1993 in Athens, Greece; the remaining data were used for verification. The solar irradiance forecast tool has 28 input neurons, one hidden layer of 16 and another of eight neurons, and one output neuron. The inputs are the last and six previous values of the following parameters: daily-normalized time, ambient temperature, day of the year and solar irradiance. The output is the one-step-ahead prediction of the difference of the global solar irradiance. The prediction of solar irradiance was satisfactory. The difference exhibits a correlation with the hour of the day, being larger at times close to sunrise and sunset, therefore their effects are not so important.

Solar resource estimation

Reddy and Ranjan (2003) used ANN-based models for the estimation of monthly mean daily and hourly values of solar global radiation. Solar radiation data from 13 stations spread over India around the year were used for training and testing the ANN. The solar radiation data from 11 locations (6 from South India and 5 from North India) were used for training the neural networks and data from the remaining 2 locations (1 from South India and 1 from North India) were used for testing the estimated values. The results of the ANN model were compared with other empirical regression models. The solar radiation estimations by ANN were in good agreement with the actual values and are superior to those of other available models. The maximum mean absolute relative deviation of predicted hourly global radiation tested was 4.07 per cent. The results indicate that the ANN model shows promise for evaluating solar global radiation possibilities in places where monitoring stations are not established.

Sozen et al (2005) presented a new formula based on meteorological and geographical data developed to determine the solar energy potential in Turkey using ANNs. Scaled conjugate gradient (SCG) and Levenberg–Marquardt (LM) learning algorithms and a logistic sigmoid transfer function were used in the network. Meteorological data for four years (2000–2003) from 18 cities spread over Turkey were used as training data.

Geographical and meteorological data (latitude, longitude, altitude, month, mean sunshine duration and mean temperature) were used in the input layer of the network. Solar radiation is the output parameter. One-month test data for each city was used, which were not used for training. The ANN models show good accuracy for evaluating solar resource possibilities in regions where a network of monitoring stations has not been established in Turkey. This study confirms the ability of the ANN to predict solar radiation values precisely.

Sozen et al (2004a) presented another study, the main focus of which was to estimate the solar energy potential in Turkey using ANNs. In this study, the best approach was investigated for each station by using different learning algorithms and a logistic sigmoid transfer function in the neural network. In order to train the neural network, meteorological data for three years (2000–2002) from 17 stations spread over Turkey were used as training (11 stations) and testing (6 stations) data. Meteorological and geographical data (latitude, longitude, altitude, month, mean sunshine duration and mean temperature) were used in the input layer of the network. Solar radiation was in the output layer. The maximum mean absolute percentage error was found to be less than 6.74 per cent and R^2-values were found to be about 99.89 per cent for the testing stations. For the training stations, these values were found to be 4.40 per cent and 99.97 per cent. The trained and tested ANN models show greater accuracy for evaluating the solar resource possibilities in regions where a network of monitoring stations has not been established in Turkey.

In a similar study, Sozen et al (2004b) used ANNs to determine the solar energy potential in Turkey. SCG, Pola-Ribiere conjugate gradient (PCG) and LM learning algorithms and a logistic sigmoid transfer function were used in the network. In order to train the neural network, meteorological data and network topology similar to the one presented above were used. The results indicate that the ANN model seems promising for evaluating solar resource possibilities at the places where there are no monitoring stations in Turkey. The results on the testing stations indicate a relatively good agreement between the observed and the predicted values.

In another study by the same authors (Sozen et al, 2004c) the predicted solar potential values from the ANN were given in the form of monthly maps. These maps are of prime importance for different working disciplines, like architects, meteorologists and solar engineers. The predictions from the ANN models could enable scientists to locate and design solar energy systems and determine the appropriate solar technology.

Diffuse solar radiation
Soares et al (2004) presented a perceptron neural network technique used to estimate hourly values of diffuse solar radiation in the city of Sao Paulo, Brazil, using as input the global solar radiation and other meteorological parameters measured from 1998 to 2001. The neural network verification was performed using the hourly measurements of diffuse solar radiation obtained during the year 2002. The neural network was developed based on both feature determination and pattern selection techniques. It was found that the inclusion of the atmospheric long-wave radiation as input improves the neural network performance. On the other hand traditional meteorological parameters, like air temperature and atmospheric pressure, are not as important as long-wave radiation which

acts as a surrogate for cloud-cover information on the regional scale. An objective evaluation has shown that diffuse solar radiation is better reproduced by a neural network synthetic series than by a correlation model.

Forecasting minimum temperature

Forecasting of temperature is one of the regular operational practices carried out by meteorological services worldwide. The great interest in developing methods for more accurate site-specific predictions of the temperature regime has led to the invention of several such methods based on ANNs.

One of the first efforts to employ ANNs in meteorology refers to the forecasting of minimum temperature as a result of nocturnal cooling (Schizas et al, 1991, 1994). The set of raw data involved in this work was extracted from meteorological observations recorded at three-hour intervals at Larnaca Airport in Cyprus. From each of the eight sets of observations available daily, the following elements were extracted, preprocessed and used as input to the ANN; total cloud amount (TC), visibility (VI), present weather condition (PW), atmospheric pressure (AP), dry-bulb temperature (DB), wet-bulb temperature (WB) and low cloud amount (LC). Using the wind direction and the wind speed, the eastward (W1) and northward components (W2) of the wind vector were calculated. Besides these records, the astronomical day length (DL) and the observed minimum temperature of the previous night (PM) were used as additional inputs. All of the above inputs were chosen after consultation with local weather forecasters and are considered as having an impact on nocturnal cooling.

In their study, Schizas et al (1991, 1994) used a total of 133 days in winter and spring of 1984 for training and evaluating the forecasting models. Of these days, 103 were randomly selected and used for training while the other 30 days were used for testing the models constructed.

There are many possible ways to implement ANNs and determine the optimal network and this is not an easy task. In each case, the modeller has to consider the uniqueness of each problem and, in most cases, has no other choice but to resort to a systematic testing of some of the available possibilities based on experience. Some of the available choices refer to the input and output selection, the network architecture and the training algorithm. Table 4.3 displays how the various input parameters were used with different models.

The back-propagation training algorithm was used for building the investigated ANN models. Table 4.4 shows the set of ANN models that were trained and evaluated with the above mentioned data. For all the examined models, the learning rate and the momentum were chosen to be 0.01 and 0.9, respectively. For each model, the number of inputs (the first number shown by the architecture) is determined by the number of parameters that are used for training the model. The number of units in the first and the second hidden layers of the model are represented by the second and the third number of the architecture, respectively. These numbers were chosen empirically. The number of outputs (last number in the architecture) was chosen to be 40 for all the models. The 40 outputs represent the temperature range 0°C to +20°C with 0.5°C intervals that form the available classes of the models.

Most of the models learned the training set very well, as suggested by the 'total sum of squares' which is defined as the sum of the squared difference between the expected

TABLE 4.3 Input parameters to various models (from Schizas et al, 1991; 1994)

MODEL	INPUT PARAMETERS										
	TC	W1	W2	VI	PW	AP	DB	WB	LC	DL	PM
1	✓	✓	✓	✓	✓	✓	✓	✓	✓	✓	✓
2	✓	✓	✓	✓	✓	✓	✓	✓	✓	✓	✓
3	✓	✓	✓	✓	✓	✓	✓	✓	✓	✓	✓
4					✓						
5					✓		✓	✓			✓
6		✓	✓				✓	✓			✓
7	✓	✓	✓				✓	✓	✓		
8		✓	✓				✓	✓			✓
9		✓	✓				✓	✓		✓	✓
10		✓	✓				✓	✓			✓
11		✓	✓		✓		✓	✓			✓
12		✓	✓				✓	✓			✓
13		✓	✓		✓		✓	✓			✓

TABLE 4.4 Various models' architectures and forecasting yield (from Schizas et al, 1991; 1994)

MODEL	ARCHITECTURE	EPOCHS	TOTAL SUM OF SQUARES	FORECASTING YIELD (%)			
				±0.5°C	±1.0°C	±2.0°C	±3.0°C
1	74–120–160–40	290	0.9	21	32	45	53
2	74–80–120–40	410	0.9	16	29	45	50
3	64–40–60–40	670	2.5	18	37	60	68
4	9–16–32–40	8060	30.6	24	332	39	53
5	25–40–60–40	2460	7.3	21	29	47	66
6	32–40–80–40	4070	5.0	24	29	47	50
7	48–60–100–40	431	0.9	18	26	37	42
8	33–40–80–40	4020	2.0	18	29	53	58
9	34–40–80–40	587	0.9	16	32	47	55
10	41–50–100–40	4180	3.0	18	29	50	60
11	41–50–100–40	453	0.9	24	26	39	55
12	29–40–80–40	1330	2.2	21	39	50	58
13	36–50–100–40	1100	0.9	21	32	42	53

and the calculated output for all cases (days) in the training cycle. The forecasting yield, also shown in Table 4.4, is a measure of the success of a model during the testing (evaluation) phase. The percentage of days in the testing set of data, for which the minimum temperature was successfully forecast, is defined as the forecasting yield of a model. This forecasting yield is given for temperature confidence ranges ±0.5, ±1.0, ±2.0 and ±3.0°C, as shown in Table 4.4. It is obvious that an increase of the temperature confidence range will increase the forecasting yield, at the expense of accuracy.

Arca et al (1998) also developed a model based on ANN for short-term prediction of the daily minimum temperature of air, analysed its performance under different meteorological conditions and evaluated its capabilities as a tool for decision making related to frost occurrence. The database used consisted of measurements collected from December 1996 to March 1998 at an agrometeorological station in Florence, Italy. The measured variables were:

- air temperature;
- soil temperature;
- relative humidity of air;
- temperatures of a black body in the sun and in the shade;
- global solar radiation;
- diffuse radiation;
- net radiation;
- wind speed and direction;
- rain intensity; and
- atmospheric pressure.

The data were acquired every 2 minutes and a mean value of each variable was calculated and stored every 20 minutes. In addition, at sunset, the form and the total amount of cloud were recorded. For each quantity, the value at sunset and the daily minimum, maximum and mean value were calculated. The data collected were divided in two parts: the first, consisting of 90 days refer to the period December 1996 to February 1997 and were used for training the ANN; the second, composed of 28 records, collected in February 1998, were used for testing. Seven different feed-forward ANNs with three layers were developed. In all of the seven ANNs, both the input layer, which receives the values of selected quantities, and the hidden layer, which performs the data computations, have a variable number of neurons. The output layer has a unique neuron yielding the estimate of minimum air temperature. The input variables were chosen on the basis of physical laws describing the radiative cooling of soil and atmosphere.

The best performing architecture yields predictions of minimum air temperatures with a low standard error of estimate (1.63°C) and a good correlation with the observed values (0.91). The accuracy of estimates provided by the neural model developed in this study is high in comparison with other statistical or semi-empirical methods. In addition, the ANN models require few, easily measurable atmospheric parameters whereas other models prove to be more demanding.

Ambient temperature prediction

To cover a basic need regarding temperature forecasting, Tasadduq et al (2002) have used ANNs for the prediction of hourly mean values of ambient temperature 24 hours in advance. One year hourly values of ambient temperature were used to train a neural network model for a coastal location in Saudi Arabia. The ANN had one input, one output and one hidden layer. The neural network was trained offline using back-propagation and a batch learning scheme. The model developed required only one temperature value as input to predict the temperature for the following day for the same hour. The trained neural network was tested with temperatures for three years other than the ones used for training and it was found that the model was able to forecast hourly temperatures 24 hours in advance with sufficient accuracy.

Argiriou et al (2000) developed a predictive tool for ambient temperature by applying the feed-forward neural network model and supervised training with the method of back-propagation with momentum term. The input to this tool was the measured ambient temperature at regular time intervals of 15 minutes. The ANN developed was trained using part of real meteorological data for 1993 in Athens, Greece; the remaining data were used for verification. The forecasting of the ambient (outdoor) temperature used a simple ANN with ten input neurons, one hidden layer of eight and another of four neurons and one output neuron. The inputs to this ANN were the (daily normalized) time value for the next interval, the last and three previous values of the outdoor temperature, the (yearly normalized) day number for the next and the last and three previous values of the solar irradiance. Hourly values were normalized by dividing them by 24 and the day number values by 365. Also all other parameters were normalized to fit in the interval [0,1]. The output was the one-step-ahead prediction of the outdoor temperature difference. The ANN forecasting of the one-step-ahead outdoor temperature performed satisfactorily with differences from the recorded values up to one degree in just a few cases.

Wind speed prediction

Modelling the wind is a highly non-linear issue due to the numerous factors that can influence both the wind speed and its direction (terrain configuration and characteristics, elevation, prevailing weather pattern, etc.). Forecasting the wind is traditionally sought with dynamical modelling (i.e. with numerical weather prediction models in which the wind at various levels is routinely predicted); also, stochastic analysis and modelling have been used extensively and related methodologies include time series modelling. ANNs have recently been employed in wind forecasting. They can be an alternative to conventional techniques and in many respects can be more advantageous for tackling such non-linear problems.

A suitable artificial neural network was trained to predict the mean monthly wind speed in regions of Cyprus where data are not available (Kalogirou et al, 1999a). Data for the period 1986–1996 (11 years) were used to train the network and data for the year 1997 were used for validation. Both learning and prediction were performed with an acceptable accuracy. Two multilayered artificial neural network architectures of the same type were tried, one with 5 neurons in the input layer (month, wind speed at 2 m and 7 m for two stations) and one with 11. The additional input data for the 11-input network were the x and y coordinates of the meteorological stations. The 5-input network proved to be more successful in the prediction of the mean wind speed.

A comparison of the mean wind speed at the two levels (2 m and 7 m) for the two networks is shown in Table 4.5. As can be seen, the network using only five input parameters is more successful, giving a maximum percentage difference of only 1.8 per cent.

The two networks can be used for different types of jobs, i.e. the network having 5 inputs can be used to fill missing data from a database, whereas the one having 11 inputs

TABLE 4.5 Maximum percentage differences of the annual results of the two networks

NETWORK	MEAN WIND SPEED (ACTUAL)		MEAN WIND SPEED (ANN PREDICTED)		% DIFFERENCE	
	H=2m	H=7m	H=2m	H=7m	H=2m	H=7m
11 input neurons			2.43	3.52	1.2	5
	2.4	3.35				
5 input neurons			2.4	3.41	0	1.8

can be used for predicting mean wind speed in other nearby locations. In the former, the station can be located within the area marked by the three stations (interpolation) or outside (extrapolation).

Mohandes et al (1998b) describe the utilization of neural networks in predicting the monthly and daily mean wind speed for Jeddah in Saudi Arabia. These predictions are subsequently compared with those of autoregressive models. For the training of the multilayer neural network adopted in this study, the back-propagation algorithm has been used.

The observed data used in this study cover a period of 12 years, from 1970 to 1982. For the mean monthly wind speed prediction study, the data were first normalized and were subsequently split into two sets; one for training (120 months) and the other for testing (24 months). A similar procedure was adopted for the mean daily wind speed prediction study, in which case the data were also divided into two sets; a training set (730 days) and a testing set. The testing data were used in either defining the coefficients of the autoregressive models or for adjusting the weights of the neural networks. The same training and testing data sets were used for both the autoregressive and the neural network approaches.

Two types of models were built and tested:

1 single-step models (one month ahead for the monthly models and one day ahead for the daily models) by using one input and one output;
2 multi-step models (several days ahead) by feeding the output of the system as input.

Several networks with different numbers of hidden neurons were trained. Based on the root mean square error over the training data, Mohandes et al (1998b) concluded that, for the daily predictions, the best network was one with 24 hidden units. The results on testing data indicate that for both the single-step and the multi-step models, the neural network approach outperforms the autoregressive approach.

Wind classification and prediction

For most applications, users of wind speed predictions are interested in ranges or classes of wind values rather than in an exact wind values. Kretzschmar et al (2004) addressed the prediction of average hourly local ground wind speed and hourly local ground wind gust maximum values for prediction lead times between 1 and 24 hours; in their efforts, the local wind prediction task is formulated as a classification task in order to meet the aforementioned need.

The wind speed and gust classes were defined on the basis of the widely used Beaufort scale. The neural networks employed were conventional feed-forward neural networks with two variants; directly (as classifiers) and indirectly (as function approximators that predict the exact wind values). The classifiers presented exhibited a superior quality when compared with persistence with respect to performance as well as hit and false alarm rates for most of the prediction lead times for wind speed and wind gusts. For the prediction of strongest winds, satisfying results could only be obtained for the 1 hour lead time.

The inputs selected for the classifiers were several lags of the local wind speed, wind gust and wind direction time series, time and date. To test the potential of additional input from numerical weather prediction models, model data from the European Centre for Medium-Range Weather Forecasts (ECMWF) analyses were also used. Except for the model features, all inputs were obtained from a single wind observation device at the site of interest. The chosen model input led to improvements for wind speed prediction for 12 and 24 hour lead times but the improvement was smaller than expected.

Renewable energy systems
A summary of the ANN applications in renewable energy systems applied to buildings is shown in Table 4.6.

Solar domestic water heating (SDWH) systems
An ANN was trained based on 30 known cases of systems, varying from collector areas between 1.81 m^2 and 4.38 m^2 (Kalogirou et al, 1999b). Open and closed systems were considered, both with horizontal and vertical storage tanks. In addition to the above, an attempt was made to consider a large variety of weather conditions. In this way, the network was trained to accept and handle a number of unusual cases. The data presented as input were the collector area, storage tank heat loss coefficient (U-value), tank type, storage volume, type of system, and 10 readings from real experiments of total daily solar radiation, mean ambient air temperature and the water temperature in the storage tank at the beginning of a day for each system. The network output was the useful energy extracted from the system and the stored water temperature rise. Unknown data were used to investigate the accuracy of prediction. Predictions within 7.1 per cent and 9.7 per cent were obtained, respectively (Kalogirou et al, 1999b). These results indicate that the

TABLE 4.6 Summary of applications of ANNs in building renewable energy systems

SYSTEM	REMARKS	REFERENCES
Solar water heating systems	Solar domestic water heating systems, long-term performance prediction, identification of collector parameters, fault diagnostic systems.	Kalogirou et al (1999b); Kalogirou et al (1999c); Kalogirou and Panteliou (2000); Kalogirou (2000b); Kalogirou and Panteliou (1999); Lalot (2000); Kalogirou et al (2008).
Photovoltaics	Peak power tracking.	Veerachary and Yadaiah (2000).

proposed method can successfully be used for the estimation of the useful energy extracted from the system and the stored water temperature rise. The advantages of this approach compared with the conventional algorithmic methods are the speed, simplicity and capacity of the network to learn from examples. This is done by embedding experiential knowledge in the network. Additionally, actual weather data were used for the training of the network, which leads to more realistic results compared with other modelling programmes which rely on typical meteorological year (TMY) data that are not necessarily similar to the actual environment in which a system operates.

An ANN was trained using performance data for four thermosyphon types of systems, all employing the same collector panel under varying weather conditions (Kalogirou et al, 1999c). The output of the network is the useful energy extracted from the system and the stored water temperature rise. Predictions with maximum deviations of 1 MJ and 2.2°C were obtained for the two output parameters, respectively. Random data were also used both with the performance equations obtained from the experimental measurements and with the artificial neural network to predict the above two parameters. The predicted values thus obtained were very comparable. These results indicate that the proposed method can successfully be used for the estimation of the performance of the particular thermosyphon system at any of the different types of configurations used.

Solar domestic water heating systems long-term performance prediction

30 thermosyphon SDWH systems were tested and modelled according to the procedures outlined in the standard ISO 9459-2 at three locations in Greece (Kalogirou and Panteliou, 2000). From these, data for 27 systems were used for training and testing the network while data for the remaining 3 were used for validation. Two ANNs were trained using the monthly data produced by the modelling programme supplied with the standard. Different networks were used due to the nature of the required output, which was different in each case. The first network was trained to estimate the solar energy output of the system (Q) for a draw-off quantity equal to the storage tank capacity, and the second one to estimate the solar energy output of the system (Q) and the average quantity of hot water per month (V_d) at demand temperatures of 35°C and 40°C. The input data in both networks were similar to the ones used in the programme supplied with the standard. These were the size and performance characteristics of each system and various climatic data. In the second network, the demand temperature was also used as input. The statistical coefficient of multiple determination (R^2-value) obtained for the training data set was equal to 0.9993 for the first network and 0.9848 and 0.9926 for the second for the two output parameters, respectively. Unknown data were subsequently used to investigate the accuracy of prediction. Predictions with R^2-values equal to 0.9913 for the first network and 0.9733 and 0.9940 for the second were obtained (Kalogirou and Panteliou, 2000).

A similar approach was followed for the long-term performance prediction of three forced circulation-type SDWH systems (Kalogirou, 2000b). The maximum percentage differences obtained when unknown data were used, were 1.9 per cent and 5.5 per cent for the two networks, respectively.

The performance of a solar hot water thermosyphon system was tested with the dynamic system method according to standard ISO/CD/9459.5. The system is of closed circuit type and consists of two flat plate collectors with total aperture area of 2.74 m^2 and of a 170 litre hot water storage tank. The system was modelled according to the procedures outlined in the standard with the weather conditions encountered in Rome. The simulations were performed for hot water demand temperatures of 45°C and 90°C and volume of daily hot water consumption varying from 127 to 200 litres. These results were used to train a suitable neural network to perform long-term system performance prediction (Kalogirou and Panteliou, 1999). The input data were learned with adequate accuracy with correlation coefficients varying from 0.993 to 0.998, for the four output parameters. When unknown data were used to validate the network, satisfactory results were obtained. The maximum percentage difference between the actual (simulated) and predicted results is 6.3 per cent. These results prove that artificial neural networks can be used successfully for this type of prediction.

Identification of the time parameters of solar collectors
Lalot (2000) used ANNs for the identification of time parameters of solar collectors. Two parameters fully described the static behaviour whereas two other parameters were necessary to fully describe the dynamic behaviour of a flat plate collector. The discrimination ability of the network, however, was not very high when a second-order system was considered. It was shown that collectors may be considered as third-order systems. A radial basis function (RBF) neural network was used to accurately identify pure third-order systems. The neural network was validated by the computation of the Euclidean distance between the collectors and their models, depending on the number of learning steps. Finally, it was shown that the neural networks were able to discriminate collectors that have close parameters within 2 per cent.

Solar systems fault diagnostic system
Kalogirou et al (2008) developed an automatic solar water heater (SWH) fault diagnosis system (FDS). The FDS consists of a prediction module, a residual calculator and the diagnosis module. A data acquisition system measures the temperatures at four locations of the SWH system and the mean storage tank temperature. In the prediction module, a number of ANNs were used, trained with values obtained from a TRNSYS model of a fault-free system operated with the typical meteorological year for Nicosia, Cyprus and Paris, France. Thus, the neural networks were able to predict the fault-free temperatures under different environmental conditions. The input data to the ANNs were various weather parameters, the incidence angle, flow condition and one input temperature. The residual calculator received both the current measurement data from the data acquisition system and the fault-free predictions from the prediction module. The system can predict three types of faults – collector faults and faults in insulation of the pipes connecting the collector with the storage tank (two circuits) – and these are indicated with suitable labels. The system was validated by using input values representing various faults of the system.

Peak power tracking for PV-supplied DC motors
Veerachary and Yadaiah (2000) applied an ANN for the identification of the optimal operating point of a PV water pumping system. A gradient descent algorithm was used to

train the ANN controller for the identification of the maximum power point of a solar cell array and gross mechanical energy operation of the combined system. The input parameter to the neural network was solar insolation and the output parameter was the converter chopping ratio corresponding to the maximum power output of the PV cells or gross mechanical energy output of the combined PV system. The error in the ANN predictions was less than 2 per cent for centrifugal and 7 per cent for volumetric pump loads, respectively. According to the authors, the ANN provides a highly accurate identification/tracking of optimal operating points even with stochastically varying solar insolation.

Naturally ventilated buildings

The air flow distribution inside a lightweight test room, which was naturally ventilated, was predicted using artificial neural networks (Kalogirou et al, 2001a). The test room was situated in a relatively sheltered location and was ventilated through adjustable louvres. Indoor air temperature and velocity were measured at four locations and six different levels. The outside local temperature, relative humidity, wind velocity and direction were also monitored. The collected data were used to predict the air flow across the test room. Experimental data from a total of 32 trials were collected. Data for 28 of these were used for the training of the neural network while the data for 4 trials were used for validation of the network. The data were recorded at 2 minute intervals and the length of each trial varied but were generally 12 hours. A multilayer feed-forward neural network was employed with three hidden slabs. Satisfactory results for the indoor temperature and combined velocity were obtained when unknown data were used as input to the network. A comparison between the actual and the ANN-predicted data for the indoor air temperature is shown in Figure 4.6.

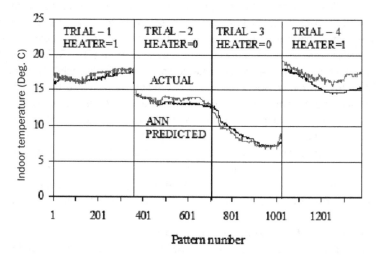

FIGURE 4.6 Comparison between actual and ANN-predicted data for indoor air temperature

Energy consumption and conservation

A summary of the ANN applications in energy consumption and conservation is shown in Table 4.7.

Modelling of the space and hot-water heating energy consumption in the residential sector

Aydinalp et al (2004) used two methods to model residential end-use energy consumption at the national or regional level – the engineering method and the conditional demand-analysis method. It was shown that the neural network method is capable of accurately modelling the behaviour of the appliances, lighting and space-cooling energy consumption in the residential sector. As a continuation of the work on the use of the ANN method for modelling residential end-use energy consumption, two ANN-based energy-consumption models were developed to estimate the space and domestic hot-water heating energy consumptions in the Canadian residential sector. Both models achieved a very high prediction performance with R^2-values equal to 0.91 and 0.87, respectively.

Energy conservation in buildings through efficient air-conditioning control

Ben-Nakhi and Mahmoud (2002) used a GRNN to optimize the air-conditioning setback scheduling in public buildings. To save energy, the temperature inside these buildings is allowed to rise after business hours by setting back the thermostat. The objective is to predict the time of the thermostat end of setback (EoS) such that the design temperature inside the building is restored in time for the start of business hours. State-of-the-art building simulation software, ESP-r, was used to generate a database that covered the past five years. The software was used to calculate the EoS for two office buildings using the climate records in Kuwait. The EoS data for 1995 and 1996 were used for training and testing the neural networks. The robustness of the trained ANN was tested by applying them to a 'production' data set (1997–1999) that the networks had never 'seen' before. A parametric study showed that the optimum GRNN design is one that uses a genetic adaptive algorithm, a so-called City Block distance metric, and a linear scaling function for the input data. External hourly temperature readings were used as network inputs, and the

TABLE 4.7 Summary of applications of ANNs in energy consumption and conservation

SYSTEM	REFERENCES
Modelling of the space and hot-water heating energy consumption in the residential sector	Aydinalp et al (2004)
Energy conservation in buildings through efficient air-conditioning control	Ben-Nakhi and Mahmoud (2002)
Cooling load prediction in buildings	Ben-Nakhi and Mahmoud (2004)
Prediction of the indoor air temperature of a building	Mechaqrane and Zouak (2004)
Energy consumption of residential buildings	Michalakakou et al (2002)
Energy demand controller	Argiriou et al (2000)
Heating and cooling load estimation	Kalogirou et al (1997, 2001b)

thermostat EoS is the output. The ANN predictions were improved by developing a neural control scheme. This scheme is based on using the temperature readings as they become available. Six ANNs were designed and trained for this purpose. The performance of the ANN analysis was evaluated using the coefficient of multiple determination (R^2-value) and by examination of the error patterns. The results show that the neural control scheme is a powerful instrument for optimizing air-conditioning setback scheduling based on external temperature records.

Cooling load prediction in buildings

Ben-Nakhi and Mahmoud (2004) used GRNNs which were designed and trained to investigate the feasibility of using this technology to optimize HVAC thermal energy storage in public buildings as well as office buildings. Building simulation software, ESP-r, was used to generate a database covering the years 1997–2001. The software was used to calculate hourly cooling loads for three office buildings using climate records in Kuwait. The cooling load data for 1997–2000 were used for training and testing the neural networks, while robustness of the trained ANN was tested by applying them to a 'production' data set (2001 data) that the networks had never 'seen' before. Three buildings of various densities of occupancy and orientational characteristics were investigated. Parametric studies were performed to determine optimum GRNN design parameters that best predict cooling load profiles for each building. External hourly temperature readings for a 24 hour period were used as network inputs and the hourly cooling load for the next day was the output. The performance of the ANN analysis was evaluated using the coefficient of multiple determination and by statistical analysis of the error patterns, including confidence intervals of regression lines, as well as by examination of the error patterns. The results show that a properly designed ANN is a powerful instrument for optimizing thermal energy storage in buildings based only on external temperature records.

Prediction of the indoor air temperature of a building

Mechaqrane and Zouak (2004) used a neural network autoregressive with exogenous input (NNARX) model to predict the indoor temperature of a residential building. First, the optimal regressor of a linear ARX model was identified by minimizing Akaike's final prediction error (FPE). This regressor was then used as the input vector of a fully connected feed-forward neural network with one hidden layer of 10 units and one output unit. The results showed that the NNARX model outperformed the linear model considerably, i.e. the sum of the squared error (SSE) was 15.05 with the ARX model and 2.06 with the NNARX model. The optimal network topology was subsequently determined by pruning the fully connected network according to the optimal brain surgeon (OBS) strategy. With this procedure, nearly 73 per cent of connections were removed and, as a result, the performance of the network was improved. The final SSE obtained was equal to 0.906.

Energy consumption of residential buildings

Michalakakou et al (2002) used a neural network approach for modelling and estimating the energy consumption time series for a residential building in Athens, using as inputs several climatic parameters. The hourly values of the energy consumption, for heating and cooling the building, were estimated for several years using feed-forward back-propagation neural networks. Various neural network architectures were designed and trained for the output estimation, which is the building's energy consumption. The results were tested with extensive sets of non-training measurements and it was found that they corresponded well with the actual values. Furthermore, 'multi-lag' output predictions of ambient air temperature and total solar radiation were used as inputs to the neural network models for modelling and predicting the future values of energy consumption with sufficient accuracy.

Energy demand controller

Argiriou et al (2000) demonstrated how ANNs can be usefully applied in optimizing the energy demand of buildings and systems that maximize the use of solar energy for space heating. An ANN model developed for this purpose plays the role of a controller which is able to forecast the energy demand leading to a simple On/Off decision of the electrical heating system maintaining acceptable indoor conditions. The controller developed is of a modular structure. The relevance to meteorological ANN forecasting lies in the inclusion of two weather modules that make use of and forecast the solar irradiance and ambient temperature by using the values of these parameters at the previous time steps and parameters characterizing the specific time interval (hour and day). Other modules comprising the ANN controller predict the On/Off operation status of the heating system, the indoor temperature and the final output from the controller to the heating system, that is, the On/Off switch position.

Heating and cooling load estimation

Kalogirou et al (1997) used an ANN to predict the required heating load of buildings with the minimum of input data. An ANN was trained based on 250 known cases of heating load, varying from very small rooms (1–2 m^2) to large spaces of 100 m^2 floor area. The type of rooms varied from small toilets to large classroom halls, while the room temperatures varied from 18°C to 23°C. In addition to the above, an attempt was made to use a large variety of room characteristics. In this way, the network was trained to accept and handle a number of unusual cases. The data presented as input were, the areas of windows, walls, partitions and floors, the type of windows and walls, the classification on whether the space has a roof or ceiling, and the design room temperature. The network output was the heating load. Preliminary results on the training of the network showed that the accuracy of the prediction could be improved by grouping the input data into two categories, one with floor areas up to 7 m^2 and another with floor areas from 7 to 100 m^2. The statistical R^2-value for the training data set was equal to 0.988 for the first case and 0.999 for the second. Unknown data were subsequently used to investigate the accuracy of prediction. Predictions within 10 per cent for the first group and 9 per cent for the second were obtained. These results indicate that the proposed method can successfully be used for the estimation of the heating load of a building. The advantages of this approach compared with the conventional algorithmic methods are the speed of calculation, the simplicity and the capacity of the network to learn from examples and thus gradually improve its performance. This is done by embedding experiential knowledge in the network and thus the appropriate U-values are considered. Such an approach is very useful for countries where accurate thermal properties of building materials are not readily available.

Kalogirou et al (2001b) also used ANNs for the estimation of the daily heating and cooling loads. The daily loads of nine different building structures were estimated using the TRNSYS programme and a typical meteorological year of Cyprus. This set of data was used to train a neural network. For each day of the year, the maximum and minimum loads were obtained from which heating or cooling loads could be determined. All the buildings considered had the same areas but different structural characteristics. Single and double walls were considered as well as a number of different roof insulations. A multi-slab feed-forward architecture having three hidden slabs was employed. Each hidden slab was comprised of 36 neurons. For the training data set, the R^2-values obtained were 0.9896 and 0.9918 for the maximum and minimum loads, respectively. The method was validated by using actual (modelled) data for one building, for all days of the year, which the network had not seen before. The R^2-values obtained in this case were 0.9885 and 0.9905 for the two types of loads, respectively. The results indicate that the proposed method can be used for the required predictions for buildings of different constructions.

Heating, ventilating and air-conditioning systems

A summary of the ANN applications in HVAC systems is shown in Table 4.8.

Models of discharge air temperature system

Zaheer-uddin and Tudoroiu (2004a) developed non-linear neuro-models for a discharge air temperature (DAT) system. Experimental data gathered in an HVAC test facility was used to develop multi-input multi-output (MIMO) and single-input single-output (SISO) neuro-models. Several different network architectures were explored to build the models. Results show that a three-layer second-order neural network structure is necessary to achieve good accuracy of the predictions. Results from the developed models are compared and some observations on sensitivity and standard deviation errors are presented.

TABLE 4.8 Summary of applications of ANNs in HVAC systems

SYSTEM	REFERENCES
Discharge air temperature	Zaheer-uddin and Tudoroiu (2004a; 2004b).
Fault detection and diagnosis	Wang and Jiang (2004).
Prediction of the optimal start time of a heating system in buildings	Yang et al (2003)
Subsystem level fault diagnosis of a building's air-handling unit	Lee et al (2004)
Modelling the thermal dynamics of a building's space and its heating system	Gouda et al (2002)
Prediction of the COP of existing rooftop units	Zmeureanu (2002)

The same authors (Zaheer-uddin and Tudoroiu, 2004b) also explored the problem of improving the performance of a DAT system using a PID (proportional–integral–derivative) controller and augmenting it with neural network-based tuning and tracking functions. The DAT system is modelled as a SISO system. The architecture of the real-time neuro-PID controller and simulation results obtained under realistic operating conditions are presented. The neural network-assisted PID tuning method is simple to implement. Results show that the network-assisted PID controller is able to track both constant and variable setpoint trajectories efficiently in the presence of disturbances acting on the DAT system.

Fault detection and diagnosis
Wang and Jiang (2004) presented a method for monitoring and diagnosing the degradation in the performance of heating/cooling coil valves, which may result in serious energy waste, without requiring the valves to be demounted or adding additional sensors. A recurrent cerebellar model articulation controller (RCMAC) was developed to learn the normal characteristics of the valve. When degradation in the performance of the valve occurred, the response of the RCMAC deviated from the normal. Two characteristic variables were defined as the degradation index and the waveform index for analysing the residual errors. A strategy was developed to identify the type of degradation and estimate the severity of the degradation. Tests on a typical valve with five faulty cases in an air-handling unit (AHU) demonstrated the effectiveness and robustness of the strategy.

Prediction of the optimal start time of a heating system in buildings
Yang et al (2003) presented an application of an ANN in a building control system. The objective of this study was to develop an optimized ANN model to determine the optimal start time for a heating system in a building. For this, programmes for predicting the room air temperature and the learning of the ANN model based on back-propagation learning were developed. Learning data for various building conditions were collected through programme simulation for predicting the room air temperature. Then the optimized ANN model was presented through the learning of the ANN, and its performance to determine the optimal start time was evaluated.

Subsystem level fault diagnosis of a building's air-handling unit
Lee et al (2004) described a scheme for online fault detection and diagnosis (FDD) at the subsystem level in an air-handling unit. The approach consists of process estimation, residual generation, and fault detection and diagnosis. The schematic of the system is shown in Figure 4.7. Residuals are generated using general regression neural network models. The GRNN is a regression technique and uses a memory-based feed-forward network to produce estimates of continuous variables. The main advantage of a GRNN is that no mathematical model is needed to estimate the system. Also, the inherent parallel structure of the GRNN algorithm makes it attractive for real-time fault detection and diagnosis. Several abrupt and performance degradation faults were considered. Because performance degradations are difficult to introduce artificially in real or experimental systems, simulation data were used to evaluate the method. The simulation results show

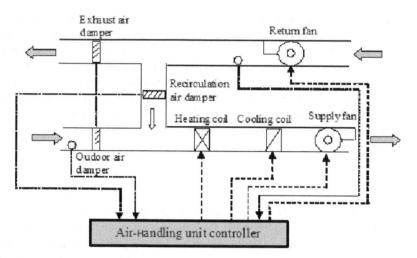

FIGURE 4.7 AHU system diagrammatic

that the GRNN models are accurate and reliable estimators of highly non-linear and complex AHU processes, and demonstrate the effectiveness of the proposed method for detecting and diagnosing faults in an AHU.

Modelling the thermal dynamics of a building's space and its heating system

Gouda et al (2002) used ANNs for modelling the thermal dynamics of a building's space, its water heating system and the influence of solar radiation. A multilayered feed-forward neural network using the Levenberg-Marquardt back-propagation training algorithm was applied to predict the future internal temperature. Real weather data for a number of winter months, together with a validated model (based on the building construction data), were used to train the network in order to generate a mapping between the easily measurable inputs (outdoor temperature, solar irradiance, heating valve position and the building indoor temperature) and the desired output, i.e. the predicted indoor temperature. The objective of this work was to investigate the potential of using an ANN with singular value decomposition (SVD) method, as shown in Figure 4.8, to predict the indoor temperature which may allow early shutdown of the heating system controller with a consequent reduction in energy consumption for heating inside the building.

Prediction of the COP of existing rooftop units

Zmeureanu (2002) proposed a new approach for evaluating the coefficient of performance (COP) of existing rooftop units, using general regression neural networks. This approach reduces the installation cost of monitoring equipment since only a minimum number of sensors are needed and it also reduces the costs for recalibration or replacement of sensors during the operation. The new approach was developed and tested using measurements taken on two existing rooftop units in Montreal, Canada.

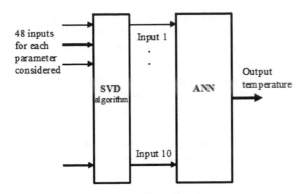

FIGURE 4.8 Feed-forward neural network with SVD algorithm

APPLICATIONS OF GENETIC ALGORITHMS IN BUILDINGS
Only three applications of GAs in buildings were identified. A summary of these applications is shown in Table 4.9.

Optimization of building thermal design and control
Wright et al (2002) showed that the design of buildings is a multi-criterion optimization problem where there is always a trade-off that needs to be made between capital expenditure, operating cost and occupant thermal comfort. This chapter investigates the application of a multi-objective genetic algorithm (MOGA) search method in the identification of the optimum pay-off characteristic between the energy cost of a building and the occupant thermal discomfort. Results are presented for the pay-off characteristics between energy cost and zone thermal comfort, for three design days and three building weights. Inspection of the solutions indicates that the MOGA is able to find the optimum pay-off characteristic between the daily energy cost and zone thermal comfort. It can be concluded that multi-criterion genetic algorithm search methods offer great potential for the identification of the pay-off between the elements of building thermal design and as such can help the building design process.

Energy input estimation
Ozturk et al (2004) developed the energy input estimation equations for the residential–commercial sector (RCS), in order to estimate the future projections based on the genetic algorithm notion, and examined the effect of design parameters on the energy input of the sector. For this purpose, the Turkish RCS is given as an example. The GA

TABLE 4.9 Summary of applications of GAs in buildings

SYSTEM	REFERENCES
Optimization of building thermal design and control	Wright et al (2002)
Energy input estimation	Ozturk et al (2004)
Determination of heat transfer coefficient of exterior wall surface	Zhang et al (2004)

Energy Input Estimation Model (GAEIEM) is used to estimate Turkey's future residential–commercial energy input based on gross domestic product (GDP), population, import, export, house production, cement production and basic house appliance consumption figures. It may be concluded that the three forms of the models proposed here can be used as alternatives to the available solution and estimation techniques to available estimation techniques. It is also expected that this study will be helpful in developing highly applicable and productive planning for energy policies.

Determination of heat transfer coefficient of exterior wall surface

Zhang et al (2004) proposed a new experimental procedure based on genetic algorithms for analysing the exterior wall surface heat transfer process and deducing exterior wall surface heat transfer coefficients under actual conditions. Tests were carried out under a wide range of conditions and the heat transfer coefficient was found to vary from 14.315 to 24.412 W/m²K while wind speed ranged from 1.04 to 7.36 m/s, and correlation was obtained in terms of the wind speed. Tests revealed that the commonly used correlation for predicting the heat transfer coefficient was overestimated, especially at high wind speeds.

CONCLUSIONS

From the research work described in this chapter, one can see that ANNs and GAs have been applied in a wide range of fields for modelling and prediction in building energy systems. What is required for setting up such systems is data that represent the past history and performance of the real system and a suitable selection of an ANN or GA model. The selection of this model is done empirically and after testing various alternative solutions. The performance of the selected models is tested with the data of the past history and performance of the real system.

The number of applications presented here is neither complete nor exhaustive but merely a sample that demonstrate the usefulness of ANN and GA models. ANNs and GAs, like all other approximation techniques, have relative advantages and disadvantages. There are no rules as to when this particular technique is more or less suitable for an application. Based on the work presented here, it is believed that they offer an alternative method that should not be underestimated.

AUTHOR CONTACT DETAILS

Soteris A. Kalogirou: Department of Mechanical Engineering and Materials Sciences and Engineering, Cyprus University of Technology, PO Box 50329, Limassol 3603, Cyprus; Soteris.kalogirou@cut.ac.cy

REFERENCES

Alawi, S. M. and Hinai, H. A. (1998) 'An ANN-based approach for predicting global radiation in locations with no direct measurement instrumentation', *Renewable Energy*, vol 14, no 1–4, pp199–204

Arca, B., Benincasa, F., De Vincenzi, M. and Fasano, G. (1998) 'A neural model to predict the daily minimum of air temperature', in *7th CCTA – International Congress for Computer Technology in Agriculture, Computer Technology in Agricultural Management and Risk Prevention*, Firenze, pp1–9

Argiriou, A. A., Bellas-Velidia, I. and Balaras, C. A. (2000) 'Development of a neural network heating controller for solar buildings', *Neural Networks*, vol 13, pp811–820

Aydinalp, M., Ugursal, V. I. and Fung, A. S. (2004) 'Modeling of the space and domestic hot-water heating energy-consumption in the residential sector using neural networks', *Applied Energy*, vol 79, no 2, pp159–178

Ben-Nakhi, A. E. and Mahmoud, M. A. (2002) 'Energy conservation in buildings through efficient A/C control using neural networks', *Applied Energy*, vol 73, no 1, pp5–23

Ben-Nakhi, A. E. and Mahmoud, M. A. (2004) 'Cooling load prediction for buildings using general regression neural networks', *Energy Conversion and Management*, vol 45, no 13–14, pp2127–2141

Cao, S. and Cao, J. (2005) 'Forecast of solar irradiance using recurrent neural networks combined with wavelet analysis', *Applied Thermal Engineering*, vol 25, no 2–3, pp161–172

Davis, L. (1991) *Handbook of Genetic Algorithms,* Van Nostrand, New York

Goldberg, D. E. (1989) *Genetic Algorithms in Search Optimization and Machine Learning,* Addison-Wesley, Reading, MA

Gouda, M. M., Danaher, S. and Underwood, C. P. (2002) 'Application of artificial neural network for modelling the thermal dynamics of a building's space and its heating system', *Mathematical and Computer Modelling of Dynamical Systems,* vol 8, no 3, pp333–344

Haykin, S. (1994) *Neural Networks: A Comprehensive Foundation,* Macmillan, New York

Kalogirou, S. A. (2000a) 'Application of artificial neural networks for energy systems', *Applied Energy,* vol 67, no 1–2, pp17–35

Kalogirou, S. A. (2000b) 'Forced circulation solar domestic water heating systems long-term performance prediction using artificial neural networks', *Applied Energy,* vol 66, no 1, pp63–74

Kalogirou, S. A. (2001) 'Artificial neural networks in renewable energy systems: A review', *Renewable & Sustainable Energy Reviews,* vol 5, no 4, pp373–401

Kalogirou, S. A. (2003) 'Artificial intelligence for the modelling and control of combustion processes: A review', *Progress in Energy and Combustion Science,* vol 29, no 6, pp515–566

Kalogirou, S. A. and Panteliou, S. (1999) 'Dynamic system testing method and artificial neural networks for solar water heater long-term performance prediction', in *Proceedings of the European Symposium on Intelligent Techniques ESIT'99 on CD-ROM,* Crete, Greece

Kalogirou, S. A. and Panteliou, S. (2000) 'Thermosyphon solar domestic water heating systems long-term performance prediction using artificial neural networks', *Solar Energy,* vol 69, no 2, pp163–174

Kalogirou, S., Neocleous, C. and Schizas, C. (1997) 'Artificial neural networks used for estimation of building heating load', in *Proceedings of the CLIMA 2000 International Conference,* Brussels, Belgium, Paper Number P159

Kalogirou, S. A., Neocleous, C., Paschiardis, S. and Schizas, C. (1999a) 'Wind speed prediction using artificial neural networks', in *Proceedings of the European Symposium on Intelligent Techniques ESIT'99* on CD-ROM, Crete, Greece

Kalogirou, S. A., Panteliou, S. and Dentsoras A. (1999b) 'Modelling of solar domestic water heating systems using artificial neural networks', *Solar Energy,* vol 65, no 6, pp335–342

Kalogirou, S. A., Panteliou, S. and Dentsoras, A. (1999c) 'Artificial neural networks used for the performance prediction of a thermosyphon solar water heater', *Renewable Energy,* vol 18, no 1, pp87–99

Kalogirou, S.A., Eftekhari, M. and Pinnock, D. (2001a) 'Artificial neural networks used for the prediction of air flow in a single-sided naturally ventilated test room', *Building Services Engineering Research and Technology Journal,* vol 22, no 2, pp83–93

Kalogirou, S., Florides, G., Neocleous, C. and Schizas, C. (2001b) 'Estimation of the daily heating and cooling loads using artificial neural networks', in *Proceedings of CLIMA 2000 International Conference on CD-ROM,* Naples, Italy

Kalogirou, S. A., Lalot, S., Florides, G. and Desmet, B. (2008) 'Development of a neural network-based fault diagnostic system for solar thermal applications', *Solar Energy,* vol 82, no 2, pp164–172

Kemmoku, Y., Orita, S., Nakagawa, S. and Sakakibara, T. (1999) 'Daily insolation forecasting using a multi-stage neural network', *Solar Energy*, vol 66, no 3, pp193–199

Kolokotsa, D. (2007) 'Artificial intelligence in buildings: A review of the application of fuzzy logic', *Advances in Building Energy Research*, vol 1, no 1, pp29–54

Kretzschmar, R., Eckert, P. and Cattani, D. (2004) 'Neural network classifiers for local wind prediction', *Journal of Applied Meteorology*, vol 43, pp727–738

Lalot, S. (2000) 'Identification of the time parameters of solar collectors using artificial neural networks', in *Proceedings of Eurosun 2000 on CD-ROM*, Copenhagen, Denmark

Lee, W., House, J. M. and Kyong, N. (2004) 'Subsystem level fault diagnosis of a building's air-handling unit using general regression neural networks', *Applied Energy*, vol 77, no 2, pp153–170

Lopez, G., Batlles, F. J. and Tovar-Pescador, J. (2005) 'Selection of input parameters to model direct solar irradiance by using artificial neural networks', *Energy*, vol 30, no 9, pp1675–1684

Mechaqrane, A. and Zouak, M. (2004) 'A comparison of linear and neural network ARX models applied to a prediction of the indoor temperature of a building', *Neural Computing and Applications*, vol 13, no 1, pp32–37

Michalakakou, G., Santamouris, M. and Tsagrassoulis, A. (2002) 'On the energy consumption in residential buildings', *Energy and Buildings*, vol 34, no 7, pp727–736

Michalewicz, Z. (1996) *Genetic Algorithms + Data Structures = Evolution Programs*, 3rd edn, Springer, Berlin

Mohandes, M., Rehman, S. and Halawani, T. O. (1998a) 'Estimation of global solar radiation using artificial neural networks', *Renewable Energy*, vol 14, no 1–4, pp179–184

Mohandes, M., Rehman, S. and Halawani, T. O. (1998b) 'A neural network approach for wind speed prediction', *Renewable Energy*, vol 13, no 3, pp345–354

Nannariello, J. and Fricke, F. R. (2001) 'Introduction to neural network analysis and its applications to building services engineering', *Building Services Engineering Research & Technology*, vol 22, no 1, pp58–68

Ozturk, H. K., Canyurt, O. E., Hepbesli, A. and Utlu, Z. (2004) 'Residential-commercial energy input estimation based on genetic algorithm (GA) approaches: An application for Turkey', *Energy and Buildings*, vol 36, no 2, pp175–183

Reddy, K. S. and Ranjan, M. (2003) 'Solar resource estimation using artificial neural networks and comparison with other correlation models', *Energy Conversion and Management*, vol 44, no 15, pp2519–2530

Ripley, B. D. (1996) *Pattern Recognition and Neural Networks*, Cambridge University Press, Cambridge

Rumelhart, D. E., Hinton, G. E. and Williams, R. J. (1986) *Learning Internal Representations by Error Propagation, Parallel Distributed Processing: Explorations in the Microstructure of Cognition*, vol 1, Chapter 8, MIT Press, Cambridge, MA

Schizas, C. N., Michaelides, S. C., Pattichis, C. S. and Livesay, R. R. (1991) 'Artificial neural networks in forecasting minimum temperature', in *Proceedings of Second International Conference on Artificial Neural Networks*, Bournemouth, UK, Institution of Electrical Engineers, Publication no 349, pp112–114

Schizas, C. N., Pattichis, C. S. and Michaelides, S. C. (1994) 'Forecasting minimum temperature with short time-length data using artificial neural networks', *Neural Networks World*, vol 2, pp219–230

Soares, J., Oliveira, A. P., Boznar, M. Z., Mlakar, P., Escobedo, J. F. and Machado, A. J. (2004) 'Modelling hourly diffuse solar-radiation in the city of Sao Paulo using artificial neural-network technique', *Applied Energy*, vol 79, no 2, pp201–214

Sozen, A., Arcaklioglu, E. and Ozalp, M. (2004a) 'Estimation of solar potential in Turkey by artificial neural networks using meteorological and geographical data', *Energy Conversion and Management*, vol 45, no 18–19, pp3033–3052

Sozen, A., Ozalp, M. , Arcaklioglu, E. and Kanit, E. G. (2004b) 'A study for estimating solar resources of Turkey using artificial neural networks', *Energy Resources*, vol 26, no 14, pp1369–1378

Sozen, A., Arcaklioglu, E., Ozalp, M. and Kanit, E. G. (2004c) 'Use of artificial neural networks for mapping of solar potential of Turkey', *Applied Energy*, vol 77, no 3, pp273–286

Sozen, A., Arcaklioglu, E., Ozalp, M. and Kanit, E. G. (2005) 'Solar-energy potential in Turkey', *Applied Energy*, vol 80, no 4, pp367–381

Tasadduq, I., Shafiqur Rehman, S. and Bubshait, K. (2002) 'Application of neural networks for the prediction of hourly mean surface temperatures in Saudi Arabia', *Renewable Energy*, vol 25, pp545–554

Tsoukalas, L. H. and Uhrig, R. E. (1977) *Fuzzy and Neural Approaches in Engineering*, John Wiley & Sons, New York

Veerachary, M. and Yadaiah, N. (2000) 'ANN based peak power tracking for PV supplied DC motors', *Solar Energy*, vol 69, no 4, pp343–350

Wang, S. and Jiang, Z. (2004) 'Valve fault detection and diagnosis based on CMAC neural networks', *Energy and Buildings*, vol 36, no 6, pp599–610

Werbos, P. J. (1974) *Beyond Regression: New Tools for Prediction and Analysis in the Behavioural Science*, PhD thesis, Harvard University, Cambridge, MA

Wright, J. A., Loosemore, H. A. and Farmani, R. (2002) 'Optimisation of building thermal design and control by multi-criterion genetic algorithm', *Energy and Building*, vol 34, no 9, pp959–972

Yang, I., Yeo, M. and Kim, K. (2003) 'Application of artificial neural network to predict the optimal start time for heating system in building', *Energy Conversion and Management*, vol 44, no 17, pp2791–2809

Zaheer-uddin, M. and Tudoroiu, N. (2004a) 'Neuro-models for discharge air temperature', *Energy Conversion and Management*, vol 45, no 6, pp901–910

Zaheer-uddin, M. and Tudoroiu, N. (2004b) 'Neuro-PID tracking control of a discharge air temperature system', *Energy Conversion and Management*, vol 45, no 15–16, pp1405–2415

Zalzala, A. and Fleming, P. (1997) *Genetic Algorithms in Engineering Systems*, The Institution of Electrical Engineers, London, UK

Zhang, L., Zhang, N., Zhao, F. and Chen, Y. (2004) 'A genetic-algorithm-based experimental technique for determining heat transfer coefficient of exterior wall surface', *Applied Thermal Engineering*, vol 24, no 2–3, pp339–349

Zmeureanu, R. (2002) 'Prediction of the COP of existing rooftop units using artificial neural networks and minimum number of sensors', *Energy*, vol 27, no 9, pp889–904

earthscan

publishing for a sustainable future

5

Decision support methodologies on the energy efficiency and energy management in buildings

D. Kolokotsa, C. Diakaki, E. Grigoroudis, G. Stavrakakis and K. Kalaitzakis

Abstract

The aim of the present chapter is to analyse the decision support processes towards energy efficiency and improvement of the environmental quality in buildings. The main criteria in the decision analysis of buildings are categorized. The decision alternatives which may formulate specific actions, or group of actions (strategies) for buildings' sustainability are analysed. The decision methodologies presented are separated to online (based on real-time operation of buildings) and offline decision approaches. Both approaches are supported by simulation, multi-objective programming optimization techniques, multi-criteria decision analysis techniques and their combinations in order to reach optimum solution, rank alternatives or provide trade-offs between the criteria. The advantages and drawbacks of the various methods are discussed and analysed.

■ *Keywords* – energy efficiency; indoor environment; multi-criteria decision analysis; multi-objective decision support; energy management systems

INTRODUCTION

The building sector has a substantial share of the primary energy supply being a major contributor to conventional fuels consumption, thus creating a significant environmental burden through materials production and global warming gas releases. Buildings account for about 40 per cent of the global energy use. To save a significant portion of this energy, the International Energy Agency (IEA, 2008) recommends action on:

● building codes for new buildings;
● passive energy houses and zero energy buildings;
● policy packages to promote energy efficiency in existing buildings;
● building certification schemes;
● energy efficiency improvements in windows.

ADVANCES IN BUILDING ENERGY RESEARCH ■ 2009 ■ VOLUME 3 ■ PAGES 121–146

doi:10.3763/aber.2009.0305 ■ © 2009 Earthscan ■ ISSN 1751-2549 (Print), 1756-2201 (Online) ■ www.earthscanjournals.com

The European energy policy has a clear orientation towards the preservation of energy and the improvement of indoor environmental quality in buildings through the adoption of the European Commission's (EC) Energy Performance of Buildings Directive (EPBD) 2002/91/EC (EC, 2003). To support EPBD, the Comité Européen de Normalisation (CEN) introduced several new CEN standards (e.g. CEN 2005; CEN 2006). The EC, on January 2008, unveiled the Climate Action package to fight climate change and promote renewable energy in line with European Union (EU) commitments. This builds on the many targets that the EU set itself in 2007 for 2020 as part of the Energy Policy for the EU, including 20 per cent reduction in greenhouse gases, 20 per cent increase in energy efficiency, and increasing renewable energy use to 20 per cent of the total energy consumption.

To this end, in the past decades, there have been significant efforts towards designing, operating and maintaining energy efficient and environmentally conscious buildings.

Optimal energy management and energy efficiency in the buildings sector is a valuable tool for natural resources conservation. Moreover, the financial benefits from using energy efficient technologies constitute the major motivation for building owners. Buildings' energy efficiency and environmental burden are treated differently in the design and operational phase. In selecting the most appropriate actions or strategies for energy efficiency and reduction of buildings' environmental impact, either in the design or in the operational phase, various methodologies are utilized.

The aim in the buildings' design or renovation phase usually is:

- compliance of the building with regulations in the design or in the operational phase;
- improvement of the building's energy efficiency in a sector (i.e. heating or cooling) or overall improvement;
- improvement of the indoor environment (i.e. improvement of indoor thermal comfort, visual comfort or indoor air quality or a combination of these);
- environmental impact of the building for global warming, etc.;
- reduction of the energy-related costs.

More specifically, in the design phase, the objective is to achieve the best equilibrium between the essential design parameters versus a set of criteria that are subject to specific constraints. The essential design variables that contribute to the energy and environmental burden of a building and influence the occupants' comfort, the heating and cooling energy demand as well as the lighting demand can be (Gero et al, 1983):

- building shape;
- orientation;
- building mass;
- type of glazing and glazing ratio;
- shading.

Usually, the designer uses simulation building modelling to assess predefined design aspects and solutions, subject to the building owner's subjective preferences (cost of the building, energy efficiency, aesthetics, etc.).

In the operational stage of a building, decisions towards energy efficiency are usually undertaken with the support of energy audit and survey procedures (Krarti, 2000). Energy auditing of buildings can range from a short walk-through survey to a detailed analysis with hourly computer simulation. Any actions in the building undertaken during its operational stage can be either refurbishment or retrofit. The term refurbishment implies the necessary modifications in order to return a building to its original state, while retrofit includes the necessary actions that will improve the building's energy and/or environmental performance.

The state of practice procedure for the improvement of a building's energy efficiency in its operational phase follows four steps:

- Step 1: Buildings analysis. The main purpose of this step is to evaluate the characteristics of the energy systems and the patterns of energy use for the building. The building characteristics can be collected from the architectural/mechanical/ electrical drawings and/or from discussions with building operators. The energy use patterns can be obtained from a compilation of utility bills over several years. Analysis of the historical variation of the utility bills allows the energy auditor to determine if there are any seasonal and weather effects on the building energy use.
- Step 2: Walk-through survey. Potential energy saving measures are identified in this part. The results of this step are important since they determine whether the building warrants any further energy auditing work. Some of the tasks involved in this step are:
  - identification of the customer concerns and needs;
  - checking of the current operating and maintenance procedures;
  - determination of the existing operating conditions of major energy use equipment (lighting, heating ventilation and air-conditioning systems, motors, etc.);
  - estimation of the occupancy, equipment and lighting (energy use density and hours of operation).
- Step 3: Creation of the reference building. The main purpose of this step is to develop a base-case model, using energy analysis and simulation tools, that represents the existing energy use and operating conditions of the building. This model is to be used as a reference to estimate the energy savings incurred from appropriately selected energy conservation measures.
- Step 4: Evaluation of energy saving measures. In this step, a list of cost-effective energy conservation measures is determined using both energy saving and economic analysis. A predefined list of energy conservation measures is prepared. The energy savings due to the various energy conservation measures pertinent to the building using the baseline energy use simulation model are evaluated. The initial costs required to implement the energy conservation measures are estimated. The cost-effectiveness of each energy conservation measure using an economic analysis method (simple payback or life cycle cost analysis) is assessed.

Regardless of the dwelling's phase (design or operational), energy efficiency and sustainability in buildings is a complex problem. This is attributed mainly to the fact that buildings consist of numerous subsystems that interrelate with each other. The subsystems are:

● structural system and materials;
● building systems, like heating, ventilation, air-conditioning and lighting;
● building services, such as support of indoor comfort requirements and management.

Therefore buildings' sustainability is reached by taking the necessary decisions that are optimum for the overall system. This implies a decision support approach with the following steps:

● identification of the overall goal in making a decision, subsidiary objectives and the various indices or criteria against which option performance may be measured (objective function);
● identification of the alternative options or strategies;
● assessment of each option and/or strategy performance against the defined criteria;
● weighting of objectives or criteria;
● evaluation of the overall performance;
● evaluation and ranking of options;
● sensitivity analysis.

The purpose of the present chapter is to analyse the state of the art of the methodologies for decision support in energy management and sustainability for the building sector. The steps usually followed in this procedure and listed above are depicted in Figure 5.1. The first major step is to find the overall goal and criteria based on which the process will be assessed. The main criteria and objectives in the state of the art are analysed in 'Criteria in Decision Support for Energy Efficiency and Energy Management' below. The actions usually undertaken by the designers, architects, building scientists, etc. are presented in 'Alternative Options and Strategies'. 'Assessment methodologies' analyses the assessment methods for buildings' energy and environmental improvement, while the decision support methodologies are discussed in the following section. 'Conclusion and Future Prospects' summarizes the main conclusions of the chapter and highlights further development needs.

CRITERIA IN DECISION SUPPORT FOR ENERGY EFFICIENCY AND ENERGY MANAGEMENT

The criteria for energy efficiency and energy management in a new construction or retrofit can be either quantitative or qualitative and can be divided into the categories depicted in Figure 5.2.

More specifically, regarding *energy use* (primary or delivered), the following indices have been utilized in chronological order:

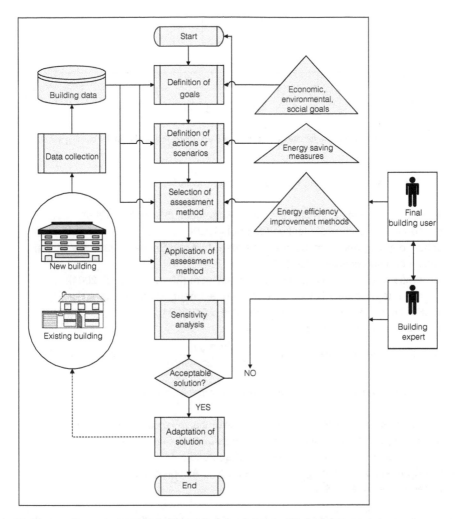

FIGURE 5.1 The methodology for buildings' design and operational improvement

- heating and cooling load for conditioned buildings (D'Cruz and Radford, 1987; Bouchlaghem, 2000);
- normalized annual energy consumption and energy use for heating in kWh/m^2 (Rey, 2004; Zhu, 2006);
- annual electricity use in kWh/m^2 (Rey, 2004);
- embodied energy (Chen et al, 2006);
- energy and time consumption index (ETI) (Chen et al, 2006);

- energy savings by retrofitting expressed by $\left(1 - \dfrac{Energy}{Energy\,baseline}\right)\%$ (Gholap and Khan, 2007).

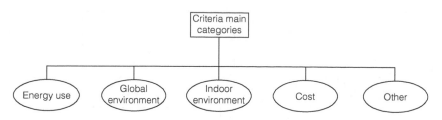

FIGURE 5.2 The main criteria for energy efficiency and environmental quality in the building sector

Regarding *costs*, the following indices have been found in the literature:

● direct costs and initial investment costs (Rosenfeld and Shohet, 1999);
● economic life span (Rosenfeld and Shohet, 1999);
● annual ongoing maintenance charges (Rosenfeld and Shohet, 1999; Rey, 2004);
● annual ongoing charges (Rey, 2004);
● net present value (NPV) of the energy investment (Martinaitis et al, 2004);
● internal rate of return (IRR) of the energy investment (Martinaitis et al, 2004);
● cost of conserved energy (CCE) (Martinaitis et al, 2004);
● life cycle cost (LCC) (Wang et al, 2005);
● cost of retrofitting expressed by $\left(1-\dfrac{Energy}{Energy\,baseline}\right)\%$ (Gholap and Khan, 2007).

As far as *global environment* is concerned, the criteria usually set are:

● annual emissions GWP (global warming potential in $kg_{eq}CO_2/m^2$) (Rey, 2004);
● reduction potential of global warming emissions (Alanne, 2004);
● life cycle environmental impact (Wang et al, 2005);
● acidification potential in $kg_{eq}SO_2/m^2$ (Rey 2004; Alanne et al, 2007);
● water use (Alanne et al, 2007).

Indoor environmental quality and *comfort* have subcategories for the evaluation of thermal sensation, visual comfort, indoor air quality and acoustic comfort. More specifically, regarding *thermal comfort*, the following criteria and indicators have been used:

● predicted mean vote based on ISO 7730 standard (ISO, 1984);
● dry resultant temperature for unconditioned buildings (Bouchlaghem, 2000);
● indoor temperature and humidity (Jaggs and Palmer, 2000);
● discomfort hours during summer or winter (Roulet et al, 2002);
● daily overheating in K (Rey, 2004);
● percentage of people dissatisfied index (Blondeau et al, 2002; Rutman et al, 2005);
● effective draught temperature index (Rutman et al, 2005);
● summer thermal discomfort severity index, which indicates the summer severity of excessive mean radiant temperature during summer (Becker et al, 2007).

For *visual comfort*, the assessment criteria can be:

● daylight availability (Radford and Gero, 1980);
● lighting and visual comfort (e.g. EPIQR method, see Bluyssen, 2000; Rey, 2004);
● daylight factor (Rey, 2004);
● discomfort glare severity indicator, which indicates the annual severity of excessive discomfort glare (Becker et al, 2007).

Indoor air quality can be assessed via:

● CO_2 concentration index (Kolokotsa et al, 2001; Doukas et al, 2007);
● maximum ratio between the mean concentration of a contaminant over the occupancy period and the contaminant's threshold limit value for short-term or long-term exposure (Blondeau et al, 2002);
● ventilation rates (Blondeau et al, 2002).

Acoustic comfort criteria include:

● noise level in dB at workplace (Rey, 2004);
● noise rating index (Rutman et al, 2005).

Other criteria for assessing buildings' performance in conjunction with energy efficiency can be:

● construction duration (Rosenfeld and Shohet, 1999);
● level and ranking of the service according to state of the art (Rosenfeld and Shohet, 1999);
● uncertainty factors (Rosenfeld and Shohet, 1999);
● communication cost between spaces (Homoud, 2001);
● allocation of activities within spaces (Homoud, 2001);
● functionality (Alanne, 2004) influenced by:
  ● easiness of implementation of the retrofit
  ● effect on comfortability
  ● space requirements
  ● adaptability to existing structure
  ● usability
  ● serviceability;
● work efficiency of building energy management systems (BEMS) and intelligent buildings, such as reliability (Wong and Li, 2008);
● security, i.e. fire and safety features, alarm systems, etc. (Wong and Li, 2008).

Most of the aforementioned criteria are competitive. As a consequence, it is impossible to find a global solution to satisfy all of them simultaneously. For this reason, several decision support techniques are used in both the design and operational phases to enable reaching a solution that will be satisfactory enough according to the preferences and priorities of the building user/owner.

ALTERNATIVE OPTIONS AND STRATEGIES

The different actions that may be undertaken include more than 400 alternatives (Wulfinghoff, 1999) and may be grouped in categories as tabulated in Table 5.1. Moreover these actions may be accomplished separately or combined in groups thus formulating a strategy (see Table 5.1 and Figure 5.3). Additionally, all actions may be applied to both the design (Becker et al, 2007) and the operational phase (Alanne 2004).

The selection of actions for each building depends upon its characteristics, the scope of the retrofit or design and the criteria that the actions will serve. For example, a decision support approach for selecting energy-saving building components in the building design is proposed by Wilde and van der Voorden (2004). The alternative actions for this specific work include solar walls, advanced glazing, sunspaces and photovoltaic arrays. Additionally, the influence of building design variables and improvement of the ventilation schemes on the thermal performance, indoor air quality and energy efficiency is studied by Becker et al (2007). The alternatives selected are on the one hand related to the building's design, i.e. building orientation, size of windows, thermal insulation, mass, colour and facades and, on the other hand, to various ventilation schemes.

Lists of retrofit actions are provided by various researchers (Alanne, 2004; Doukas et al, 2008). Indicatively, these lists include:

- lighting improvements such as replacement of lamps and use of lighting control systems;
- heating and cooling improvements such as installation of extra monitoring devices, introduction of natural gas or solar energy systems for absorption cooling, etc.;
- electromechanical equipment improvements such as load factor corrections, etc.;
- general improvements, i.e. installation of renewables, insulation, etc.

An example of a set of alternatives in a strategic form is provided by Rosenfeld and Shohet (1999) including full retrofit, partial retrofit actions, demolition and reconstruction of the building and construction of a new building nearby.

Moreover, regarding strategies, i.e. combined actions that lead to a specific result, Rey (2004) proposes the following:

- the stabilization strategy (STA) which consists of a set of actions that do not alter the building's appearance;
- the substitution strategy (SUB) which consists of a set of actions that alter the building's appearance to a great extent;
- the double-skin facade strategy (DSF) which corresponds to adding a new glass skin.

As a general comment, all efforts for energy efficiency and improvement of building performance are focusing on specific actions or action categories without the adoption of a global and holistic approach mainly due to the problem's complexity.

TABLE 5.1 The different actions for improving buildings' energy efficiency

BUILDING COMPONENT	STRATEGY	ACTIONS
Building envelope and design aspects	Increase of insulation	Roofs: Increase the quantity of attic insulation; add rigid insulation to the top surface of roofs; apply sprayed foam insulation to the top surface of roofs; Install a suspended insulated ceiling, etc.
		Walls: Insulate wall cavities; insulate the inside surfaces of walls; insulate the outside surface of walls, etc.
		Glazing: Install high-efficiency glazing; install storm windows or supplemental glazing; reduce the areas of glazing; install thermal shutters; use window films that reflect heat back into the building, etc.
	Reduction of air leakage	Install appropriate weather-stripping on exterior doors; install high-efficiency doors; maintain the fit, closure, and sealing of windows; install weather-stripping on openable windows; install supplemental ('storm') windows; install high-efficiency windows; seal gaps in the envelope structure, etc.
	Use of advanced building envelope technologies	Spectrally selective glasses; chromogenic glazings; cool materials, etc.
	Control and exploitation of sunlight	Reduction of cooling loads: Install external shading devices appropriate for each exposure of the glazing; install internal shading devices; install high-efficiency glazing; install solar control films on existing glazing; reduce the area of glazing (with insulating panels), etc.
		Daylight: Install skylight or light pipes; install diffusers for existing clear skylights; install translucent roof and wall sections for daylight; install diffusers to make windows more effective for daylight; install a system of light shelves and shading, etc.
		Passive solar heating: Keep open the window shades of unoccupied spaces that need heating; install combinations of sunlight absorbers and reflectors inside windows and skylights; install solar enclosures over areas that can benefit from heating, etc.

TABLE 5.1 The different actions for improving buildings' energy efficiency (Cont'd)

BUILDING COMPONENT	STRATEGY	ACTIONS
Building services	Heating Ventilation and Air-Conditioning Systems (HVAC)	Setting up/back thermostat temperatures; regular retrofit of constant air volume systems, etc.; installation of heat recovery systems; retrofit of control heating plants, etc.
	Mechanical Equipment	Boiler types and ratings maintenance; use of Compression cooling; absorption cooling; variable-speed motors and drives, etc.
	Office equipment	Control of operating time, use of high-efficiency office equipment.
	Motors	Reduction of operating time; optimized control; Installation of energy efficient motors.
	Electric systems	High-efficiency lamps & ballasts; addition of reflecting devices; light pipe technologies; replace fluorescent lamps with high-efficiency or reduced-wattage types; replace ballasts with high-efficiency or reduced-wattage types, or upgrade both ballasts and lamps; install current limiters; install fluorescent dimming equipment; consider retrofit 'reflectors' for fluorescent fixtures, etc.
Energy management tools	Monitor and control of the building during its operation	Sensors; clock controls and programmable thermostats; measurement of liquid, gas, and heat flow; control signal pooling; energy analysis computer programmes; advanced control systems; decision support systems; energy management control systems, infrared thermal scanning, etc.

FIGURE 5.3 The actions and strategies tree

ASSESSMENT METHODOLOGIES

The decision problems are generally based on the description of the set of alternatives and can be (Ehrgott, 2005):

- Problems with a large but finite set of alternatives that are explicitly known. Usually, the goal is to select a most preferred one.
- Discrete problems where the set of alternatives is described by constraints in the form of mathematical functions.
- Continuous problems – the set of alternatives is generally given through constraints.

Finding the optimum decision is usually an optimization procedure. Optimization is a technique of maximizing or minimizing specific objective functions under constraints. The objective functions are formulated to represent the decision criteria. Therefore, the selection of the optimization procedure depends on the problem's particularities.

In the building sector, the assessment phase involves the evaluation of the predefined actions or strategies analysed in 'Alternative Options and Strategies' versus the selected criteria that have been pinpointed in 'Criteria in Design Support for Energy Efficiency and Energy Management'. Difficulties and drawbacks exist due to buildings' complexity (Homoud, 2001).

The major issues that arise are:

- The criteria are usually more than one and are conflicting (indoor comfort and energy, energy consumption and investment cost, etc.).
- The actions discussed in 'Alternative Options and Strategies' involve a number of decision variables that are not negligible.

Therefore, the assessment procedure is an iterative procedure strongly influenced by the criteria, the alternatives, actions and strategies and finally by the end user. This iterative procedure is illustrated in Figure 5.4.

The approaches towards the improvement of energy efficiency in buildings that are met in the relevant literature may be distinguished according to their different characteristics (Figure 5.5).

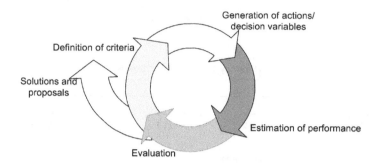

FIGURE 5.4 The iterative decision support process (Alanne, 2004)

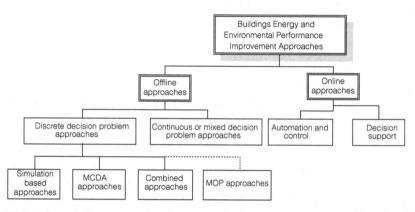

FIGURE 5.5 Categorization of methodological approaches for the improvement of energy efficiency in buildings

Depending on how an approach is implemented, they may be distinguished in offline and online approaches. Offline approaches aim to identify particular measures such as insulation materials, wall construction, boiler type, etc. that are expected to lead to improved building energy and environmental performance. These approaches may be applied either during the design phase or in the frame of a refurbishment or retrofit during the operational phase of the building. The offline approaches are not interacting with the building in real time. Online approaches, on the other hand, aim to identify particular parameters such as setpoints, control strategies (night setback, compensation, etc.) based on real-time measurements collected through a BEMS that improves the energy performance of the building during its real-time operation. Both online and offline approaches are analysed in the following sections.

OFFLINE APPROACHES

Offline approaches may be further divided into approaches that are applied to a decision problem formulated as a discrete one, and approaches applied to a decision problem formulated as a continuous or mixed (i.e. both continuous and discrete) one. This separation is performed according to the way the different available alternatives (actions or strategies) are considered.

Discrete decision problem approaches

In this case, approaches are met where a possibly large but in any case discrete and definite set of potential actions or strategies – in the sense described in 'Alternative Options and Strategies' – is considered. The final assessment and selection under this category may be based on various sub-approaches as depicted in Figure 5.5. These are as follows.

Simulation based approaches

The simulation-based approaches can be either simplified (analytical methods) or detailed (numerical methods) using powerful simulation programmes (Clarke, 2001). Indicatively, the simplified methods are the degree-day method, the variable-base degree-day method, the bin method and the modified bin method (Homoud, 2001; Kreider et al, 2002).

In the simulation-based process, a basic model is developed using simulation tools. Then, through an iterative procedure, a series of recommendations are defined, using the simulation energy analysis in order to 'move' the building from a typical construction to a best practice construction (Horsley et al, 2003). These recommendations may include increase of insulation, use of innovative highly efficient glazing, change of building shape and aspect ratios, etc. (Table 5.1).

The detailed simulation programmes are analysed by many authors (see for example, Hong et al, 2000; Homoud, 2001). An overview of the computational support for energy efficient building components is provided by Wilde and van der Voorden (2004). The main simulation tools for energy analysis are TRNSYS, DOE-2, EnergyPlus, BLAST, ESP-r, etc.

TRNSYS is used by a number of researchers. Florides et al (2002) used TRNSYS to examine measures such as natural and controlled ventilation, solar shading, various types of glazing, orientation, shape of buildings, and thermal mass aiming to reduce the thermal load, while Zurigat et al (2003) used TRNSYS to evaluate different passive measures aiming to reduce the peak cooling load of school buildings.

EnergyPlus is used by Becker et al (2007) to assess specific factors of building design elements (window orientation, glazing type, thermal resistance of walls, etc.) and 20 ventilation strategies for schools' energy consumption and efficiency.

Visual DOE is used by Tavares and Martins (2007) to perform a sensitivity analysis that results in energy efficient design solutions for a specific case study. A significant number of predefined solutions are modelled and evaluated.

By concluding the simulation-based approach, the most appropriate energy tool is selected based upon:

- type of criteria to be assessed;
- the required accuracy;
- easiness;
- availability of required data;
- building's phase (design or operational).

Multi-criteria decision analysis approaches

Multi-criteria decision analysis (MCDA) is used by many researchers to support the synthesis of the potential actions and includes areas such as multi-attribute utility theory (Keeney and Raiffa, 1993) and outranking methods (ELECTRE method, see Roy, 1991). Moreover, MCDA supports the inclusion of subjective aspects through the decision makers' preferences that influence the decision process.

In the building design process, MCDA is utilized by:

- Gero et al (1983) to assess the optimum orientation, window fraction, etc. versus capital cost, area that is used and total thermal load ration; and
- Jedrzejuk and Marks (2002) to find the optimum solution for building shape, internal partitions and optimization of heat sources for blocks of flats, using an iterative procedure.

In the operational and retrofit stage, MCDA is introduced by various researchers. Combinatorial and outranking methods are used by Blondeau et al (2002) to assess indoor air quality, thermal comfort and energy consumption. With the combinatorial method, each potential action is assessed via a utility function that fits the decision process. Then each action is ranked according to its utility function. In the outranking method, the potential actions are compared in pairs and final ranking of the predefined actions is extracted after ascending and descending ranking. The utility functions used for each action are I_{area}, which is the surface area of the triangle shaped when joining the coordinates of each considered action in the three-dimensional criteria space (Figure 5.6) and I_{norm}, which corresponds to the norm of the vector joining the origin to the action point.

A similar approach is followed by Rutman et al (2005) who use the multi-criteria ELECTRE method to evaluate the thermal comfort, acoustic comfort and indoor air

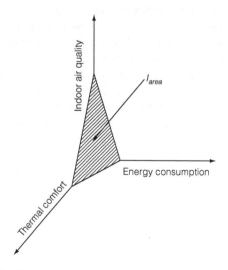

FIGURE 5.6 The utility function to assess the indoor thermal comfort, air quality and energy consumption (Blondeau et al, 2002)

distribution in an office air-conditioned room. The purpose of the procedure is to extract design rules for air-conditioning systems that satisfy indoor comfort requirements.

Outranking methods with criterion by criterion comparison are also used by Rey (2004). ELECTRE III method is used to rank the three strategies discussed in 'Alternative Options and Strategies' after ascending and descending ranking. Each criterion is weighted according to its importance. The retrofitting strategies are placed in a final ranking if they have the same position in the ascending and descending ranking.

Moreover, the Office Rating Methodology (ORME) proposed by Roulet et al (2002) uses ELECTRE algorithms to rank office buildings based on comfort, waste and energy consumption criteria. The ORME method introduces the energy efficient retrofit score (EERS) which is defined as:

$$EERS = 1 - \frac{\text{Distance of scenario i from the target}}{\text{Distance of base case from the target}} \qquad [5.1]$$

Different predefined scenarios are evaluated using the EERS.

Another simplified MCDA approach considering the economic benefits resulting from energy efficiency investments is the two-factor method proposed by Martinaitis et al (2007). This method, which uses only two criteria in order to overcome the problem of using complex MCDA techniques, is applied to building retrofits or renovations. The investments are differentiated between those relating to energy efficiency and those relating to building renovation. The costs and benefits of measures incorporating energy efficiency and building renovation are then separated by using the building rehabilitation coefficient (Martinaitis et al, 2004). The cost of conserved energy index is used to evaluate the energy efficiency investments. Homogeneous building renovation measures are evaluated using standard tools for the assessment of investments in maintenance, repair and rehabilitation.

Kaklauskas et al (2005) used multivariant design and MCDA for the refurbishment of a building envelope to prioritize and rank the alternative solutions. The alternatives' significance, utility degree and priority are extracted using this methodology and, as a consequence, the strongest and weakest points of the refurbishment are revealed.

Pasanisi and Ojalvo (2008) developed an MCDA tool called REFLEX (Effective Retrofitting of an Existing Building Tool) for building refurbishments. The multi-criteria approach ranks the retained solutions according to the end users' point of view and energy suppliers' satisfaction.

Other tools that are used in the literature for energy efficiency and indoor environment using multi-criteria decision aid are the EPIQR (Energy Performance Indoor Environmental Quality Retrofit Method for Apartment Building Refurbishment) for residential buildings (Jaggs and Palmer, 2000; Flourentzou and Roulet, 2002) and the TOBUS (Tool for Selecting Office Building Upgrading Solutions) for office buildings (Caccavelli and Gugerli, 2002).

Combined approaches
Combinatorial multi-criteria decision aid in a knapsack model is proposed by Alanne (2004) to support building retrofit and renovation. The knapsack problem is a well-known

problem in combinatorial optimization. It derives its name from the maximization problem of the best choice of essentials that can fit into one bag to be carried on a trip. In the approach, proposed by Alanne, MCDA is used to extract the utilities of the renovation actions proposed, as well as the total utility versus the selected criteria. The obtained utility scores are then used as weights in a knapsack optimization model to identify which actions should be undertaken, through the maximization of the following objective function, subject to budget constraints (Alanne, 2004):

$$ObjF = max \sum_{i=1}^{n} a_i S_i \qquad [5.2]$$

where a_i is a binary decision variable corresponding to the renovation action i ($a_i = 1$ if action i is carried out, else $a_i = 0$) and S_i is the utility score of the renovation action i. This model is suitable for use when the number of renovation actions is large and their combinations lead to a great number of options that cannot be assessed manually.

Multi-objective programming (MOP) approaches

Multi-objective decision aid deals with mathematical models including more than one objective function (Mavrotas et al, 2008). As a consequence of vector optimization, multi-objective problems do not provide a single optimal solution. The main reason is that an 'ideal' solution optimizing all the objective functions, simultaneously, is seldom feasible. Multi-objective optimization algorithms are used by various researchers to find the Pareto optimal solutions that correspond to the Pareto frontier (Figure 5.7), which is used to facilitate the determination and understanding of the trade-off relationships between the optimization objectives (Kalyanmoy, 2001). The scope of applying multi-objective optimization is twofold: first to find solutions as close as possible to the true Pareto optimal frontier and, second, to find solutions as diverse as possible in the Pareto frontier (Fonseca and Flemming, 1998).

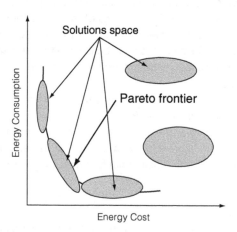

FIGURE 5.7 The Pareto frontier for two criteria

Although MOP approaches can be utilized for the specific problem and under the conditions described in the beginning of this section, the authors did not find any MOP discrete approach in the literature.

Continuous or mixed decision problem approaches

In the discrete decision problem approaches, the whole process, as well as the final decisions, are significantly affected by the experience and the knowledge of the corresponding building expert or the decision maker. Although this experience and knowledge are certainly significant and irreplaceable elements in the whole process, it is necessary to develop practical tools that will assist him/her in taking into account as many feasible alternatives and decision criteria as possible, without the restrictions imposed by predefined actions or strategies. To overcome this shortcoming of the discrete decision problem approaches, the continuous or mixed decision problem approaches have emerged. These latter approaches are mainly based upon the concepts of MOP techniques, a scientific area that offers a wide variety of methods with great potential for the solution of complicated decision problems. The main characteristic of this approach is that the whole possible set of feasible measures is taken into account. Some examples of continuous or mixed decision problem approaches are given below:

- The thermal design of a building is optimized using the Nelder-Mead (downhill simplex) method (Nelder and Mead, 1965) and the non-random complex method (Mitchell and Kaplan, 1968) by Bouchlaghem (2000). The Nelder-Mead method minimizes an objective function in a multidimensional space. The non-random complex method is similar to the downhill simplex method, except that, when the decision variables fall outside the constraints or boundaries, they are replaced by the boundaries. The decision support procedure follows three steps: (a) specification of the building, (b) simulation and (c) optimization, with a continuous interaction between the simulation and optimization steps in order to extract the optimum values of the decision variables that satisfy the comfort criterion. The methodology provides the optimum elevations and angles of the building that minimize discomfort.
- More recently, building envelope improvement during either the design or operational phase using MOP has been proposed by Diakaki et al (2008). The developed decision problem is a mixed-integer multi-objective combinatorial optimization problem with two competitive criteria (cost and energy efficiency). The problem is solved using three different MOP techniques: the compromise programming; the global criterion method; and the goal programming. This approach is still at an early stage of development.

The complexity of the decision problem resulting from the consideration of multiple criteria and its formulation as a continuous or mixed decision problem has led various researchers to use genetic algorithms (GAs), usually coupled with simulation tools. The notable characteristic of GAs compared with other optimization approaches is that they facilitate the exploration of the decision space in the search for optimal or near optimal solutions. Applications of multi-objective genetic algorithms in buildings include the following:

- Wright et al (2002) use GAs to find the optimum Pareto set of solutions to trade-off between the HVAC system energy cost and occupant discomfort. The solutions evaluation is performed using a simplified model.
- Wang et al (2005) use GAs for building design. The two objectives that are minimized are the life cycle cost and the life cycle environmental impact. GAs generate simultaneously a group of solutions and search the state space based on the evolution principle of mutation and crossover (Michalewicz, 1994).The Pareto frontier extracted by the specific optimization procedure provides the best solution for each criterion, the target for each criterion at the design stage and the trade-off relationship between the life cycle cost and life cycle environmental impact.
- Verbeeck and Hens (2007) proposed one more multi-objective GA approach. The design alternatives of the buildings are merged in a chromosome. The fitness function represents the three objectives, i.e. primary energy consumption, net present value and global warming potential. During the assessment process, each solution is ranked according to its fitness for the three objectives. The assessment process is supported also by simulation using TRNSYS. The ranking is based on the Pareto score of the solution, being the number of solutions in the population by which the solution is dominated. The Pareto concept combined with GAs does not result in one single optimum but in the trade-off between the objectives.

ONLINE APPROACHES

In the operational phase of a building, energy efficiency is achieved by data collection and BEMS, thus improving the building's intelligence. The architecture of a BEMS incorporating energy efficiency techniques is illustrated in Figure 5.8.

BEMS can contribute to a significant reduction of the energy consumption of buildings and improvement of the indoor comfort through advanced control techniques (Kolokotsa et al, 2005). Modern control systems provide an optimized operation of the energy systems while satisfying indoor comfort. A comparison of the various control schemes for buildings (Proportional–Integral–Derivative, On–Off, fuzzy, etc.) is provided by Kolokotsa (2003). Recent technological developments based on artificial intelligence techniques (neural networks, fuzzy logic, GAs, etc.) offer several advantages compared with classical control systems. A review of the fuzzy logic contribution in indoor comfort regulation as well as HVAC control and energy efficiency is performed by Kolokotsa (2007). The role of neural networks in buildings is analysed by Kalogirou (2006). A model-based supervisory control strategy for online control and operation of a building is presented by Ma et al (2008).

Control rules concerning the energy efficiency via BEMS can be (Mathews et al, 2000; Doukas et al, 2007):

- Start/stop optimization: Rules about model starting and ending, according to each area or room working hours, including pre-warming and smooth power-down procedures for accomplishing the possible energy savings.
- Setpoint-related energy management strategies: Rules to reduce the energy consumption by regulation setpoints (temperature reset, zero energy band control, enthalpy economizer, adaptive comfort control).

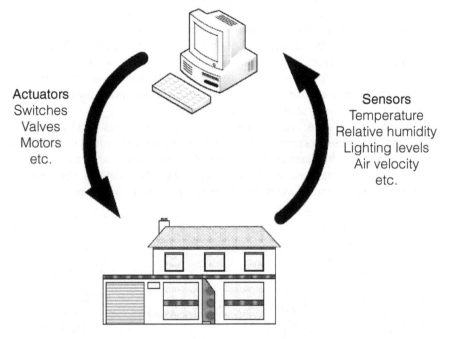

BEMS
Database and monitoring
Decision support
Advanced building automation and control

Actuators
Switches
Valves
Motors
etc.

Sensors
Temperature
Relative humidity
Lighting levels
Air velocity
etc.

FIGURE 5.8 The building energy management system and its role in energy efficiency

- Procedural hierarchy: Rules concerning the intervention hierarchy for the temperature, relative humidity, air quality and luminance adjustment for their optimum operation.
- Energy management optimization: Rules that allocate high-consumption periods in order to propose actions for peak shaving and shifting (demand limiting, duty cycling, load resetting, etc.).

Using BEMS, the required data that in the design phase are extracted by simulation, are taken by the BEMS database. Usually, the energy efficiency measures proposed are pinpointed by a specific list of actions based on energy experts' experience and the criteria assessed can differ according to the application's scope (Doukas et al, 2008).

BEMS and intelligent buildings can also be combined with advanced decision support methodologies to monitor and improve building performance during real-time operation. In terms of online advanced decision support, the following approaches have been found in the literature:

- Analytic hierarchy process (AHP) survey is proposed by Wong and Li (2008) to weight the most critical selection criteria for intelligent buildings. In their survey, the work efficiency criterion (e.g. system's reliability, compliance, etc.) of a BEMS is the most critical one, followed by indoor comfort, safety and cost-effectiveness.
- Analytic network process (ANP), an MCDA approach, is implemented by Chen et al (2006) for lifespan energy efficiency assessment of intelligent buildings. The MCDA is implemented in four steps: model construction; paired comparisons between each two clusters or nodes; supermatrix calculation based on results from paired comparisons; and finally result analysis for the assessment (Figure 5.9). Two alternatives are assessed. The alternative with the highest rate is the one that regulates the building performance of lifespan energy efficiency with the best solutions for building services systems, least energy consumption, lowest ratio of wastage, and lower adverse environmental impacts.
- Gradient-based multi-objective decision support algorithm is proposed by Atthajariyakul and Leephakpreeda (2004) for the real-time optimization of the HVAC system with criteria the thermal comfort, the CO_2 concentration and the energy consumption. The criteria are weighted and summed to formulate the objective function of the optimization and decision support procedure. The weights are set by

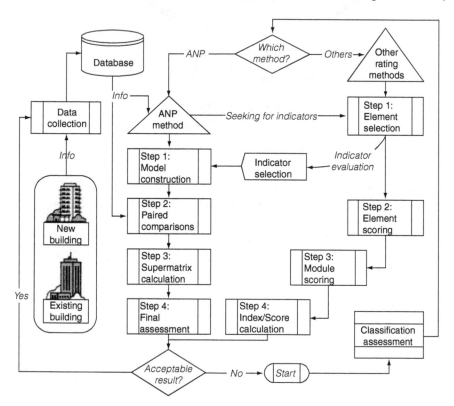

FIGURE 5.9 The ANP model (Chen et al, 2006)

the decision maker. The authors achieved optimum setpoints for the HVAC system by simultaneously maximizing the thermal comfort and the indoor air quality and the energy consumption.

DISCUSSION

Decision problems about a building's sustainability are usually unstructured and ill defined. Moreover, they are characterized by competitive objectives, they involve multiple stakeholders and key actors, dynamic and uncertain procedures and limited timeframes to make significant decisions. Another important issue is the fact that with technological developments in the field, the solutions and decision alternatives are steadily increasing.

The decision makers in the domain are the end users, the design team, the maintenance and operational team, organizational teams, etc. that usually have conflicting subjective preferences and fragmented expertise. Therefore, the role of decision support is critical. Although, as stated in the literature, decision support systems (DSS) should be used to aid decision makers in their work, in practice, DSS can be used to restrict or expand the decision maker's options and to facilitate or provide redirected options. Provision of only a limited set of operations or criteria leads to restrict the techniques and solutions that can be applied and consequently restricts the decision-making process. On the other hand, the inclusion of many objectives and the permitting of user specification of input data, system parameters and models, generally increases system flexibility and increases decision support freedom.

Based on the overall analysis of the present chapter and since the energy efficiency and environmental performance measures can be of a discrete or a continuous nature, the decision variables and consequently the corresponding decision problem can be either discrete, continuous or mixed. For this reason, the approaches employed to analyse and solve this problem can be:

- MCDA approaches where an optimum solution is selected or ranking is performed for a finite set of alternatives based on a set of criteria. This methodology is mainly used for discrete problems, but it can be used for continuous problems as well, if the decision variables are discretized. The advantage of the MCDA approaches is that the decision maker can easily understand the problem and express his/her preferences, while their main drawback is usually the large number of alternative options to be considered. In many cases, researchers perform a preliminary screening phase to eliminate those with a limited or negligible contribution to the criteria (Blondeau et al, 2002; Becker et al, 2007).
- MOP approaches, which target the optimization of a set of objective functions subject to a set of linear or non-linear constraints. Although these approaches are mainly utilized for continuous decision problems, they can also be used for discrete problems. In MOP approaches, the problem is formulated as an integer or a mixed integer mathematical programme. The advantage of such a treatment is that the decision problem is structured in a holistic way. Moreover, the number of alternatives, i.e. the decision space that can be studied, is not limited as it is not necessary to enumerate the set of actions to be considered. The main drawbacks of

the MOP approaches are on the one hand the difficulty in modelling the objective functions and constraints with mathematical equations and on the other hand the complexity and the problem size.

- The third approach is a combination of the above two approaches in two ways:
  - Initially MOP is performed to evaluate the Pareto front and afterwards MCDA is used for the selection of the optimum solution(s).
  - Initially MCDA is utilized in order to evaluate the alternatives on specific criteria and estimate an aggregated 'score', which is then introduced in a mathematical programming problem with a single objective function for the evaluation of the optimum solution.

A critical issue that the above analysis revealed is the role of simulation in the building's sustainability. The approaches used for energy efficiency and minimization of the environmental burden in the building sector are in most cases based on simulation analysis. Even when advanced decision support (i.e. MCDA, MOP) is introduced, the evaluation and the assessment of the decision support process are performed via dynamic or simplified models. Usually, a first evaluation of the alternative options is initially performed through, for example, simulation and then an MCDA technique is employed to aggregate the simulation results and provide a trade-off between the alternative options that is expected to enable the researcher to reach a better decision. In past decades, many efforts were dedicated to developing simulation tools in order to model the dynamics of energy and the indoor environment. Although valuable for detailed design and analysis, few methods are able to simultaneously handle the thermal, visual and indoor air quality aspects in conjunction with energy demands and environmental burden prediction. Therefore, the decision support, if multiple criteria are set, should come from different simulation models that sometimes are not compatible with each other. New simplified tools that integrate several aspects of the building sustainability sector are needed in order to facilitate the decision support applicability.

An additional critical point in the analysis of the present chapter is the fact that in the building sector predefined actions and decision variables are evaluated against a set of decision criteria. This procedure is met in both the design decision phase and in the operational decision phase. Since the predefined actions and options are selected by building experts, the selection procedure is influenced by their relevant expertise. Consequently, to some extent the decisions are biased by the building experts' knowledge and subjective preferences. The end user's criterion is viewed only as the economic aspect of the overall procedure. The economic aspect though, in most cases, is the overriding criterion. Conventional economic indices such as net present value and internal rate of return are not explicitly representative for buildings' sustainability. The cost of conserved energy introduced by Martinaitis et al (2004) provides an alternative view of buildings' related costs by separating investments into those that are absolutely necessary for building renovation and those that contribute to energy efficiency.

CONCLUSION AND FUTURE PROSPECTS

Energy efficiency, the indoor environment and reduction of buildings' environmental impact are the major priorities of the energy and environmental policy worldwide.

In the decision support process, several measures are available, and the decision maker has to take into consideration environmental, energy, financial and social aspects in order to make an optimum design or operational choice. The problem of the decision maker is characterized by the existence of multiple and in several cases conflicting objectives each of which should be optimized against a set of realistic and available alternatives that is influenced by a set of parameters and constraints that should be taken into account.

Consequently, the decision maker is facing a multi-objective optimization problem that up to now is usually dealt with by using either simulation or multi-criteria decision-making techniques that concentrate on particular aspects of the problem.

Thus, there is a need for further development of the decision aid systems to support building experts in the application of their expertise, and assist less-experienced decision makers to approach decisions in the same way as experts would, while taking into account the continuous development of technological expertise in energy efficient solutions.

AUTHOR CONTACT DETAILS

D. Kolokotsa (corresponding author) and **C. Diakaki**: Technological Educational Institute of Crete, Department of Natural Resources and Environment, 3, Romanou str, 73133, Chania, Crete, Greece; dekol@otenet.gr

E. Grigoroudis: Technical University of Crete, Department of Production Engineering and Management, University Campus, Kounoupidiana, 73100 Chania, Greece

G. Stavrakakis and **K. Kalaitzakis**: Technical University of Crete, Department of Electronics and Computer Engineering, University Campus, Kounoupidiana, 73100 Chania, Greece

REFERENCES

Alanne, K. (2004) 'Selection of renovation actions using multi-criteria 'knapsack' model', *Automation in Construction*, vol 13, no 3, pp377–391

Alanne, K., Salo, A., Saari, A. and Gustafsson, S. I. (2007) 'Multi-criteria evaluation of residential energy supply systems', *Energy and Buildings*, vol 39, pp1218–1226

Atthajariyakul, S. and Leephakpreeda, T. (2004) 'Real-time determination of optimal indoor-air condition for thermal comfort, air quality and efficient energy usage', *Energy and Buildings*, vol 36, no 7, pp720–733

Becker, R., Goldberger, I. and Paciuk, M. (2007) 'Improving energy performance of school buildings while ensuring indoor air quality ventilation', *Building and Environment*, vol 42, no 9, pp3261–3276

Blondeau, P., Sperandio, M. and Allard, F. (2002) 'Multi-criteria analysis of ventilation in summer period', *Building and Environment*, vol 37, no 2, pp165–176

Bluyssen, P. M. (2000) 'EPIQR and IEQ: Indoor environment quality in European apartment buildings', *Energy and Buildings*, vol 31, no 2, pp103–110

Bouchlaghem, N. (2000) 'Optimising the design of building envelopes for thermal performance', *Automation in Construction*, vol 10, no 1, pp101–112

Caccavelli, D. and Gugerli, H. (2002) 'TOBUS – a European diagnosis and decision-making tool for office building upgrading', *Energy and Buildings*, vol 34, no 2, pp113–119

CEN EN 15217:2005 (2005) *Energy Performance of Buildings – Methods for Expressing Energy Performance and for Energy Certification of Buildings*, Comité Européen de Normalisation

CEN EN15251:2006 (2006) *Indoor Environmental Input Parameters for Design and Assessment of Energy Performance of Buildings – Addressing Indoor Air Quality, Thermal Environment, Lighting and Acoustics*, Comité Européen de Normalisation

Chen, Z., Clements-Croome, D., Hong, J., Li, H. and Xu, Q. (2006) 'A multi-criteria lifespan energy efficiency approach to intelligent building assessment', *Energy and Buildings*, vol 38, no 5, pp393–409

Clarke, J. A. (2001) *Energy Simulation in Building Design*, 2nd edn, Butterworth-Heinemann, London

D'Cruz, N. A. and Radford, A. D. (1987) 'A multi-criteria model for building performance and design', *Building and Environment*, vol 22, no 3, pp167–179

Diakaki, C., Grigoroudis, E. and Kolokotsa, D. (2008) 'Towards a multi-objective optimization approach for improving energy efficiency in buildings', *Energy and Buildings*, vol 40, no 9, pp1747–1754

Doukas, H., Patlitzianas, K.D., Iatropoulos, K., Psarras, J. (2007) 'Intelligent building energy management system using rule sets', *Building and Environment*, vol 42, no 10, pp3562–3569

Doukas, H., Nychtis, C. and Psarras, J. (2008) 'Assessing energy-saving measures in buildings through an intelligent decision support model', *Building and Environment*, vol 44, no 2, pp290–298

EC (European Community) (2003) *Directive 2002/91/EC of the European Parliament and of the Council of 16 December 2002 on the Energy Performance of Buildings*, L1/65, Official Journal of the European Communities

Ehrgott, M. (2005) *Multi-criteria Optimization*, 2nd edn, Springer, Berlin

Florides, G. A., Tassou, S. A., Kalogirou, S. A. and Wrobel, L. C. (2002) 'Measures used to lower building energy consumption and their cost effectiveness', *Applied Energy*, vol 73, no 3–4, pp299–328

Flourentzou, F. and Roulet, C. A. (2002) 'Elaboration of retrofit scenarios', *Energy and Buildings*, vol 34, no 2, pp185–192

Fonseca, C. M. and Flemming, P. J. (1998) 'Multiobjective optimization and multiple constraint handling with evolutionary algorithms – Part 1: A unified formulation', *IEEE Transactions on Systems, Man and Cybernetics, Part A*, vol 28, no 1, pp26–37

Gero, J. S., Neville, D. C. and Radford, A. D. (1983) 'Energy in context: A multi-criteria model for building design', *Building and Environment*, vol 18, no 3, pp99–107

Gholap, A. K. and Khan, J. A. (2007) 'Design and multi-objective optimization of heat exchangers for refrigerators', *Applied Energy*, vol 84, no 12, pp1226–1239

Homoud Al, M. S. (2001) 'Computer-aided building energy analysis techniques', *Building and Environment*, vol 36, no 4, pp421–433

Hong, T., Chou, S. K. and Bong, T. Y. (2000) 'Building simulation: An overview of developments and information sources', *Building and Environment*, vol 35, no 4, pp347–361

Horsley, A., France, C. and Quatermass, B. (2003) 'Delivering energy efficient buildings: A design procedure to demonstrate environmental and economic benefits', *Construction Management and Economics*, vol 21, no 4, pp345–356

IEA (International Energy Agency) (2008) *Towards a Sustainable Energy Future*, IEA programme of work on climate change, clean energy and sustainable development

ISO (International Organization for Standardization) (1984) *Moderate Thermal Environments – Determination of PMV and PPD Indices and Specification of the Conditions for Thermal Comfort*, ISO Standard 7730-84, ISO, Geneva

Jaggs, M. and Palmer, J. (2000) 'Energy performance indoor environmental quality retrofit – a European diagnosis and decision making method for building refurbishment', *Energy and Buildings*, vol 31, no 2, pp97–101

Jedrzejuk, H. and Marks, W. (2002) 'Optimization of shape and functional structure of buildings as well as heat source utilisation example', *Building and Environment*, vol 37, no 12, pp1249–1253

Kaklauskas, A., Zavadskas, E. K. and Raslanas, S. (2005) 'Multivariant design and multiple criteria analysis of building refurbishments', *Energy and Buildings*, vol 37, no 4 pp361–372

Kalogirou, S. A. (2006) 'Artificial neural networks in energy applications in buildings', *International Journal of Low Carbon Technologies*, vol 1, no 3, pp201–216

Kalyanmoy, D. (2001) *Multi-Objective Optimization Using Evolutionary Algorithms*, Wiley and Sons, New York

Keeney, R. and Raiffa, H. (1993) *Decisions with Multiple Objectives: Preferences and Tradeoffs*, Cambridge University Press, Cambridge, UK

Kolokotsa, D. (2003) 'Comparison of the performance of fuzzy controllers for the management of the indoor environment', *Building and Environment*, vol 38, no 12, pp1439–1450

Kolokotsa, D. (2007) 'Artificial intelligence in buildings: A review on the application of fuzzy logic', *Advances in Building Energy Research*, vol 1, pp29–54

Kolokotsa, D., Tsiavos, D., Stavrakakis, G., Kalaitzakis, K. and Antonidakis, E. (2001) 'Advanced fuzzy logic controllers design and evaluation for buildings' occupants' thermal–visual comfort and indoor air quality satisfaction', *Energy and Buildings*, vol 33, no 6, pp531–543

Kolokotsa, D., Niachou, K., Geros, V., Kalaitzakis, K., Stavrakakis, G. and Santamouris, M. (2005) 'Implementation of an integrated indoor environment and energy management system', *Energy and Buildings*, vol 37, no 1, pp93–99

Krarti, M. (2000) *Energy Audit of Building Systems*, CRC Press, Boca Raton, London, New York

Kreider, J. F., Curtiss, P. and Rabl, A. (2002) *Heating and Cooling for Buildings, Design for Efficiency*, McGraw-Hill, New York

Ma, Z., Wang, S., Xu, X. and Xiao, F. (2008) 'A supervisory control strategy for building cooling water systems for practical and real time applications', *Energy Conversion and Management*, vol 49, no 8, pp2324–2336

Martinaitis V., Rogoža, A. and Bikmaniene, I. (2004) 'Criterion to evaluate the "twofold benefit" of the renovation of buildings and their elements', *Energy and Buildings*, vol 36, no 1, pp3–8

Martinaitis, V., Kazakevicius, E. and Vitkauskas, A. (2007) 'A two-factor method for appraising building renovation and energy efficiency improvement projects', *Energy Policy*, vol 35, no 1, pp192–201

Mathews, E. H., Arndt, D. C., Piani, C. B. and van Heerden, E. (2000) 'Developing cost efficient control strategies to ensure optimal energy use and sufficient indoor comfort', *Applied Energy*, vol 66, no 2, pp135–159

Mavrotas, G., Diakoulaki, D., Florios, K. and Georgiou, P. (2008) 'A mathematical programming framework for energy planning in services' sector buildings under uncertainty in load demand: The case of a hospital in Athens', *Energy Policy*, vol 36, no 7, pp2415–2429

Michalewicz, Z. (1994) *Genetic Algorithms + Data Structures = Evolution Programs*, 2nd edn, Springer-Verlag, New York

Mitchell, R. A. and Kaplan, J. L. (1968) 'Non-linear constrained optimization by a non-random complex method', *Journal of Research of the National Bureau of Standards*, C7, pp249–258

Nelder, J. A. and Mead, R. (1965) 'A simplex method for function minimization', *Computer Journal*, vol 7, no 15, pp308–313

Pasanisi, A. and Ojalvo, J. (2008) 'A multi-criteria decision tool to improve the energy efficiency of residential buildings', *Foundations of Computing and Decision Sciences*, vol 33, no 1, pp71–82

Radford, A. D. and Gero, J. S. (1980) 'Tradeoff diagrams for the integrated design of the physical environment in buildings', *Building and Environment*, vol 15, no 2, pp3–15

Rey, E. (2004) 'Office building retrofitting strategies: Multi-criteria approach of an architectural and technical issue', *Energy and Buildings*, vol 36, no 4, pp367–372

Rosenfeld, Y. and Shohet, I. M. (1999) 'Decision support model for semi-automated selection of renovation alternatives', *Automation in Construction*, vol 8, no 4, pp503–510

Roulet, C. A., Flourentzou, F., Labben, H. H., Santamouris, M., Koronaki, I., Dascalaki, E. and Richalet, V. (2002) 'ORME: A multi-criteria rating methodology for buildings', *Building and Environment*, vol 37, no 6, pp579–586

Roy, B. (1991) 'The outranking approach and the foundations of ELECTRE methods', *Theory and Decision*, vol 31, pp49–73

Rutman, E., Inard, C., Bailly, A. and Allard, F. (2005) 'A global approach of indoor environment in an air-conditioned office room', *Building and Environment*, vol 40, no 1, pp29–37

Tavares, P. F. A. F. and Martins, A. M. O. G. (2007) 'Energy efficient building design using sensitivity analysis – A case study', *Energy and Buildings*, vol 39, no 1, pp23–31

Verbeeck, G. and Hens, H. (2007) 'Life cycle optimization of extremely low energy dwellings', *Journal of Building Physics*, vol 31, no 2, pp143–177

Wang, W., Zmeureanu, R. and Rivard, H. (2005) 'Applying multi-objective genetic algorithms in green building design optimization', *Building and Environment*, vol 40, no 11, pp1512–1525

Wilde, P. and van der Voorden, M. (2004) 'Providing computational support for the selection of energy saving building components', *Energy and Buildings*, vol 36, no 8, pp749–758

Wong, J. K. W. and Li, H. (2008) 'Application of the analytic hierarchy process (AHP) in multi-criteria analysis of the selection of intelligent building systems', *Building and Environment*, vol 43, no 1, pp108–125

Wright, J. A., Loosemore, H. A. and Farmani, R. (2002) 'Optimization of building thermal design and control by multi-criterion genetic algorithm', *Energy and Buildings*, vol 34, no 9, pp959–972

Wulfinghoff, D. R. (1999) *Energy Efficiency Manual*, Energy Institute Press, Wheaton, Maryland, US

Zhu, Y. (2006) 'Applying computer-based simulation energy auditing: A case study', *Energy and Buildings*, vol 38, no 5, pp421–428

Zurigat, Y. H., Al-Hinai, H., Jubran, B. A. and Al-Masoudi, Y. S. (2003) 'Energy efficient building strategies for school buildings in Oman', *International Journal of Energy Research*, vol 27, no 3, pp241–253

earthscan

publishing for a sustainable future

6

Progress in Numerical Modelling for Urban Thermal Environment Studies

Isaac Lun, Akashi Mochida and Ryozo Ooka

Abstract

Urbanization is progressing rapidly in many Asian cities. The process of urbanization has modified the land use from natural environment into built environment. It alters not only the surface energy balance of the urban canopy, but also brings about a great quantity of anthropogenic sources of waste heat through air-conditioning, cars, etc. In addition, the effect of urbanization on urban wind environment is likewise significant. Thus, the primary precondition is to understand how the urban environment affects the physical and climatic pattern in and around the city resulting from urban encroachments. Commonly, wind-tunnel measurements and observational campaigns enable us to understand the physical processes that take place with the morphology of urban areas. This understanding is then used to represent these processes within numerical models of different urban scales. The ever-increasing computational power together with high-resolution computational fluid dynamic models has now become a useful tool to gain significant insight into detailed processes occurring within the urban context. This chapter gives an overview of the latest simulation studies for mesoscale and microscale climates, and also the assessment tools used in urban climate research. Various assessment tools are introduced and classified according to corresponding modelling scales. Next, the chapter addresses recent achievements in urban climate research for urban thermal environment studies. Examples of numerical results obtained by researchers of Japan are presented.

■ *Keywords* – microclimate assessment tools; urban thermal climates; urban heat island; subgrid scale flow obstacles

INTRODUCTION

The world population now numbers more than 6 billion, with more than half of these people living in urban areas; the urban population is expected to swell to almost 5 billion by 2030 (UNFPA, 2007). In line with population growth, rapid urbanization is expected to

doi:10.3763/aber.2009.0306 ■ © 2009 Earthscan ■ ISSN 1751-2549 (Print), 1756-2201 (Online) ■ www.earthscanjournals.com

take place in most developing countries. As a result, occurrence of urban environmental problems is inevitable. In this respect, the urban thermal environment is one of the major urban environmental issues, which has led to extensive research into this topic. As a city grows, the heat of the city builds. This hot city phenomenon has far-reaching environmental sustainability and human liveability implications, ranging from the aggravation of health problems such as hyperthermia, increasing the intensity of urban air pollution, and contributing to extreme heat waves (National Weather Service, 2005). The impact of urbanization and industrialization on the quality of the environment also multiplies. There has been a lot of discussion in the media and among the public about the effect of urban climate change on urbanites (Smith, 1997; NASA/Goddard Space Flight Center, 2002; Swanson, 2007; Earth Observatory, 2008). This urban localized weather is a condition that scientists refer to as the urban heat island (UHI) effect. Due to urbanization, concentration of the population is taking place, the area covered by the city is expanding and natural ground surfaces are modified. As a result, energy consumption and city metabolism heat increase significantly and eventually change the heat balance mechanism of urban space.

The urban thermal environment is closely related to the UHI phenomenon. The UHI phenomenon was first discovered in the early 1800s in London by the British meteorologist Luke Howard. Not long after, he documented the temperature difference between an urban area and its rural environment (Howard, 1833a, 1833b, 1833c). This urban–rural temperature contrast was later termed the 'urban heat island' by Manley (1958). Since then the term has been widely used across different disciplines and various UHI-related urban climate research has been carried out worldwide (Chandler, 1960; Nieuwolt, 1966; Sham, 1972; Oke and Maxwell, 1975; Oke, 1981, 1987, 1988; Lars et al, 1985; Karl and Jones, 1989; Yinka, 1990; Lee et al, 2002; Li and Lin, 2002; Lee et al, 2003; Lin and Lee, 2004; Li et al, 2004). The UHI is now regarded as one of the most serious urban environmental problems all over the world (see Figure 1 of Ojima (1991) and Figure 2 of Ooka (2007)). The major causes of the heat island phenomenon are alternations of the land surface covering, increased anthropogenic exhaustion heat, and low wind velocity due to the high density of urban structures (see Figure 3 of Sera (2006)). For instance, in the Tokyo metropolitan area, about half of the land is occupied by buildings and about half of the anthropogenic exhaustion heat generated in the summer in this area comes from the buildings' facilities (see Figure 4 of Murakami (2006)).

Figure 6.1 illustrates the development of urbanization in Tokyo since the Meiji era (1868–1912) (Ojima, 1991) and Figure 6.2 shows the increase in air temperature during the period 1880–1980 in Tokyo. There has been a 2°C increase in average annual temperature within the capital area over that time which is greater than the rise recorded across the rest of the globe. The existence of an urban heat island around Tokyo is clearly and remarkably indicated from these figures.

In Japan, the urban heat island effect has caused various problems such as heat stroke and large electric power demand for cooling devices, etc. In recent years, various countermeasures for reducing the urban heat island effect, e.g. urban tree planting, high-albedo building surfaces and the introduction of sea breezes into urban areas, have been proposed and examined as well as implemented. It should be noted that the urban climate of each region is influenced by regional characteristics such as urban scale, geographical

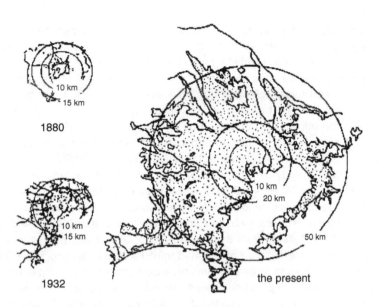

FIGURE 6.1 Development of urbanization in Tokyo (Ojima, 1991)

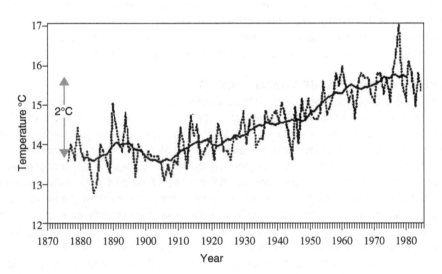

FIGURE 6.2 Increase of air temperature in Tokyo (height of about 1.5 m) (Ooka, 2007)

features, land use, sea breeze and anthropogenic heat releases, etc. Thus, effective countermeasures vary from region to region according to the regional climatic characteristics.

Inhabitants become more sensitive to their environmental conditions as a result of urbanization. Thus, the micro- and meso-climatology of urban areas have attracted much attention from urban climatologists. Many studies examined the dispersion and chemistry of urban-derived air pollutants, while others investigated urban canyon energy budgets such as storage heat flux (Arnfield and Grimmond, 1998) and modelled thermal environments of street canyons (Herbert et al, 1998). Kitada et al (1998) simulated the interacting effects of topography and built-up areas on local winds and temperatures. Simulations at the interface between urban climatology and urban planning (the use of green space within cities) were undertaken by Palomo Del Barrio (1998) and Sailor (1998). Simulations of plant canopy processes and climates were presented by Leuning et al (1998), Su et al (1998), Wang and Leuning (1998) and Wilson et al (1998), while Patton et al (1998) employed large-eddy simulation (LES) to model the flow around windbreaks set within a wheat canopy. Vu et al (2002) used k-ε turbulence closure in a boundary layer model for urban atmospheric simulation. Another popular area of numerical simulation is the modelling of the surface energy budget and surface temperature (Avissar and Schmidt, 1998; Bélair et al, 1998; Best, 1998; Bünzli and Schmid, 1998; Flerchinger et al, 1998).

A variety of environmental problems related to UHI are now affecting our entire world. Various counter measures to these problems have been proposed and implemented such as roof greening, use of high albedo (highly reflective) paints, etc. However, tools to assess the urban climate are very important to estimate the effects of these measures. The urban climate is composed of various spatial scale phenomena in and around the urban area (Figure 6.3) and hence various urban climate assessment tools that correspond to different scales are required. This chapter describes various assessment tools for urban climate study according to the modelling scales, and presents examples with results achieved from these models, in particular those involving urban thermal modelling.

URBAN CLIMATE ASSESSMENT TOOLS

In the field of urban climate research, the study of wind flow and pollutant transport within and over urban street canyons has attracted great research interest since the late 20th century. In the past, the conventional approach was to take field measurements and use scale physical modelling. However, computational power has increased dramatically over the past decade, together with advances in computer software, allowing engineers/ researchers to more accurately simulate many types of specific case within complex urban building environments. It is, however, worth noting that much of the effort seems focused on integrating computer modelling techniques that currently exist at divergent scales. For instance, while numeric models allow weather predictions at the larger 'synoptic' scale and 'mesoscale', these models do not allow an understanding of climate phenomena at resolutions finer than several kilometres. On the other hand, computational fluid dynamic (CFD) techniques are the common tools used to study fluid flow, and engineering-type CFD models allow a better understanding of the behaviour of wind at the scale of an individual building or cluster of buildings within street canyons (Jeong and Andrews, 2002; Calhoun et al, 2004; Cui et al, 2004; Cheng and Hu, 2005; Lien and Yee, 2005; Lien et al, 2005), but do not accurately simulate the complexity of actual climate and wind behaviour. The development and use of CFD models is a very active area of scientific enquiry. They are now widely used in both the industrial and research communities and can provide a relatively economic approach to urban climate study. In addition, the models

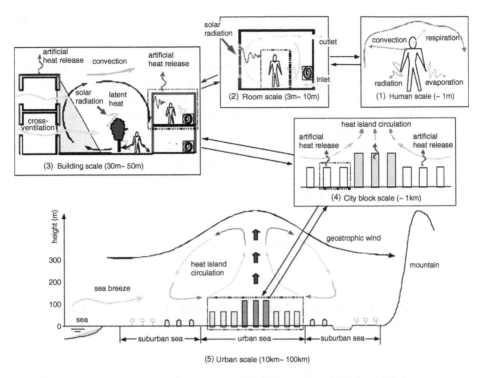

FIGURE 6.3 Various scales of phenomena concerning with urban climate (Murakami, 2004)

are becoming more sophisticated in terms of numerical methods, mesh structures and turbulence modelling approaches (direct numerical simulation (DNS), large-eddy simulation and Reynolds averaged Navier-Stokes (RANS) simulation) (e.g. Lien and Yee, 2004; Liu et al, 2004; Hamlyn and Britter, 2005).

Urban climate assessment tools based on various modelling scales, can be attributed to four categories:

1 mesoscale meteorological model;
2 microclimate model;
3 building model;
4 human thermal model.

These models correspond respectively to various scales: the meteorological model to an urban and city block scale, the microclimate model to a city block and building scale, the building model to a building and room scale, and the human thermal model to a human body scale (Ooka, 2007).

The mesoscale meteorological model is also subclassified as a mesoscale meteorological model in the narrow sense and a one-dimensional urban canopy model. Originally, the one-dimensional urban canopy model was developed as a surface sublayer model for the mesoscale meteorological model. However, thanks to its ease of use, the one-dimensional urban canopy model is often used independently to estimate the effect of urban heat island countermeasures. As the calculation load is very small, many case studies of various measures to mitigate the heat island effect can be handled. With regard to the mesoscale meteorological model, it is well suited to predicting phenomena in the 100-km order (mesoscale), such as land and sea breezes. However, the smallest applicable mesh size for the mesoscale meteorological model is about 1 km, so it is unsuitable for evaluating pedestrian level outdoor thermal environments.

The microclimate model was developed in the engineering field as well as the meteorological field and is usually based on a three-dimensional CFD model, often being coupled with radiation and conduction calculations. This model is useful for predicting detailed spatial distributions of flow, temperatures and scalar fields inside complex urban areas.

These models are sometimes used separately and sometimes simultaneously depending on the targeted scale and resolution level required. The present situation and application of meteorological and microclimate models for urban heat island analysis are described in 'Mesoscale meteorological model and its application'.

MESOSCALE METEOROLOGICAL MODEL AND ITS APPLICATION

Bornstein (1972, 1975) and Lee and Olfe (1974) were possibly the first to simulate the heat island phenomena using a meteorological model. In Japan, Kikuchi et al (1981) conducted a numerical simulation of the sea breeze over the Kanto plain. From that point, a number of organizations developed their own models. For example, the National Center for Atmospheric Research (NCAR) and Pennsylvania State University developed their 5th Generation Mesoscale Model (MM5), while Colorado State University developed the Colorado State University Mesoscale Model (CSUMM) and later the Regional Atmospheric Modelling System (RAMS). Yamada and Bunker (1988) developed a Higher Order Turbulence Model for Atmospheric Circulation (HOTMAC). Recently, a Weather Research and Forecasting model (WRF) was developed by NCAR. Many urban climate researchers employ these models. Table 6.1 shows a comparison of the various mesoscale meteorological models that are used for urban climate and heat island research.

Mochida et al (1999) analysed the flow and temperature fields in summer in the greater Tokyo area using HOTMAC. Their results are shown in Figure 6.4 and compared with actual observations. The measurements and predictions demonstrate fairly good correspondence. Figure 6.5 compares the temperature distributions at ground level at 3 pm in early August under land-use conditions from the present back to the Edo era (about 200 years ago). The current surface temperature in central Tokyo is about 4°C higher than that of the Edo era. This indicates that urbanization as evidenced by a decrease in greenery and an increase in anthropogenic heat has advanced the progress of this urban heat island. Ichinose et al (2000) evaluated the impact of anthropogenic heat and greenery coverage ratio on the urban climate across the Tokyo metropolitan area using CSUMM. The air temperature without anthropogenic heat would be about 1.5°C lower at 10 pm in Otemachi than it is at present. The air temperature in the case where all of the study area is assumed to be grassland would be about 2.5°C lower at

TABLE 6.1 Comparison of various mesoscale meteorological models for urban climate and heat island study

NAME	DEVELOPER	EQUATION	TURBULENCE MODEL	SURFACE SUBLAYER	MAIN USER FOR HEAT ISLAND STUDY
LSM	Kimura and Arakawa (1983)	Hydro	0 equation	Monin Obukhov	F. Kimura, H. Kusaka
FITNAH	Wippermann and Gross (1985) Gross (1987)	Non-hydro, Hydro +dynamic	1 equation	Non-slip for momentum & Heat Balance	Gross, G
AIST*1-MM	Kondo (1995)	Hydro	0 equation Gambo (1978)	Monin Obukhov AIST-CM	H. Kondo, Y. Genchi, Kikegawa
NEDO*2 Software Platform	Murakami et al (2000)	Hydro	k-l two equation Yamada and Bunker (1988)	Monin Obukhov urban canopy	S. Murakami, A. Mochida, R. Ooka
BRI*3 UCSS	Ashie et al (1999) Ashie and VuThanh Ca (2004).	Non-hydro	k-two equation VuThanh Ca et al (2000, 2002)	urban canopy	Y. Ashie
Osaka-Univ. OASIS	Oh et al (2000) Kondo et al (2001)	Non-hydro optional	k-l two equation Yamada and Bunker (1988)	urban canopy	D. Narumi, A. Kondo
YSA*4 HOTMAC	Yamada and Bunker (1988)	Non-hydro optional	k-l two equation Yamada and Bunker (1988)	Monin Obukhov forest canopy	S. Murakami, A. Mochida
CSU*5 CSUMM	Pielke (1974)	Hydro	0 equation Pielke's model	Monin Obukhov	T. Ichinose, I. Uno
RAMS	Pielke et al (1992)	Non-hydro optional	k-l two equation, LES optional	Monin Obukhov	M. Kanda, A. Velazques-Lozada, C. Sarrat
NCAR*6 MM5	Grell et al (1994)	Non-hydro	k-l two equation	Monin Obukhov	A. Kondo, H. Fan
MC2	Laprise (1995) Benoit et al (1997)	Non-hydro	0 equation	Monin Obukhov	E. Krayenhoff
NCAR*6 WRF	Skamarock et al (2005)	Non-hydro	k-l two equation	Monin Obukhov urban canopy	H. Kusaka

*1 Advanced Institute of Science and Technology
*2 New Energy Development Organization
*3 Building Research Institute, Japan
*4 Yamada Science and Art Co.
*5 Colorado State University
*6 National Center for Atmospheric Research

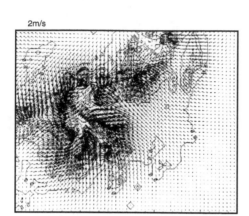

(1) Numerical prediction (early August)

(2) Measurement (1985 summer, 50 days)

FIGURE 6.4 Comparison between numerical prediction and measurement (horizontal distribution of velocity vectors at a height of 100 m, at 3 pm) (Mochida et al, 1999)

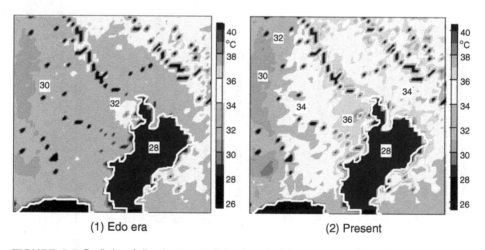

(1) Edo era

(2) Present

FIGURE 6.5 Prediction of climatic change in Tokyo from the Edo era to present (ground temperature at 3 pm in early August) (Mochida et al, 1999)

9 pm in Otemachi than at present. Fan and Sailor (2005) also evaluated the impact of anthropogenic heat on the urban climate of Philadelphia using MM5. Kondo et al (2001) investigated the effects of high albedo paint on the road on the urban heat island in Osaka using OASIS, a simulation model for evaluating the thermal environment in an urban area. The air temperature at pedestrian level in the case where the albedo value of the road surface is 0.45 is about 0.15°C lower at 12 am in the central part of Osaka than

it would be without the high albedo paint (albedo value = 0.15). Taha (1996, 1997) evaluated the impacts of increased urban vegetation and albedo change on the air quality in California's South Coast Air Basin. Although increased urban vegetation and increased albedo reduces the air temperature in the area, the increase in urban vegetation causes more biogenic hydrocarbon emissions and a higher ozone concentration through photochemical reactions. Kanda et al (2001) simulated a small-scale cloud over a main street in the Tokyo metropolitan area using RAMS. Murakami et al (2003a) proposed the concept of a heat balance model in which heat balance within a virtual control volume in urban space is estimated from the calculation results of the mesoscale meteorological model. This model enables quantitative consideration of the various factors that form the urban thermal environment.

ONE-DIMENSIONAL URBAN CANOPY MODEL AND ITS APPLICATION
There are three kinds of surface sublayer models for surface boundary conditions in mesoscale meteorological models:

1 a surface-layer scheme;
2 a single-layer model;
3 a multilayer model.

Here, the single-layer and multilayer models are called the canopy model. However, the urban canopy model is often used independently because of its ease of use as described above. Sometimes the urban canopy model is coupled with the building energy model in order to investigate the interaction between urban climate and building energy use. Table 6.2 shows various urban canopy models.

Kikegawa et al (2001) and Genchi (2001) used the Advanced Industrial Science & Technology – Canopy Model (AIST-CM) to evaluate the effect of heat release from building air-conditioning units in Tokyo. Three scenarios were studied: an external air-conditioning unit placed on the roof of a building (case 1), placed at a height of 3 m above the ground (case 2), and with heat from the air-conditioning system being injected into the ground (case 3). The daily average temperatures at a height of 3 m for cases 1, 2 and 3 were 30.37, 30.99 and 29.09°C, respectively. Thus, the inclusion of heat release from air-conditioning systems is very important. Hagishima et al (2002) carried out sensitivity analyses on the various factors closely related to the urban heat island effect, such as building density, ground coverage conditions, building roof and wall surface conditions, use and type of air-conditioning systems, and so on. They clarified that the highest temperature inside the urban canopy is most strongly affected by the air-conditioning systems, internal heat generation from buildings and the conditions of ground coverage. Recently, the urban canopy model has sometimes been coupled with the mesoscale meteorological model interactively in order to estimate the relationship between human activity and mesoscale climate (e.g. Narumi et al, 2002; Kusaka and Kimura, 2004; Ooka et al, 2004; Kondo et al, 2005). Figure 6.6 shows a schematic view of incorporating the urban canopy model into the mesoscale meteorological model. Figure 6.7 presents a comparison of ground surface temperatures at 1 pm in the Tokyo area between

TABLE 6.2 Comparison of various urban canopy models for urban climate and heat island study

NAME/DEVELOPER	TURBULENCE MODEL	BUILDING AREA DENSITY	DRAG	RADIATION
AIST-CM Kondo et al (1998, 2005)	0 equation Gambo (1978)	$a = BP_w(z)/((B + W)^2 - B^2P_w(z))$	Coefficient Fixed value Cdrag = 0.4	Calculation Kondo et al (1998, 2005)
RAUSSSM Hagishima et al (2001)	0 equation Gambo (1978)	$a = B/((B + W)^2 - B^2)$	Maruyama (1991)	Radio City
UCSS Ashie et al (1999) Ashie and VuThanh Ca (2004)	k-ε two equation VuThanh Ca et al (2000, 2002)	$a = \sqrt{bldr}$	Fixed value	Ashie et al (1999) Ashie and VuThanh Ca (2004)
Kusaka et al (2001)	Single layer	Single layer	Simple layer	Sakakibara (1995)
SUMM Kanda et al (2005a, 2005b)	Single layer	Single layer	Single layer	Kanda et al (2005a)
Martilli et al (2002)	k-l two equation Bougeault and Lacarrere (1989)	$a = 4\xi B/(B + W)^2$	Fixed value Cdrag = 0.4	Martilli et al (2002)
Hiraoka et al (1989)	k-ε two equation	$a = 2B/((B + W)^2 - B^2)$	–	–
Ooka et al (2004)	k-l two equation Yamada et al (1988)	$a = 4B/(B + W)^2$	Fixed value Cdrag = 0.1	Kondo et al (1998)
TEB Masson (2000)	Single layer	Single layer	Single layer	Masson (2000)
Arnfield (2000)	Single layer	Single layer	Single layer	Arnfield (2000)
Best (2005)	Single layer	Single layer	Single layer	Heat balance in urban canopy

B: building width, W: building interval, bldr: building area ratio (gross)

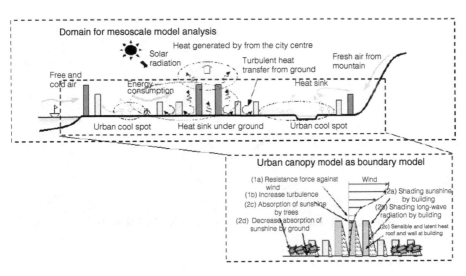

FIGURE 6.6 Schematic view of incorporating urban canopy model into mesoscale meteorological model

observations from a satellite and simulations both with and without the urban canopy model calculated by the author (Ooka et al, 2004). While the results of the conventional mesoscale meteorological model without the urban canopy model show poor agreement with the observations, those of the mesoscale model with the urban canopy model afford excellent agreement. This is because the urban canopy model defines the building and ground surfaces clearly, whereas the heat balance model used as the boundary condition in the conventional meteorological model does not define such surfaces clearly.

MICROCLIMATE MODEL
AN OVERVIEW OF MICROCLIMATE MODELS
The microclimate model is composed of CFD, radiation transfer and heat conduction models. CFD has been developed on the basis of recent developments in computational technology. Quantitative analyses of flow fields within building complexes have been conducted using wind engineering expertise (e.g. Mochida and Lun, 2008). The article provides the most informative source on recent developments in computational wind engineering (CWE) research for predicting the pedestrian level wind and thermal environments in urban areas. Mochida and Lun (2008) suggested that wind environment is one of the most important factors to be considered in UHI study as it has a significant influence on UHI effect and outdoor thermal comfort. The conventional urban wind environment assessment methods only took into consideration the influence of topographic features and geometry of buildings (Figure 6.8(1)). In modern urban climate prognostic studies, these methods may be inadequate to reflect the real conditions of

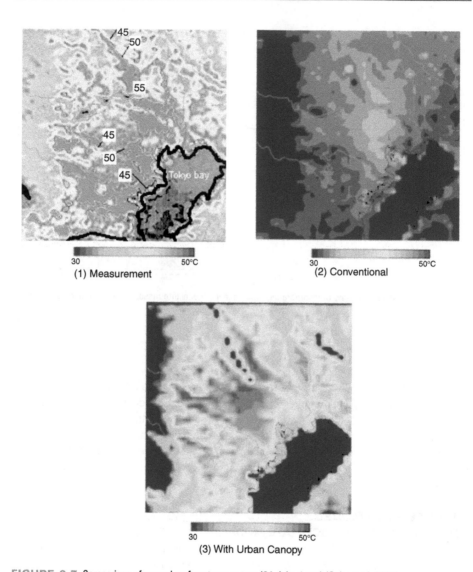

(1) Measurement

(2) Conventional

(3) With Urban Canopy

FIGURE 6.7 Comparison of ground surface temperature (24 July, 1 pm) (Ooka et al, 2004)

cities where there are objects of various scales within the street canyons. The new concept should include stationary objects (Figure 6.8(2)) such as trees and telephone boxes, and non-stationary objects (Figure 6.8(3)) such as moving vehicles and swinging hanging-signboards (commonly seen in Asian countries), because these obstacles can alter the surface roughness to a certain extent. In order to obtain accurate quantitative data for urban wind environment assessment, these objects must not be overlooked.

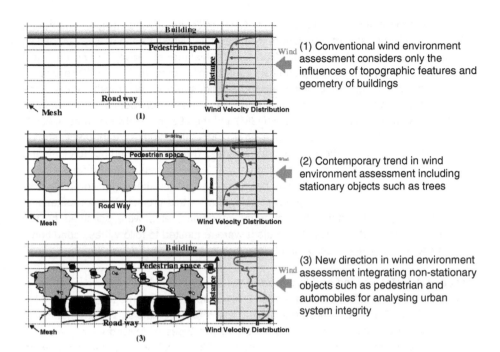

(1) Conventional wind environment assessment considers only the influences of topographic features and geometry of buildings

(2) Contemporary trend in wind environment assessment including stationary objects such as trees

(3) New direction in wind environment assessment integrating non-stationary objects such as pedestrian and automobiles for analysing urban system integrity

FIGURE 6.8 Concept of wind environment assessment: from conventional approach to contemporary tactics (Mochida and Lun, 2008)

4.2 AIJ COLLABORATIVE RESEARCH PROJECT

Cross-comparisons of CFD results for wind environment at pedestrian level

Recently, the prediction of the wind environment around a high-rise building, using CFD technique, was carried out at the practical design stage. There are a large number of previous studies of the wind environment around actual buildings using CFD (Stathopoulos and Baskaran, 1996; Timofeyef, 1998; Richards et al, 2002; Westbury et al, 2002; Blocken et al, 2003; Yang et al, 2006). However, the influences of the computational conditions – grid discretization, domain sizes and boundary conditions, for instance – on the prediction accuracy of the velocity distribution at pedestrian level near the ground have not been systematically investigated. For the aspect of industrial CFD applications, some published policy statements and guidelines have provided valuable information on the application of CFD for pedestrian wind environment around buildings (Roache et al, 1986; AIAA, 1998; ERCOFTAC, 2000), but the guideline oriented to the application of this area is somehow different. Recently, the recommendations on the use of CFD in predicting pedestrian wind environment were proposed by the COST C14 group (Franke et al, 2004; Franke, 2006), these recommendations were mainly based on the results published elsewhere.

In view of this point, a working group for CFD prediction of wind environment around buildings, which consisted of researchers from several universities and private companies (see Note), was organized by the AIJ (Architectural Institute of Japan). During the initial stage of the project, the working group carried out cross-comparisons of CFD results for flow around a single high-rise building, around a building block placed within the surface boundary layer and flow within a building complex in urban area, obtained from various k-ε models, i.e. Differential Stress Model (DSM) and the LES model (Mochida et al, 2002; Tominaga et al, 2004, 2005, 2008a; Yoshie et al, 2005, 2006, 2007). Figure 6.9 illustrates the six test cases (A–F) for these cross-comparisons. They were carried out to clarify the major factors that affect the prediction accuracy. In order to assess the performance of turbulence models, the results should be compared under the same computational conditions. Special attention was given to this point in this project. The computational conditions, i.e. grid arrangements, boundary conditions, etc., were specified by the organizers, and the participants in this project were requested to follow these conditions. Two representative cases, cases A and E, are discussed next.

Test case A (2:1:1 shaped building model)

A summary of computations carried out for 2:1:1 shaped building model (test case A) is provided in Table 6.3. Nine groups have submitted a total of 18 data sets of computational results. The performance of the standard k-ε and five types of revised k-ε models was examined. Furthermore, DSM (Murakami et al, 1993), DNS with third-order upwind scheme (Kataoka and Mizuno, 2002) and LES using the Smagorinsky Subgrid-scale Model (Tominaga et al, 2008b) were also included in the comparison.

The predicted reattachment lengths on the roof, X_R, and that behind the building, X_F, were determined for all the cases, as shown in Table 6.3. It can be seen that the reverse flow on the roof, clearly observed in the experiment, was not reproduced using the standard k-ε (KE1~8). On the other hand, the reverse flow on the roof appeared in the

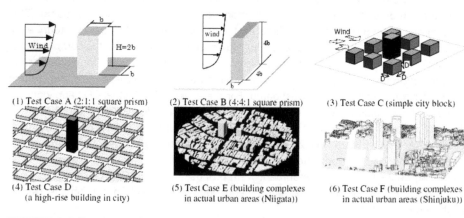

(1) Test Case A (2:1:1 square prism) (2) Test Case B (4:4:1 square prism) (3) Test Case C (simple city block)

(4) Test Case D (a high-rise building in city) (5) Test Case E (building complexes in actual urban areas (Niigata)) (6) Test Case F (building complexes in actual urban areas (Shinjuku))

FIGURE 6.9 Six test cases for cross comparisons by the CFD Working Group of AIJ

TABLE 6.3 Computed cases for 2:1:1 shaped building model (test case A; Tominaga et al, 2004)

AFFILIATION	CASE	SOFTWARE	TURBULENCE MODEL	SCHEME FOR CONVECTION TERMS	COMPUTATIONAL METHOD AND TIME INTEGRAL SCHEME	X_R^*/b	X_F^*/b
A	KE1	STREAM ver.2.10	k-ε(standard)	QUICK	SIMPLE, steady solution	–	2.54
B	KE2	STREAM ver.2.10	k-ε (standard)	QUICK (1st-order upwind for k and ε)	SIMPLE, steady solution	–	1.66
C	KE3	STREAM ver.2.10	k-ε (standard)	QUICK	SIMPLE, steady solution	–	2.00
	LK1		k-ε (LK)[3]			0.87	2.98
D	KE4	STREAM ver.2.10	k-ε (standard)	QUICK	SIMPLE, steady solution	–	2.00
	RNG1		k-ε (RNG)[10]	QUICK		0.50	2.80
E	KE5	STAR-LT ver.2.0	k-ε (standard)	QUICK	SIMPLE, steady solution	–	2.20
F	MMK1	Homemade	k-ε (MMK)[11]	QUICK	MAC, unsteady solution with implicit scheme	0.65	2.72
	KE6	FLUENT ver.5.0	k-ε (standard)	Central	SIMPLE, steady solution	–	2.41
	RNG2		k-ε (RNG)[10]			0.58	3.34
	KE8		k-ε (standard)			–	2.70
	LK2		k-ε (LK)[3]			0.58	3.19
	LK3		k-ε (modified LK)[12]	QUICK	HSMAC, unsteady solution with implicit scheme	0.53	3.11
	MMK2		k-ε (MMK)[11]			0.52	3.09
	DBN		k-ε (Durbin)[13]			0.63	2.70
G	DSM		DSM[7]			>1.0	4.22
		Homemade					
	LES1		LES (without inflow turbulence)[88]	Second-order centred difference	SMAC, Convection terms: Adams-Bashforth scheme Diffusion terms: Crank-Nicolson scheme	0.62	1.02
	LES2		LES (with inflow turbulence)[9]			0.50	2.10
H	KE7	Homemade	k-ε (standard)	QUICK	HSMAC, unsteady solution with implicit scheme	–	1.98
I	DNS	Homemade	DNS[8]	3rd-order upwind scheme	Artificial compressibility method, explicit	0.92	2.05
			Experiment			0.52	1.42

results of all the revised k-ε models (LK1, RNG1, MMK1, RNG1, LK2, LK3, MMK2, DBN), although its size was slightly larger than that of the experiment. In the DSM result, the prediction of the separated flow from a windward corner was too large and it did not reattach on the roof. The result of LES without inflow turbulence (LES1) can reproduce the reattachment on the roof, but X_R is somewhat overestimated in this case. On the other hand, the result of LES with inflow turbulence (LES2) shows close agreement with the experiment.

The reattachment length behind the building, X_F, computed using the standard and revised k-ε models was larger than the experimental value, in all the cases. It is surprising to see that there were significant differences in X_F values between the results obtained using the standard k-ε model. As was already noted, the grid arrangements and boundary conditions were set to be identical in all cases, and the Quadratic Upstream Interpolation for Convective Kinematics (QUICK) scheme was used for convection terms in many cases. The reason for the difference in X_F values predicted by the standard k-ε models is not clear, but it may be partly due to the difference in some details of numerical conditions, e.g. the convergence criteria, etc. The results of the revised k-ε models, except for the Durbin model, led to X_F being larger than the value obtained using the standard k-ε model. The computational-experimental discrepancy was significantly lower for LES and DNS computations. On the other hand, DSM greatly overestimated X_F. The overestimation of the reattachment length behind a three-dimensional obstacle was also reported by Lakehal and Rodi (1997).

The size of the recirculation region behind the building is strongly affected by the momentum transfer mechanism in the wake region, where vortex shedding plays an important role. Thus, the reproduction of vortex shedding is important for accurate prediction of the X_F value. However, none of the k-ε models compared here could reproduce vortex shedding. This resulted in underestimation of the mixing effects in the lateral direction causing a large recirculation region behind the building.

Test case E (building complex in actual urban area)

The prediction accuracy for wind environment within an actual building complex, located in Niigata City, Niigata Prefecture, Japan, was examined in this test case. The prediction was carried out by three different codes, namely an in-house developed CFD code and two commercial CFD codes. The computational conditions are given in Table 6.4. In order to reproduce the geometries of the surrounding building blocks, data from an identical CAD file were used in the three codes. This file was produced from the drawings of the experimental model.

Although CFD simulations have been performed for 16 different wind directions, only the wind distributions for wind directions of NNE and W, which are the prevailing wind directions in Niigata City, are shown here. Since there were no clear differences between the horizontal distributions of wind speed near ground surface, predicted by the three CFD codes, only the results obtained using Code T are presented (Figure 6.10). This figure illustrates the horizontal distributions of scalar velocity near ground surface (z = 2 m). The values in Figure 6.10 are normalized by the velocity value at the same height at inflow

TABLE 6.4 Computed cases for building complex in actual urban area (test case E)

CFD Code	Code O Homemade code	Code M Commercial code	Code T Commercial code
Computational method and time integral scheme	Overlapping structured grid Artificial compressibility	Structured grid SIMPLE, steady state	Unstructured grid SIMPLE, steady state
Turbulence model	Standard k-ε	Standard k-ε	Standard k-ε
Scheme for advection term	3rd-order Upwind	QUICK	MUSCL (2nd-order)
Grid arrangements			

boundary. Figure 6.11 shows a comparison of CFD results with wind tunnel experiments. It was found that the scalar velocity predicted by all CFD codes was smaller in the wake region compared with the experimental value. This discrepancy is mainly due to the fact that the k-ε model cannot reproduce the vortex shedding from tall buildings, as seen in test case A. This problem is absent in LES results, but the computational cost of adopting LES is still beyond the scope of practical applications. In order to overcome this drawback, the adaptation of Hybrid RANS/LES and DES (detached-eddy simulation) (Spalart, 1999;

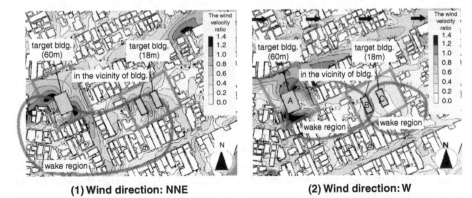

(1) Wind direction: NNE   (2) Wind direction: W

FIGURE 6.10 Distributions of scalar velocity near the ground surface around a building complex (Tominaga et al, 2004, 2005, 2008a; Yoshie et al, 2005, 2006, 2007)

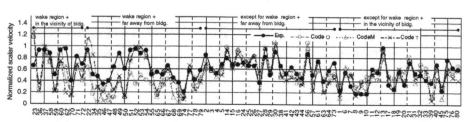

(1) Wind direction: NNE

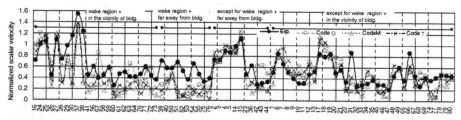

(2) Wind direction: W

FIGURE 6.11 Comparison of scalar velocity value for each measurement (Tominaga et al, 2004, 2005, 2008a; Yoshie et al, 2005, 2006, 2007)

Menter and Kunts, 2001; Squires, 2003, 2006; Kenjers and Hanjalic, 2005; Hanjalic, 2006) would be needed.

In addition, the prediction accuracies and the performances of various turbulence models are also important. Table 6.5 shows a comparison of various turbulence models used for wind engineering. Although the prediction accuracy of large-eddy simulations is better than that of the Reynolds averaged Navier-Stokes (RANS) models, such as the k-ε model, the RANS model is still generally used because of its ease of use and low computational load.

COUPLED ANALYSIS OF CONVECTION, RADIATION AND CONDUCTION

Yoshida et al (2000) and Chen et al (2004) proposed a prediction method for microclimates within city blocks, which is a coupled simulation using CFD, radiation and conduction calculations including the effect of radiation heat transfer between buildings and ground surfaces within building blocks. Figure 6.12 presents the calculation flow for the microclimate model. The human thermal sensation index, a new standard effective temperature (SET*) proposed by Gagge et al (1986), is also incorporated in this method. The prediction accuracy of this method was confirmed by Chen et al (2004) through comparisons with field measurements. Mochida et al (2001) estimated the effect of an increase in the albedo of building surfaces on the outdoor thermal environment. Although an increased albedo of the building walls decreases the air temperature at pedestrian level, the SET* increases offset these changes because of the increase in reflected radiation from the wall onto pedestrians. Thus, one advantage of the

TABLE 6.5 Relative performance of various turbulence models for practical applications

TURBULENCE MODEL		RANS			LES	
		STANDARD k-ε	NON-LINEAR k-ε	RSM	CONVENTIONAL SGS	DYNAMIC SGS
1. Simple flows (channel flow, pipe flow, etc.)		O	O	O	O	O
2. Flow around bluff body (with turbulent approaching wind, local equilibrium is not valid)	1) Impinging area	×	Δ, O	×	O	O
	2) Separated area	×	Δ, O	Δ	O**	O
	3) With oblique wind angle	×, Δ	Δ, O	Δ	O**	O
3. Transitional flow (low Reynolds number effects)	1) Near wall	O*	O*	O*	O**	O
	2) Not near wall	×	×	×	×	O
4. Unsteady flow	1) Vortex shedding	×	Δ	Δ	O	O
	2) Fluctuation over wide spectrum range	×	×	×	O	O
5. Stratified flow		×	×	Δ	Δ	O
6. CPU time required		O	O	Δ	×	×

Note: O: functions well, Δ; insufficiently functional, ×: functions poorly,

*: functions well when low Reynolds number type model is employed,

**: functions well with wall damping function

microclimate model is its ability to estimate detailed spatial distribution of various quantities at the human level.

HUMAN THERMAL MODEL

There are many indices that account for outdoor thermal environments other than air temperature. The wet bulb globe temperature (WBGT) is used for heat disorder prevention information in Japan. However, the human body heat balance and human thermal physiological response are not taken into account in this index. The human thermal model, which includes these processes, is very important for predicting human thermal sensations in the outdoor environment. Various other indices such as the predicted mean vote (PMV) by Fanger (1973), a new standard effective temperature (SET*) by Gagge et al (1986) and the predicted heat strain (PHS) model by Malchaire et al (2001) include their own human thermal models. However, it is not clear whether these models can be applied to outdoor environments or not, as these have been developed for indoor environments. For outdoor environments, the Steadman apparent temperature by Steadman (1979), the physiological equivalent temperature (PET) by Hoeppe (1999) and Matzarakis et al (1999), the Out_SET*

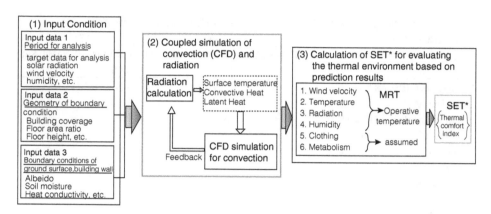

FIGURE 6.12 Flow chart for assessing outdoor human comfort based on CFD (Yoshida et al, 2000, 2006)

by Pickup and de Dear (1999) and the expected thermal sensation by Kuwabara et al (2005) have been proposed as new thermal indices. The accuracy of these models should be examined from the viewpoint of human thermal physiology for hot outdoor environments. For example, PET is almost completely insensitive to latent heat fluxes because the potential (= maximum) evaporation is employed as the measure for evaporation heat loss by sweating (Hoeppe, 1999). Minami et al (2006) compared the performance of SET* and PHS models with the results of a subject experiment in a hot environment and developed a new sweating model for SET* including the effect of the metabolic rate.

There are still many problems pertaining to human thermal models such as clothing, typical metabolic rate, etc. The accurate prediction of heat exchange between a human body and the outdoor environment is one such problem. Ono et al (2006) estimated the convective heat transfer coefficient of the human body surface with the aid of wind tunnel experiments and CFD analysis. Figure 6.13 presents the distributions of the convective heat transfer coefficient on the surface of the human body under various wind velocity conditions. In this figure, $\overline{\alpha_c}$ represents the average surface value of the convective heat transfer coefficient of a human body. In the future, it is expected that a comprehensive human thermal model integrating various factors described above will be developed.

MODELLING OF SMALL-SCALE FLOW OBSTACLES
Various subgrid-scale obstacles in real situations
The real situations of environments in street canyons are influenced by the interaction of various objects, both stationary and non-stationary. In most of the previous CFD simulations of flow around buildings, only the influences of topographical features and building geometry were considered (Figure 6.8(1)). At the pedestrian level, influences of small obstacles such as trees (stationary) and automobiles (non-stationary) are significant (Figures 6.8(2) and 6.8(3)), but their effects have been neglected in most conventional CFD predictions of wind environment. However, modelling of the effects of such subgrid-scale flow obstacles have been investigated in recent years by many researchers.

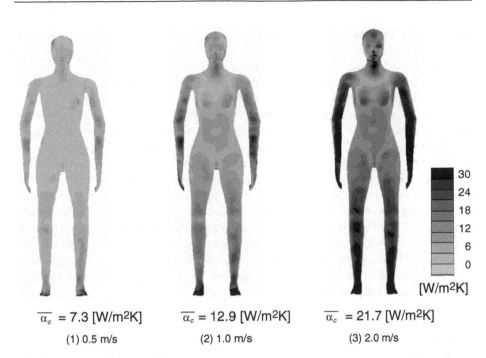

$\overline{\alpha_c} = 7.3\ [W/m^2K]$ $\overline{\alpha_c} = 12.9\ [W/m^2K]$ $\overline{\alpha_c} = 21.7\ [W/m^2K]$

(1) 0.5 m/s   (2) 1.0 m/s   (3) 2.0 m/s

FIGURE 6.13 Distribution of convective heat transfer coefficients on human body surface under various wind velocity conditions (Ono et al, 2006)

MODELLING OF AERODYNAMIC AND THERMAL EFFECTS OF THE TREE CANOPY

Outline of tree canopy model for reproducing various effects of planted trees

Tree planting is one of the most popular measures of environmental design for improving the outdoor climate. It reduces strong winds around high-rise buildings and improves outdoor thermal comfort, etc. In order to reproduce the effects of trees, a lot of research has been conducted (Yamada, 1982; Uno et al, 1989; Svensson and Häggkvist, 1990; Green, 1992; Hiraoka, 1993; Liu et al, 1996; Hiraoka, 2004; Hiraoka and Ohashi, 2006; Ohashi, 2006; Yoshida et al, 2006; Mochida et al, 2008a). Hiraoka, one of the pioneers in this research field, has developed models for expressing the aerodynamic effects of trees, stomatal conductance, radiative heat transfer, and the balance of heat, water vapour and carbon dioxide within vegetation canopy (Hiraoka, 2004). The present model developed by Hiraoka is very comprehensive and can provide a very accurate prediction of the microclimate within the vegetation canopy, provided that proper input parameters are given. From the viewpoint of engineering application, however, it is not easy to apply this model because it requires too many input parameters, some of which are not usually available. Yoshida et al (2006) developed a three-dimensional tree canopy model which can be easily applied to the practical applications related to microscale climate prediction. It includes the following effects (Figure 6.14):

- aerodynamic effects of planted trees;
- thermal effects:
  - shading effects on short-wave and long-wave radiations;
  - productions of water vapour (latent heat) and sensible heat from planted trees.

Yoshida et al (2006) predicted the effects of tree planting on outdoor thermal comfort by using coupled simulation of convection (CFD) and radiation combined with the tree canopy model. Figure 6.14 illustrates the outline of the computational approach for predicting outdoor thermal comfort. Four equations: (1) transport equation of momentum, (2) transport equation of heat, (3) transport equation of moisture, and (4) heat transfer equation by radiation are solved as a system of coupled equations.

This three-dimensional canopy model employed the k-ε turbulence model, with extra terms (Table 6.6) added into the transport equations, to simulate the aerodynamic effects of trees. The term F_i included in the i component of the momentum equation denotes the drag force. F_k and F_ε are added, respectively, into the transport equations for turbulent kinetic energy, k, and energy dissipation rate, ε, to represent the effects of trees on the turbulent flow field. These extra terms were derived by applying the spatial average to the basic

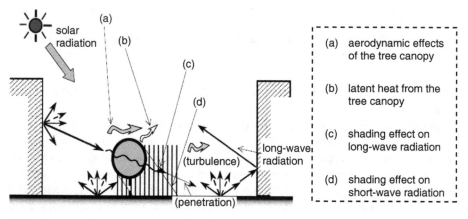

FIGURE 6.14 Effects of tree considered in the tree canopy model (Yoshida et al, 2006)

TABLE 6.6 Additional terms for tree canopy (Yoshida et al, 2006)

F_i	$-\eta C_f a(x_1,x_2,x_3)\overline{<u_i>}\sqrt{\overline{<u_j>}^2}$	η : green coverage ratio,
F_k	$\overline{<u_i>}F_i$	C_f : drag coefficient,
		$a(x_1,x_2,x_3)$: leaf surface area density,
F_ε	$\dfrac{\varepsilon}{k}C_{p\varepsilon}F_i$	$<f>$: ensemble-average
		$\overline{f}$: spatial-average or filtered quantities

equation (Hiraoka, 1993). The radiative heat transport was computed using the method based on Monte Carlo simulation (Omori et al, 1990). The solar and long-wave radiant fluxes incident to the plant canopy were calculated by assuming a decay rate of $\{1 - \exp(-k'a(x_1, x_2, x_3)/)\}$, where k' is the absorption coefficient and $/$ is the length by which radiant flux passes through the plant canopy. The mean leaf surface temperature of the plant canopy was estimated using the heat balance equations listed in Table 6.7. In the heat balance equation, i.e. Equation [27], the heat conductivity term is neglected, as the heat capacity of the leaf was negligibly small.

The results obtained using this model were presented by Yoshida et al (2006). Based on similar approaches, various studies have been conducted in recent years to investigate the effects of planted trees on the outdoor thermal environment and optimize their shapes, densities and layouts (Mochida et al, 2005; Ooka et al, 2006; Lin et al, 2008).

Optimization of tree canopy model for reproducing the aerodynamic effects

Recently, the present authors carried out a series of numerical studies to examine the accuracy of the existing canopy models in reproducing the aerodynamic effects of trees, and to optimize the model coefficients (Murakami et al, 2003b; Mochida et al, 2008a). The canopy models adopted in these studies used the revised k-ε model, which based on a 'mixed timescale' concept (S-Ω model), as a base, with extra terms added into the transport equations as shown in Table 6.6. The additional terms contained four parameters, C_{pe}, η, a and C_f. C_{pe} was regarded as a model coefficient in turbulence modelling for prescribing the timescale of the process of energy dissipation in the canopy

TABLE 6.7 Heat balance equations of plant canopy (Yoshida et al, 2006)

$S_P + R_{DP} + H_P + LE_P = 0$	(27)
$H_P = A_P \alpha_C (T_{aP} - T_P)$	(28)
$LE_P = A_P \alpha_W \beta_P L (f_{aP} - f_{sP})$	(29)

where,

S_P	: absorbed solar radiation flux on leaf surfaces	[W]
R_{DP}	: absorbed long-wave radiation flux on leaf surfaces	[W]
H_P	: sensible heat on leaf surfaces	[W]
LE_P	: latent heat on leaf surfaces	[W]
L	: latent heat of water vaporization	$(2.5 \times 10^6$ [J/kg])
A_P	: leaf area within tree crown (A_P is twice the product of the volume of tree crown and the leaf area density of tree crown)	[m²]
α_C	: convective heat transfer coefficient on leaf surfaces	[W/m²K]
T_P	: leaf surface temperature	[K]
T_{aP}	: air temperature in the mesh including tree crown	[K]
α_W	: convective moisture transfer coefficient on leaf surfaces	[kg/(m²skPa)]; ($\alpha_W = 7.0 \times 10^{-6} \alpha_C$)
β_P	: moisture availability on leaf surfaces	[–]
f_{aP}	: water vapour pressure in the mesh including tree crown	[kPa]
f_{sP}	: saturation water vapour pressure at leaf surface temperature	[kPa]

layer, while η, *a* and C_f were the parameters that should be determined according to the conditions of trees. The choice of value of C_{pe} in F_ε (see equation [26] in Table 6.6) strongly affects the prediction accuracy (Mochida et al, 2008a) and considerable differences were observed between the values adopted in the previous researches (Yamada, 1982; Uno et al, 1989; Svensson and Häggkvist, 1990; Green, 1992; Hiraoka, 1993; Liu et al, 1996; Hiraoka, 2004; Ohashi, 2006; Yoshida et al, 2006; Mochida et al, 2008a). The C_{pe} was optimized by the present authors and the predicted results were compared with experimental results as shown in, for an example, Figure 6.15. The value of 1.8 was selected for C_{pe} as a result of a series of numerical experiments. The tree canopy model developed here was incorporated in the Local Area Wind Energy Prediction System (LAWEPS) (Murakami et al, 2003b). A five-stage nesting grid system was adopted in the LAWEPS. The largest domain (1st domain) covered an approximate region of about 500 × 500 km in the horizontal plane and 10 km in vertical plane. The smallest domain (5th domain) occupied a region of about 1 × 1 × 1 km. Figures 6.16(1) and 6.16(2)

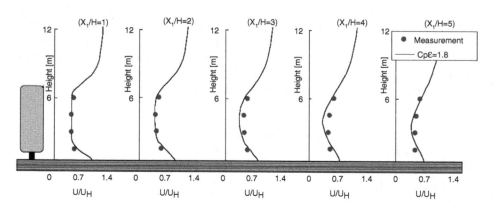

FIGURE 6.15 Comparison of normalized vertical profiles of mean streamwise velocity behind pine trees (H: a height of tree, U_H: inflow velocity at a height of H) (Mochida et al, 2008a)

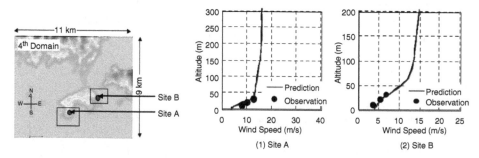

FIGURE 6.16 Comparison of vertical velocity profiles above Shionomisaki Peninsular, Wakayama, Japan (15 December 2001) (Murakami et al, 2003b)

compare the vertical velocity profiles predicted by LAWEPS with the tower observations at Shionomisaki Peninsular, Wakayama, Japan (Murakami et al, 2003b). It can be seen that the results of numerical prediction show very close agreement with the observations.

The present authors have recently carried out an extensive review of tree canopy models. Four types of canopy models were selected. The model coefficients adopted in the extra term added to the transport equation of energy dissipation rate, ε, were optimized by comparing the numerical results with field measurements. The four selected tree canopy models (Uno et al, 1989; Svensson and Häggkvist, 1990; Green, 1992; Hiraoka, 1993; Liu et al, 1996; Ohashi, 2006) are based on the k-ε model as given in Table 6.8. As seen in this table, the additional terms contain five parameters, two model coefficients, C_{pe1} and C_{pe2}, η, a and C_f. Results with type B and type C models are compared with measurements in a recent paper by Mochida et al (2008a).

MODELLING FOR AERODYNAMIC EFFECTS OF VEHICLE CANOPY

Recently, the present authors developed a simulation method named the 'vehicle canopy model' to predict the effects of moving automobiles on flow and diffusion fields within street canyons (Hataya et al, 2006; Mochida et al, 2006).

Outline of vehicle canopy model

In this model, the effects of each individual moving automobile were not directly modelled. Instead, the total effects of all the moving automobiles in the street were considered as a whole. The aerodynamic effect of the moving automobiles was modelled using the methodology of the canopy model (Mochida et al, 2008b). The proposed vehicle canopy model was based on the k-ε model, in which terms were added in the transport equations. Similarly to the tree canopy model (Yamada, 1982; Uno et al, 1989; Svensson and Häggkvist, 1990; Green, 1992; Hiraoka, 1993; Liu et al, 1996; Hiraoka, 2004; Ohashi, 2006; Yoshida et al, 2006; Mochida et al, 2008a), the extra term $-F_i$ was included in the momentum equation (see Table 6.9) to account for the effect of moving automobiles on velocity change. This $-F_i$ was defined as a function of wind velocity $<u_i>$ in the tree canopy model (see Tables 6.6 and 6.8). In modelling the vehicle canopy, $<u_i>$ was replaced by the relative velocity between the wind velocity and the moving speed of automobiles (Figure 6.17). In order to simulate the effects of moving automobiles on turbulence increase rate and energy dissipation rate, additional terms $+F_k$ and $+F_\varepsilon$ were included in the transport equations of turbulent kinetic energy, k, and energy dissipation rate, ε. Table 6.9 quantifies these extra terms (Hataya et al, 2006; Mochida et al, 2006).

CFD analyses of flow field in real situations in street canyons

By using the proposed vehicle canopy model, the flow and diffusion fields within Jozenji-Street in Sendai, Japan were predicted (Figure 6.18). The accuracy of CFD analyses was confirmed by comparing the simulation results with the field measurement results conducted by the present authors (Mochida et al, 2005; Watanabe et al, 2005). Details of the numerical settings can be obtained from Hataya et al (2006) and Mochida et al (2006).

All the test cases are shown in Table 6.10. Horizontal distributions of wind velocity vectors at a height of 1.5 m for case 1 are illustrated in Figure 6.19. A comparison of the

TABLE 6.8 Additional terms in various tree canopy models

	F_i	F_k	F_ε	SELECTED VALUES FOR NUMERICAL COEFFICIENTS
Type A		$\overline{<u_j>F_i}$	$\eta\dfrac{\varepsilon}{k}\cdot C_{p\varepsilon1}\dfrac{k^{3/2}}{L}\quad\left(L=\dfrac{1}{a}\right)$	Hiraoka (1993): $C_{p\varepsilon1}=0.8\sim1.2$
Type B		$\overline{<u_j>F_i}$	$\dfrac{\varepsilon}{k}\cdot C_{p\varepsilon1}F_k$	Yamada (1982): $C_{p\varepsilon1}=1.0$ Uno et al (1989): $C_{p\varepsilon1}=1.5$ Svensson and Häggkvist (1990): $C_{p\varepsilon1}=1.95$
Type C	$\eta C_f a\overline{<u_i>}\sqrt{<u_j>^2}$	$\overline{(u_i)F_i}-4\eta C_f a\sqrt{\overline{(u_j)}}^2$	$\dfrac{\varepsilon}{k}\left[C_{p\varepsilon1}\left(\overline{(u_i)F_i}\right)-C_{p\varepsilon2}\left(4\eta C_f a\sqrt{\overline{(u_j)}}^2\right)\right]$	Green (1992): $C_{p\varepsilon1}=C_{p\varepsilon2}=1.5$ Liu et al (1996): $C_{p\varepsilon1}=1.5$, $C_{p\varepsilon1}=0.6$
Type D		$\overline{(u_i)F_i}-4\eta C_f a\sqrt{\overline{(u_j)}}^2$	$\eta\dfrac{\varepsilon}{k}\cdot C_{p\varepsilon1}\dfrac{k^{3/2}}{L}\quad\left(L=\dfrac{1}{a}\right)$	Ohashi (2006): $C_{p\varepsilon1}=2.5$

TABLE 6.9 Additional terms for vehicle canopy model (Mochida et al, 2006; Hataya et al, 2006)

$F_i \quad \dfrac{1}{2}C_{f-car}\dfrac{A_{car}}{V_{cell}}\left(<u_i>-\overline{<u_{i-car}>}\right)\sqrt{\left(<u_j>-\overline{<u_{j-car}>}\right)^2}$ [30]	C_{f-car} : drag coefficient of automobiles A_{car} : sectional area of automobiles observed from i direction [m²]
	V_{cell} : volume of one computational mesh [m²]
$F_k \quad \left(<u_i>-<u_{i-car}>\right)F_i$ [31]	u_{i-car} : moving speed of automobiles [m/s] L : length scale for canopy layer
$F_\varepsilon \quad \dfrac{\varepsilon}{k}\dfrac{k^{3/2}}{L}C_{\varepsilon-car}$ [32]	$C_{\varepsilon-car}$: ratio of turbulence scale of automobiles

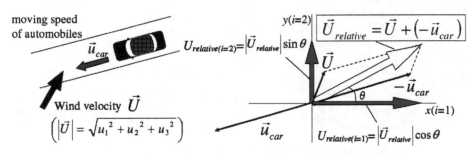

FIGURE 6.17 Relationship between wind velocity and moving speed of automobiles

results of turbulent kinetic energy, k, between field measurements and CFD analyses in Jozenji Street is given in Table 6.11. In the cases without automobiles (cases 1 and 2), k values were largely underpredicted in comparison with the measurement results. However, the magnitude of k due to turbulent diffusion generated by moving automobiles was well reproduced in case 3-2, especially on the southern sidewalk.

MODELLING FOR AERODYNAMIC AND THERMAL EFFECTS OF BUILDING CANOPY

Temperature increase due to urbanization is becoming a very serious problem in Japan. City administrators and urban planners are now willing to adopt countermeasures that can mitigate the problem of UHI effects. Various urban planning scenarios have been proposed to minimize the impact of urbanization on the urban climate. Since the mid-1990s, a lot of numerical studies of mesoscale climate in urban areas have been carried out by many researchers (Mochida et al, 1997; Murakami et al, 2003a; Ooka et al, 2004; Kondo et al, 2006; Sasaki et al, 2006; Sato et al, 2006).

Historically, a one-dimensional heat balance model was usually adopted for the ground boundary conditions in mesoscale climate analysis. In this conventional model,

FIGURE 6.18 Actual urban space; Jozenji Street, Sendai, Japan

TABLE 6.10 Test cases for vehicle canopy model (Mochida et al, 2006; Hataya et al, 2006)

	ROADSIDE TREES	EFFECTS OF MOVING AUTOMOBILES ON TURBULENT DIFFUSION PROCESS
Case 1	Without	Without automobiles
Case 2	Present situation	Without automobiles
Case 3-1	Present situation	$u_{car} = 0$ [km/h]
Case 3-2	Present situation	$u_{car} = 15$ [km/h]

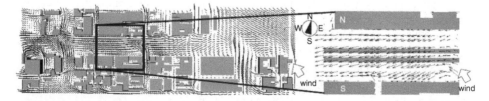

FIGURE 6.19 Horizontal distributions of wind velocity vectors (without trees and automobiles, 12 am on 3 August at a height of 1.5 m) (Hataya et al, 2006; Mochida et al, 2006)

TABLE 6.11 Comparison of turbulent kinetic energy, k [m²/s²] (Mochida et al, 2006; Hataya et al, 2006)

MEASURING POINTS	RESULT OF FIELD MEASUREMENT	RESULTS OF CFD ANALYSES			
		CASE 1	CASE 2	CASE 3-1	CASE 3-2
Northern sidewalk	0.44	0.06	0.07	0.02	0.22
Southern sidewalk	0.27	0.12	0.03	0.03	0.25

a roughness parameter was employed to represent the effect of the building complex. As the vertical grid size adjacent to the ground surface must be made several times larger than the roughness length in the conventional model, physical phenomena within the surface layer could not be estimated. Furthermore, the definition of surface temperature is vague in the conventional model because its relationship with ground, roof and wall surface temperature is unclear. Therefore, it is necessary to include the effects of the urban canopy precisely in order to analyse the thermal environment at pedestrian level, in an urban area. On the other hand, the one-dimensional urban canopy model was also commonly used to analyse urban thermal environments. Although this model can predict the thermal environment at pedestrian level easily, it does not make it possible to consider the effects of local climate, due to the limitation of the assumption of a horizontally homogeneous flow and temperature field.

Recently, Ooka et al (2004) developed a comprehensive urban canopy model that considers the following five factors in dealing with a building complex:

1 wind reduction by the building complex;
2 production of turbulence by the building complex;
3 solar radiation (short-wave) heat transfer inside and outside the building complex;
4 long-wave radiation heat transfer inside and outside the building complex; and
5 sensible and latent heat transfer from the building surfaces.

The urban canopy model developed in their study was incorporated into the meteorological mesoscale model. The effects of the plant canopy, as illustrated in Figure 6.13, were also considered. Concerning the modelling of aerodynamic effects of the building canopy, i.e. (1) and (2), Maruyama developed a building canopy model (Maruyama, 1993), which has been widely used by many researchers. He also provided a detailed database obtained by a series of wind tunnel tests to determine the model coefficients included in his model.

INTEGRATION OF URBAN CLIMATE SIMULATIONS WITH VARIOUS SCALES

Urban climate applications now cover various phenomena, at scales ranging from the microclimate around a human body to regional climate (Figure 6.3). Although scales associated with these phenomena are different, they are related and coupled to each other. Research efforts, thus, should be devoted to develop a method for integrating the sub-models into a comprehensive, total simulation system. For this purpose, it is necessary to develop a new software platform which not only can handle many subsystems for analysing each scale-dependent phenomenon, but also can integrate them for evaluating the total urban climate. Figure 6.20 illustrates the concept of the platform proposed by Murakami and Mochida (Murakami et al, 2000; Murakami, 2004; Mochida, 2005).

A number of case studies based on the proposed software platform were presented by Murakami (2004) and the present authors (Mochida, 2005). It was pointed out that CWE is now in the process of growing from a tool for analysis to a tool for environmental design.

CONCLUDING REMARKS

This chapter gives an overview of the latest microclimate simulation studies and the assessment tools used in urban climate research. The first part of the chapter introduced and classified various assessment tools, i.e. mesoscale meteorological model, microclimate model, building model and human thermal model, for urban climate studies. These are very powerful tools to estimate the mechanism of the urban heat island and the potency of countermeasures. However, validations of the prediction accuracy of these models are insufficient. Old reference data of Clarke et al (1971), for instance, are still used to validate mesoscale meteorological models. In particular, a validation database for surface boundary layers is required. The recent results obtained from a diagnostic technique by Uehara et al (2000), Christen et al (2003), Moriwaki et al (2003), Offerle et al (2003), Rotach et al (2003), Yee and Biltoft (2004), Mutoh and Narita (2005), Pearlmutter et al (2005) and Narita et al (2006) may be useful for this shortcoming. The results of the detailed simulation in the urban canyon space, such as Kanda et al (2004), are useful to develop the urban canopy model. Benchmark tests are required in order to understand the properties of the various models developed. Surface data and geographical information system (GIS) data for the boundary conditions of these models are also required. As described above, there are various spatial scale phenomena in urban climates and there are interactions between each scale. The entire structure of urban climates cannot be understood by partial analysis in each scale alone. Therefore, it is important to combine the assessment tools of each scale, from human scale to mesoscale, comprehensively.

Nevertheless, these tools have contributed to the development of urban climatology. Detailed structure of the urban climate, which cannot be observed experimentally, can be obtained using these tools. However there is a still a deep gap between researchers or specialists in urban climates and urban planners or policy-makers. This vast wealth of urban climate knowledge has not been sufficiently applied to urban design or planning. For the application of this knowledge to actual urban design, it is important to make it clearly understandable with some examples. Visualization techniques are one of the most powerful methods to express the influence of various urban heat island mitigation strategies easily and concretely for non-specialists. Moreover, for urban heat island mitigation strategies, the priorities for various countermeasures should be clearly shown. Thus, the effects of various measures should be compared with each other. It is also important to indicate the cost benefit of each countermeasure.

The latter part of the chapter described the recent and future trends of numerical modelling for predicting the wind and thermal environment, as well as turbulent diffusion

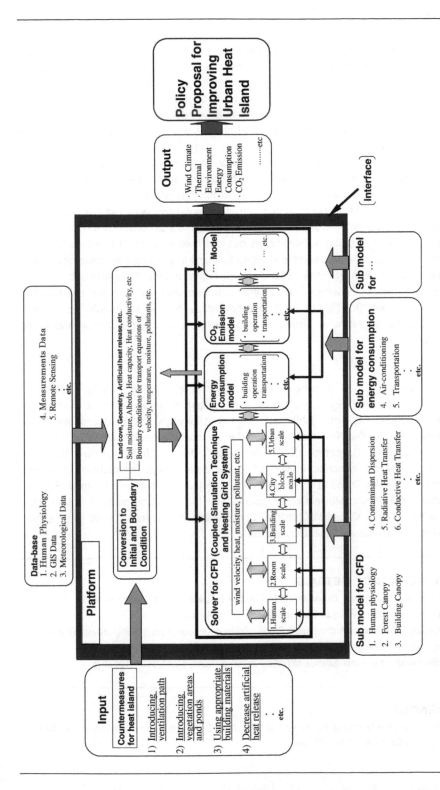

FIGURE 6.20 Prototype of software platform for the total analysis of urban climate (Murakami et al, 2000; Murakami, 2004; Mochida, 2005)

processes in real urban space in the presence of various small subgrid-scale flow obstacles. The conventional approach in numerical modelling is generally carried out based on static conditions (i.e. stationary objects such as buildings). This approach may be erroneous when applied to real situations, where dynamic conditions exist, e.g. due to non-stationary objects such as automobiles. In recent years, canopy models for reproducing the aerodynamic and thermal effects of trees/buildings/automobiles have been developed and applied to various problems related to urban climate. This chapter demonstrates some of the results regarding this subject. The research studies emphasize the significance of the effects of stationary and non-stationary objects (tree canopy and vehicle canopy, respectively) on turbulent flow fields within street canyons.

As indicated in 'Integration of urban climate simulations with various scales', the development of the total simulation system integrating the sub-model for the phenomena with various scales makes it possible to carry out total analysis of the urban climate, which is comprised of many elements and is affected by many interacting physical processes at various scales. CWE is now in the process of growing from a tool for analysis to a tool for environmental design. However, a lot of improvements and revisions are still required before the turbulent flow phenomena in urban areas can be accurately reproduced. Hence, fundamental research efforts to improve the prediction accuracy should be continued.

NOTE

The CFD Working Group members are: A. Mochida (Chairman, Tohoku University), Y. Tominaga (Secretary, Niigata Institute of Technology), Y. Ishida (Institute of Industrial Science, University of Tokyo), T. Ishihara (University of Tokyo), K. Uehara (National Institute of Environmental Studies), R. Ooka (Institute of Industrial Science, University of Tokyo), H. Kataoka (Obayashi Corporation), T. Kurabuchi (Tokyo University of Science), N. Kobayashi (Tokyo Polytechnic University), T. Shirasawa (Tokyo Polytechnic University), N. Tuchiya (Takenaka Corporation), Y. Nonomura (Fujita Corporation), T. Nozu (Shimizu Corporation), K. Harimoto (Taisei Corporation), K. Hibi (Shimizu Corporation), S. Murakami (Keio University), T. Yamanaka (Kajima Corporation), R. Yoshie (Tokyo Polytechnic University), M. Yoshikawa (Taisei Corporation).

AUTHOR CONTACT DETAILS

Isaac Lun (corresponding author): Department of Architecture & Building Science, Graduate School of Engineering, Tohoku University, Japan; lun@sabine.pln.archi.tohoku.ac.jp

Akashi Mochida: Department of Architecture & Building Science, Graduate School of Engineering, Tohoku University, Japan; mochida@sabine.pln.archi.tohoku.ac.jp

Ryozo Ooka: Institute of Industrial Science, University of Tokyo, Japan; ooka@iis.u-tokyo.ac.jp

REFERENCES

AIAA (American Institute of Aeronautics and Astronautics) (1998) *Guide for the Verification and Validation of Computational Fluid Dynamics Simulations*, AIAA G-077-1998

Arnfield, A. J. and Grimmond, C. S. B. (1998) 'An urban canyon energy budget model and its application to urban storage heat flux modeling', *Energy and Buildings*, vol 27, no 1, pp61–68

Arnfield, A. J. (2000) A simple model of urban canyon energy budget and its validation. *Physical Geography* **21**: 305–326

Ashie, Y., Vu Thanh, C., Asaeda, T. (1999) Building canopy model for the analysis of urban climate. *Journal of Wind Engineering and Industrial Aerodynamics* **81**: 237–248

Ashie, Y., Vu Thanh, CA. (2004) Development of the urban climate simulation system for urban and architectural planning Part 2, Developing a three-dimensional urban canopy model by space averaging method (in Japanese). *J. Archi. Plann. Environ. Eng., AIJ* **586**: 45–51

Avissar, R. and Schmidt, T. (1998) 'An evaluation of the scale at which ground-surface heat flux patchiness affects the convective boundary layer using large-eddy simulation', *Journal of the Atmospheric Sciences*, vol 55, no 16, pp2666–2689

Bélair, S., Lacarrère, P., Noilhan, J., Masson, V. and Stein, J. (1998) 'High-resolution simulation of surface and turbulent fluxes during HAPEXMOBILHY', *Monthly Weather Review*, vol 126, no 8, pp2234–2253

Benoit, R., Desagne, M., Pellerin, P., Pellerin, S., Chartier, Y., Desjardins, S. 1997. The Canadian MC2: A semi-Lagragian, semi-implicit widebend atmospheric model suited for finescale process studies and simulation. *Monthly Weather Review* **125**: 2382–2415

Best, M. J. (1998) 'A model to predict surface temperatures', *Boundary-Layer Meteorology*, vol 88, no 2, pp279–306

Best, M. (2005) Representing urban areas within operational numerical weather prediction model. *Boundary-Layer Meteorology* **114**: 91–109

Blocken, B., Roels, S. and Carmeliet, J. (2003) *Pedestrian Wind Conditions in Passages through Buildings – Part 1. Numerical Modeling, Sensitivity Analysis and Experimental Verification*, research report, Laboratory of Building Physics, Catholic University of Leuven

Bornstein, R. D. (1972) *Two-dimensional Non-steady Numerical Simulations of Nighttime Flows of a Stable Planetary Boundary Layer over a Rough Warm City*, PhD Thesis, Department of Meteorology and Oceanography, New York University

Bornstein, R. D. (1975) 'The two dimensional URBMET Urban Boundary Layer Model', *Journal of Applied Meteorology*, vol 14, no 8, pp1459–1477

Bünzli, D. and Schmid, H. P. (1998) 'The influence of surface texture on regionally aggregated evaporation and energy partitioning', *Journal of the Atmospheric Sciences*, vol 55, no 6, pp961–72

Calhoun, R., Gouveia, F., Shinn, J., Chan, S., Stevens, D., Lee, R. and Leone, J. (2004) 'Flow around a complex building: Comparisons between experiments and a Reynolds averaged Navier-Stokes approach', *Journal of Applied Meteorology*, vol 43, no 5, pp696–710

Chandler, T. J. (1960) 'Wind as a factor of urban temperatures: A survey in north-east London', *Weather*, vol 25, pp204–213

Chen, H., Ooka, R., Harayama, K., Kato, S. and Li, X. (2004) 'Study on outdoor thermal environment of apartment block in Shenzhen, China with coupled simulation of convection, radiation and conduction', *Energy and Buildings*, vol 36, no 12, pp1247–1258

Cheng, X. and Hu, F. (2005) 'Numerical studies on flow fields around buildings in an urban street canyon and cross-road', *Advances in Atmospheric Sciences*, vol 22, no 2, pp290–299

Christen, A., Bernhofer, C., Parlow, E., Rotach, M. and Vogt, R. (2003) 'Partitioning of turbulent fluxes over different urban surfaces', in J. Wibig and I. Gajda-Pijanowske (eds), *Proceedings of 5th International Conference for Urban Climate*, Lodz, Poland, 1–5 September, pp285–288

Clarke, R. H., Dyer, A. J., Brook, R. R., Reid, D. G. and Troup, A. J. (1971) 'The Wangara experiment: Boundary layer data', Technical Paper 19, Division of Meteorological Physics, CSIRO, Australia

Cui, Z. Q., Cai, X. M. and Baker, C. J. (2004) 'Large eddy simulation of turbulent flow in a street canyon', *Quarterly Journal of the Royal Meteorological Society*, vol 130, no 599, pp1373–1394

Earth Observatory (2008) *A Change in the Weather?*,
   http://earthobservatory.nasa.gov/Study/AncientForest/ancient_forest7.html (accessed 6 November 2008)
ERCOFTAC (2000) *Best Practices Guidelines for Industrial Computational Fluid Dynamics*, Version 1.0, January
Fan, H. and Sailor, D. (2005) 'Modeling the impacts of anthropogenic heating on the urban climate of Philadelphia: A
   comparison of implementations in two PBL schemes', *Atmospheric Environment*, vol 39, no 1, pp73–84
Fanger, P. O. (1973) *Thermal Comfort*, McGraw-Hill Education, New York
Flerchinger, G. N., Kustas, W. P. and Weltz, M. A. (1998) 'Simulating surface energy fluxes and radiometric surface temperatures
   for two arid vegetation communities using the SHAW model', *Journal of Applied Meteorology*, vol 37, no 5, pp449–460
Franke, J. (2006) 'Recommendations of the COST action C14 on the use of CFD in predicting pedestrian wind environment', in
   Proceedings of the Fourth International Symposium on Computational Wind Engineering (CWE2006), Yokohama, Japan,
   16–19 July, pp529–532
Franke, J., Hirsch, C., Jensen, A. G., Krüs, H. W., Schatzmann, M., Westbury, P. S., Miles, S. D., Wisse, J. A. and Wright,
   N. G. (2004) 'Recommendations on the use of CFD in wind engineering', in J. P. A. J. van Beeck (ed), *Proceedings of the
   International Conference on Urban Wind Engineering and Building Aerodynamics*, COST Action C14, Impact of Wind and
   Storm on City Life Built Environment, von Karman Institute, Sint-Genesius-Rode, Belgium
Gagge, A. P., Fobelets, A. P. and Berglund, P. E. (1986) 'A standard predictive index of human response to the thermal
   environment', *ASHRAE Transactions*, vol 92-1, Part 2, pp709–731
Genchi, Y. (2001) 'New countermeasures for mitigating the heat island effect – district heat supply systems, underground
   thermal energy storage systems using underground as a heat sink' (in Japanese), *Energy and Resources*, vol 22,
   pp306–315
Green, S. R. (1992) 'Modelling turbulent air flow in a stand of widely-spaced trees', *PHOENICS Journal Computational Fluid
   Dynamics and its Applications*, vol 5, pp294–312
Grell, G., Dudhia J., Satuffer D. (1994) A description of the fifth-generation Penn State/NCAR mesoscale model (MM5).
   NCAR/TN-398 + STR, NCAR Technical Note. National Center for Atmospheric Sciences, Boulder, CO.
Gross, G. (1987) Some effects of deforestation on nocturnal drainage flow and local climate – numerical study. *Boundary-
   Layer Meteorology* **38**: 315–338
Hagishima, A., Tanimoto J., Katayama T., Ohara K. (2001) An organic analysis for quantitative estimation of heat island by the
   revised architecture-urban-soil simultaneous simulation model, AUSSSM part 1, Theoretical frame of the model and
   results of standard solutions (in Japanese). *J. Archi. Plann. Environ. Eng., AIJ* **550**: 79–86.
Hagishima, A., Tanimoto, J., Katayama, T. and Ohara, K. (2002) 'An organic analysis for quantitative estimation of heat island
   by the revised architecture-urban-soil simultaneous simulation model, AUSSSM part 2, Quantitative analysis based on a
   series of numerical experiments' (in Japanese), *Journal of Architecture, Planning and Environment Engineering,
   Transactions of Architectural Institute of Japan*, vol 553, pp91–98
Hamlyn, D. and Britter, R. (2005) 'A numerical study of the flow field and exchange processes within a canopy of urban-type
   roughness', *Atmospheric Environment*, vol 39, no 18, pp3243–3254
Hanjalic, K. (2006) 'Some developments in RANS and hybrid RANS/LES for wind and environment engineering', in
   Proceedings of the Fourth International Symposium on Computational Wind Engineering (CWE2006), Yokohama, Japan,
   16–19 July, pp65–94
Hataya, N., Mochida, A., Iwata, T., Tabata, Y., Yoshino, H. and Tominaga, Y. (2006) 'Development of the simulation method
   for thermal environment and pollutant diffusion in street canyons with subgrid scale obstacles', in *Proceedings of the
   Fourth International Symposium on Computational Wind Engineering (CWE2006)*, Yokohama, Japan, 16–19 July,
   pp553–556
Herbert, J. M., Johnson, G. T. and Arnfield, A. J. (1998) 'Modelling the thermal climate in city canyons', *Environmental
   Modelling and Software*, vol 13, no 3–4, pp267–277

Hiraoka, H., Maruyama T., Nakamura Y., Katsura J. (1989) A study on modeling of turbulent flows within plant and urban canopies, formalization of turbulence model (in Japanese). *J. Archi. Plann. Environ. Eng., AIJ* **406**: 1–9

Hiraoka, H. (1993) 'Modelling of turbulent flows within plant/urban canopies', *Journal of Wind Engineering and Industrial Aerodynamics*, vol 46–47, pp173–182

Hiraoka, H. (2004) 'Modeling a microclimate within vegetation', in Y. A. Gayev (ed), *Proceedings of NATO Advanced Study Institute on Flow and Transport Processes in Complex Obstructed Geometries from Cities and Vegetative Canopies to Industrial Problems*, Kyiv, Ukraine, 5–15 May, pp106–107

Hiraoka, H. and Ohashi, M. (2006) 'A (k-ε) turbulence closure model for plant canopy flows', in *Proceedings of the Fourth International Symposium on Computational Wind Engineering (CWE2006)*, Yokohama, Japan, 16–19 July, pp693–696

Hoeppe, T. (1999) 'The physiological equivalent temperature – a universal index for the biometeorological assessment of the thermal environment', *International Journal of Biometeorology*, vol 43, no 2, pp71–75

Howard, L. (1833a) *Climate of London Deduced from Meteorological Observations*, 3rd edn, vol 1, Harvey and Darton, London

Howard, L. (1833b) *Climate of London Deduced from Meteorological Observations*, 3rd edn, vol 2, Harvey and Darton, London

Howard, L. (1833c) *Climate of London Deduced from Meteorological Observations*, 3rd edn, vol 3, Harvey and Darton, London

Ichinose, T., Shimodozono, K. and Hanaki, K. (1999) 'Impact of anthropogenic heat on urban climate in Tokyo', *Atmospheric Environment*, vol 33, no 24–25, pp3897–3909

Jeong, S. J. and Andrews, M. J. (2002) 'Application of the k-e turbulence model to the high Reynolds number skimming flow field of an urban street canyon', *Atmospheric Environment*, vol 36, no 7, pp1137–1145

Kanda, M., Inoue, Y. and Uno, I. (2001) 'Numerical study on cloud lines over an urban street in Tokyo', *Boundary-Layer Meteorology*, vol 98, no 2, pp251–273

Kanda, M., Moriwaki, R. and Kasamatsu, F. (2004) 'Large eddy simulation of turbulent organized structure within and above explicitly resolved cubic arrays', *Boundary-Layer Meteorology*, vol 112, no 2, pp343–368

Kanda, M., Kawai, T., Nakagawa, K. (2005a). Simple theoretical radiation scheme for regular building array. *Boundary-Layer Meteorology* **114**: 71–90

Kanda, M., Kawai, T., Kanega, M., Moriwaki, R., Narita, K., Hagishima, A. (2005b). Simple energy balance model for regular building array. *Boundary-Layer Meteorology* **116**: 423–443

Karl, T. and Jones, P. (1989) 'Urban biases in area-averaged surface air temperature trends', *Bulletin American Meteorological Society*, vol 70, pp265–270

Kataoka, H. and Mizuno, M. (2002) 'Numerical flow computation around aeroelastic 3D square cylinder using inflow turbulence', *Wind and Structures*, vol 5, no 2–4, pp379–392

Kenjers, S. and Hanjalic, K. (2005) 'Dynamical simulations towards optimal indoor climate and safety control', in Z. Guzovic (ed), *CD Proceedings of 3rd Dubrovnik Conference on Sustainable Development of Energy, Water and Environment System*, Dubrovnik, Croatia, June

Kikegawa, Y., Genchi, Y., Yoshikado, H. and Kondo, H. (2001) 'Development of a numerical simulation toward comprehensive assessments of urban warming countermeasures including their impacts upon the urban building's energy demands' (in Japanese), *Energy and Resources*, vol 22, pp235–240

Kikuchi, Y., Arakawa, S., Kimura, F. and Shirasaki, K. (1981) 'Numerical study on the effects of mountains on the land and sea breeze circulation in the Kanto District', *Journal of Meteorological Society of Japan*, vol 59, no 5, pp723–738

Kimura, F., Arakawa, S. (1983) A numerical experiment on the nocturnal low level jet over the Kanto plain. *Journal of the Meteorological Society of Japan* **61**: 848–878

Kitada, T., Okamura, K. and Tanaka, S. (1998) 'Effects of topography and urbanization on local winds and thermal environment in the Nohbi Plain, coastal region of central Japan: A numerical analysis by mesoscale meteorological model with a k-ε turbulence model', *Journal of Applied Meteorology*, vol 37, no 10, pp1026–1046

Kondo, H. (1995) Thermally induced local wind and surface inversion over the Kanto plain on calm winter nights. *Journal of Applied Meteorology* **34**: 1439–1448

Kondo, H., Liu Fa-Hua. (1998) A study on the urban thermal environment obtained through one-dimensional urban canopy model (in Japanese). *J. Japan Soc. Atmos. Environ.* **83**: 179–192

Kondo, A., Goda, E., Mizuma, K., Kaga, A. and Yamaguchi, K. (2001) 'Prediction of micro climate change in developed area of a few square kilometers scale by numerical method' (in Japanese), *Transactions of the Society of Heating, Air-conditioning and Sanitary Engineers of Japan*, vol 83, October, pp1–10

Kondo, H., Genchi, Y., Kikegawa, Y., Ohashi, Y., Yoshikado, H. and Komiyama, H. (2005) 'Development of a multi-layer urban canopy model for the analysis of energy consumption in a big city: Structure of the urban canopy model and its basic performance', *Boundary-Layer Meteorology*, vol 116, no 3, pp395–421

Kondo, H., Tokairin, T. and Kikegawa, Y. (2006) 'The wind calculation in Tokyo urban area with a Mesoscale Model', in *Proceedings of the Fourth International Symposium on Computational Wind Engineering (CWE2006)*, Yokohama, Japan, 16–19 July, pp235–238

Kusaka, H., Kondo, H., Kikegawa, Y., Kimura, F. (2001) A simple single-layer urban canopy model for atmospheric models: comparison with multilayer and slab models. *Boundary-Layer Meteorology* **101**: 329–358

Kusaka, H. and Kimura, F. (2004) 'Coupling a single-layer urban canopy model with a simple atmospheric model: Impact on urban heat island simulation for an idealized case', *Journal of Meteorological Society of Japan*, vol 82, no 1, pp67–80

Kuwabara, K., Horikoshi, T. and Mochida, T. (2005) 'Evaluation method for thermal and comfort sensation in outdoor environment', in *Preprints of the Third International Conference on Human-Environmental system*, Tokyo, Japan, 12–15 September, pp79–82

Lakehal, D. and Rodi, W. (1997) 'Calculation of the flow past a surface-mounted cube with two-layer turbulence models', *Journal of Wind Engineering and Industrial Aerodynamics*, vol 67–68, pp65–78

Laprise, R. (1995) The formulation of Andre Robert's MC2 (Mesoscale Compressible Community) model. *Atmos.-Ocean* **35**: 195–220

Lars, B., Jan, O. M. and Sven, L. (1985) 'Canyon geometry, street temperatures and urban heat island in Malmok, Sweden', *Journal of Climatology*, vol 5, no 4, pp433–444

Lee, R. L. and Olfe, D. B. (1974) 'Numerical calculations of temperature profiles over an urban heat island', *Boundary-Layer Meteorology*, vol 7, no 1, pp39–52

Lee, S. C., Chiu, M. Y., Ho, K. F., Zou, S. C. and Wang, X. M. (2002) 'Volatile organic compounds (VOCs) in urban atmosphere of Hong Kong', *Chemosphere*, vol 48, no 3, pp375–382

Lee, S. M., Fernando, H. J. S., Princevac, M., Zajic, D., Sinesi, M., McCulley, J. L. and Anderson, J. (2003) 'Transport and diffusion of ozone in the nocturnal and morning planetary boundary layer of the Phoenix valley', *Environmental Fluid Mechanics*, vol 3, no 4, pp331–362

Leuning, R., Dunin, F. X. and Wang, Y.-P. (1998) 'A two-leaf model for canopy conductance, photosynthesis and partitioning of available energy: II. Comparison with measurements', *Agricultural and Forest Meteorology*, vol 91, no 1–2, pp113–125

Li, C. S. and Lin, C. H. (2002) 'PM1/PM2.5/PM10 characteristics in the urban atmosphere of Taipei', *Aerosol Science and Technology*, vol 36, no 4, pp469–473

Li, C., Lau, A. and Mao, J. (2004) 'Validation of MODIS AOD products with 1-km resolution and their application in the study of urban air pollution in Hong Kong', in *Proceedings of the SPIE – The International Society for Optical Engineering*, 6 July, vol 5547, pp122–133

Lien, F. S. and Yee, E. (2004) 'Numerical modeling of the turbulent flow developing within and over a 3-D building array. Part I: a high-resolution Reynolds averaged Navier-Stokes approach', *Boundary-Layer Meteorology*, vol 112, no 3, pp427–466

Lien, F. S. and Yee, E. (2005) 'Numerical modeling of the turbulent flow developing within and over a 3-D building array. Part III: a distributed drag force approach, its implementation and application', *Boundary-Layer Meteorology*, vol 114, no 2, pp287–313

Lien, F. S., Yee, E. and Wilson, J. D. (2005) 'Numerical modeling of the turbulent flow developing within and over a 3-D building array. Part II: a mathematical foundation for a distributed drag force approach', *Boundary-Layer Meteorology*, vol 114, no 2, pp245–285

Lin, B., Zhu, Y., Li, X. and Qin, Y. (2008) 'Numerical simulation studies of the different vegetation patterns' effects on outdoor pedestrian comfort', *Journal of Wind Engineering and Industrial Aerodynamics*, vol 96, no 10–11, pp1707–1718

Lin, J. J. and Lee, L. C. (2004) 'Characterization of the concentration and distribution of urban submicron (PM1) aerosol particles', *Atmospheric Environment*, vol 38, no 3, pp469–475

Liu, C. H., Barth, M. C. and Leung, D. Y. C. (2004) 'Large-eddy simulation of flow and pollutant transport in street canyons of different building-height-to-street-width ratios', *Journal of Applied Meteorology*, vol 43, no 10, pp1410–1424

Liu, J., Chen, J. M., Black, T. A. and Novak, M. D. (1996) 'E-ε modelling of turbulent air flow downwind of a model forest edge', *Boundary-Layer Meteorology*, vol 77, no 1, pp21–44

Malchaire, J., Piette, A., Kampmann, B., Mehnert, P., Gebhardt, H., Havenith, G., Hartog, E. D., Holmer, I., Parsons, K., Alfano, G. and Griefahn, B. (2001) 'Development and validation of the predicted heat strain model', *The Annals of Occupational Hygiene*, vol 45, no 2, pp123–135

Manley, G. (1958) 'On the frequency of snowfall in metropolitan England', *Quarterly Journal of the Royal Meteorological Society*, vol 84, no 359, pp70–72

Martilli, A., Clappier A., Rotach, MW. (2002) An urban surface exchange parameterization for mesoscale models. *Boundary-Layer Meteorology* **104**: 261–304

Maruyama, T. (1993) 'Optimization of roughness parameters for staggered arrayed cubic blocks using experimental data', *Journal of Wind Engineering and Industrial Aerodynamics*, vol 46–47, pp165–171

Masson, V., (2000) A physically-based scheme for the urban energy budget in atmospheric models. *Boundary-Layer Meteorology* **94**: 357–397

Matzarakis, A., Mayer, H. and Iziomon, M. G. (1999) 'Applications of a universal thermal index: Physiological equivalent temperature', *International Journal of Biometeorology*, vol 43, no 2, pp76–84

Menter, F. R. and Kunts, M. (2001) *Development and Applications of a Zonal DES Turbulence Model for CFX-5*, ANSYS CFX Validation Report, CFX-VAL 17/0703, pp1–34

Minami, Y., Ooka, R., Tsuzuki, K., Sakoi, T. and Sawasaki, S. (2006) 'Study on human physiological models for hot environments', in *Sixth International Thermal Manikin and Modeling Meeting Preprints*, Hong Kong, China, 16–18 October, pp89–97

Mochida, A. (2005) 'Management and design of outdoor environment based on software platform for the total analysis of urban heat island', in Z. Guzovic (ed), *CD Proceedings of 3rd Dubrovnik Conference on Sustainable Development of Energy, Water and Environment System*, Dubrovnik, Croatia, June

Mochida, A. and Lun, I. Y. F. (2008) 'Prediction of wind environment and thermal comfort at pedestrian level in urban area', *Journal of Wind Engineering and Industrial Aerodynamics*, vol 96, no 10–11, October–November, pp1498–1527

Mochida, A., Murakami, S., Ojima, T., Kim, S., Ooka, R. and Sugiyama, H. (1997) 'CFD analysis of mesoscale climate in the Greater Tokyo area', *Journal of Wind Engineering and Industrial Aerodynamics*, vol 67–68, pp459–477

Mochida, A., Murakami, S., Ooka, R. and Kim, S. (1999) 'CFD study on urban climate in Tokyo – effects of urbanization on climatic change', in *Proceedings of the 10th International Conference on Wind Engineering*, Copenhagen, Denmark, 21–24 June, pp1307–1314

Mochida, A., Yoshino, H., Murakami, S., Yoshida, S., Oba, M. and Sasaki, K. (2001) 'Numerical study on effects of increased albedo for building surface on outdoor thermal comfort' (in Japanese), *Summaries of Technical Papers of Annual Meeting, Architectural Institute of Japan*, vol D-1, pp923–924

Mochida, A., Tominaga, Y., Murakami, S., Yoshie, R., Ishihara, T. and Ooka, R. (2002) 'Comparison of various k-e models and DSM applied to flow around a high-rise building: Report on AIJ cooperative project for CFD prediction of wind environment', *Wind and Structures*, vol 5, no 2–4, pp227–244

Mochida, A., Iwata, T., Hataya, N., Sasaki, K. and Watanabe, H. (2005) 'Field measurements and CFD analyses of thermal environment and pollutant diffusion in street canyon', in *Proceedings of the Sixth Asia-Pacific Conference on Wind Engineering (APCWE-VI)*, Seoul, Korea, 12–14 September, pp2681–2696

Mochida, A., Hataya, N., Iwata, T., Tabata, Y., Yoshino, H. and Watanabe, H. (2006) 'CFD analyses on outdoor thermal environment and air pollutant diffusion in street canyons under the influences of moving automobiles', in *Proceedings of the 6th International Conference on Urban Climate (ICUC6)*, Gothenburg, Sweden, 12–16 June

Mochida, A., Tabata, Y., Iwata, T. and Yoshino, H. (2008a) 'Examining tree canopy model for CFD prediction of wind environment at pedestrian level', *Journal of Wind Engineering and Industrial Aerodynamics*, vol 96, no 10–11, pp1667–1677

Mochida, A., Tabata, Y., Hagishima, A., Tanimoto, J., Maruyama, T., Kikuchi, A. and Kikuchi, Y. (2008b) 'Development of CFD model for reproducing aerodynamic effects of moving automobiles in street canyon', in C.-K. Choi et al (eds), *Proceedings of 4th International Conference on Advances in Wind and Structure (AWAS' 08)*, Jeju, Korea, 29–31 May, pp100–114

Moriwaki, R., Kanda, M. and Kimoto, Y. (2003) 'A field experiment on how atmospheric stability affects vertical profiles of momentum and heat fluxes in an urban surface layer', in J. Wibig and I. Gajda-Pijanowske (eds), *Proceedings of 5th International Conference for Urban Climate*, Lodz, Poland, 1–5 September, pp297–300

Murakami, S., Mochida, A., Kim, S., Ooka, R., Yoshida, S., Kondo, S., Genchi, Y., Shimada A. (2000) Software Platform for the total analysis of wind climate and urban heat island, -integration of CWE simulations from human scale to urban scale-. *Proceedings of Computational Wind Engineering* 2000: 23–26, Birmingham, UK, 4–7 September

Murakami, S. (2004) 'Indoor/outdoor climate design by CFD based on the software platform', *International Journal of Heat and Fluid Flow*, vol 25, no 5, pp849–863

Murakami, S. (2006) 'Technology and policy instruments for mitigating the heat island effect', in *Proceedings of 1st International Conference on Countermeasures to Urban Heat Islands*, 3–4 August, Tokyo, pp1–13

Murakami, S., Mochida, A. and Ooka, R. (1993) 'Numerical simulation of flowfield over surface-mounted cube with various second-moment closure models', in *Preprints of 9th Symposium on Turbulent Shear Flows*, Kyoto, Japan, pp13-5-1–13-5-6

Murakami, S., Mochida, A., Kim, S., Ooka, R., Yoshida, S., Kondo, H., Genchi, Y. and Shimada, A. (2000) 'Software platform for the total analysis of wind climate and urban heat island', in *Proceedings of 3rd International Symposium on Computational Wind Engineering*, Birmingham, UK, 4–7 September, pp23–26

Murakami, S., Mochida, A., Ooka, R., Yoshida, S., Yoshino, H., Sasaki, K. and Harayama, K. (2003a) 'Evaluation of the impacts of urban tree planting in Tokyo based on heat balance model', in *Preprints of 11th International Conference on Wind Engineering*, Texas, US, 2–5 June, vol 2, pp2641–2648

Murakami, S., Mochida, A. and Kato, S. (2003b) 'Development of local area wind prediction system for selecting suitable site for windmill', *Journal of Wind Engineering and Industrial Aerodynamics*, vol 91, no 12–15, pp1759–1776

Mutoh, J. and Narita, K. (2005) 'Field experiments of energy exchange process in an urban canopy layer using scale model (part 1): Outline of experiments and the measurements of transfer velocity by water evaporation method at each type of surface', *Summaries of Technical Papers of Annual Meeting, Architectural Institute of Japan*, vol D-1, pp755–756

Narita, K., Hagishima, A., Tanimoto, J., Kanda, M., Kawai, T. and Kaneda, M. (2006) 'Outdoor scale model experiments of the local bulk transfer coefficient for urban surfaces with a water evaporation method', in *6th International Conference on Urban Climate Preprints*, Gothenburg, Sweden, 12–16 June, pp72–75

Narumi, D., Otani, F., Kondo, A., Shimoda, Y. and Mizuno, M. (2002) 'Analysis of countermeasure for mitigating heat islands phenomena using a numerical model. Part 1: Effect of anthropogenic waste heat upon urban thermal environment' (in Japanese), *Journal of Architecture, Planning and Environment Engineering, Transactions of Architectural Institute of Japan*, vol 562, pp97–104

NASA/Goddard Space Flight Center (2002) 'NASA satellite confirms urban heat islands increase rainfall around cities',
    ScienceDaily, 19 June (accessed 7 November 2008) www.sciencedaily.com/releases/2002/06/020619074019.htm

National Weather Service (2005) *Heat Wave: A Major Summer Killer*, www.nws.noaa.gov/om/brochures/heat_wave.shtml
    (accessed 19 October 2008)

Nieuwolt, S. (1966) 'The urban microclimate of Singapore', *Journal of Tropical Geography*, vol 22, pp30–37

Offerle, B., Grimmond, S., Fortuniak, K. and Oke, T. (2003) 'Temporal variability in heat fluxes over a northern European
    downtown', in J. Wibig and I. Gajda-Pijanowske (eds), *Proceedings of 5th International Conference for Urban Climate*,
    Lodz, Poland, 1–5 September, pp301–304

Ohashi, M. (2006) 'A study on analysis of airflow around an individual tree' (in Japanese), *Journal of Architecture, Planning
    and Environment Engineering*, Architectural Institute of Japan, April 2004, no 578, pp91–96

Oh, E., Kondo, A., Yamaguchi, K., Mizuma, K. (2000) Transactions of the society of heating, air-conditioning and sanitary
    engineers of Japan (in Japanese). **76**: 29–39

Ojima, T. (1991) 'Changing Tokyo metropolitan area and its heat island model', *Energy and Buildings*, vol 15, pp191–203

Oke, T. R. (1981) 'Canyon geometry and the nocturnal urban heat island: Comparison of scale model and field observations',
    International Journal of Climatology, vol 1, no 3, pp237–254

Oke, T. R. (1987) *Boundary Layer Climates*, 2nd edn, Methuen, London

Oke, T. R. (1988) 'The urban energy balance', *Progress in Physical Geography*, vol 12, no 4, pp471–508

Oke, T. R. and Maxwell, G. B. (1975) 'Urban heat island dynamics in Montreal and Vancouver', *Atmospheric Environment*, vol
    9, no12, pp191–200

Omori, T., Taniguchi, H. and Kudo, K. (1990) 'Monte Carlo simulation of indoor radiant environment', *International Journal of
    Numerical Methods in Engineering*, vol 30, no 4, pp615–627

Ono, T., Murakami, S., Ooka, R., Takahashi, T., Omori, T. and Saotome, T. (2006) 'Study on convective heat transfer of a
    human body to evaluate the outdoor thermal environment', in *6th International Conference for Urban Climate Preprints*,
    Gothenburg, Sweden, 12–16 June, pp238–241

Ooka, R. (2007) 'Recent development of assessment tools for urban climate and heat-island investigation especially based on
    experiences in Japan', *International Journal of Climatology*, vol 27, no 14, pp1919–1930

Ooka, R., Harayama, K., Murakami, S. and Kondo, H. (2004) 'Study on urban heat islands in Tokyo metropolitan area using a
    meteorological mesoscale model incorporating an urban canopy model', in *Fifth Symposium on the Urban Environment*,
    Vancouver, Canada, 23–26 August, pp1–6

Ooka, R., Chen, H. and Kato, S. (2006) 'Study on optimum arrangement of trees for design of pleasant outdoor environment
    using multi-objective genetic algorithm', in *Proceedings of the Fourth International Symposium on Computational Wind
    Engineering (CWE2006)*, Yokohama, Japan, 16–19 July, pp525–528

Palomo Del Barrio, E. (1998) 'Analysis of the green roofs cooling potential in buildings', *Energy and Buildings*, vol 27,
    no 2, pp179–193

Patton, E. G., Shaw, R. H., Judd, M. J. and Raupach, M. R. (1998) 'Large-eddy simulation of windbreak flow', *Boundary-Layer
    Meteorology*, vol 87, no 2, pp275–306

Pearlmutter, D., Berliner, P. and Shaviv, E. (2005) 'Evaluation of urban surface energy fluxes using an open-air scale model',
    Journal of Applied Meteorology, vol 44, no 4, pp532–545

Pickup, J. and de Dear, R. (1999) 'An outdoor thermal comfort index (Out_SET*) – Part 1 – The model and its assumptions',
    in *Preprints of 4th International Symposium on Urban Climate*, Sydney, Australia, 8–12 November, pp279–282

Pielke, R. A. (1974) A three-dimensional numerical model of the sea breezes over south Florida. *Monthly Weather Review*
    vol 102, pp115–134

Pielke, R. A., Cotton, W. R., Walko, R. L., Tremback, C. J., Lyons, W. A., Grasso, L. D., Nicolls, M. E., Moran, M.D., Wesley, Lee, T. J.
    and Copeland. (1992) A comprehensive meteorological modeling system – RAMS. *Meteorol. Atmos. Phys.* vol 49, pp69–91

Richards, P. J., Mallinson, G. D., McMillan, D. and Li, Y. F. (2002) 'Pedestrian level wind speeds in downtown Auckland', *Wind & Structures*, vol 5, no 2–4, pp151–164

Roache, P. J., Ghia, K. and White, F. (1986) 'Editorial policy statement on the control of numerical accuracy,' *ASME Journal of Fluids Engineering*, vol 108, no 1, March, p2

Rotach, M., Christen, A. and Vogt, R. (2003) 'Profiles of turbulence statistics in the urban roughness sublayer with special emphasis to dispersion modeling', in J. Wibig and I. Gajda-Pijanowske (eds), *Proceedings of 5th International Conference for Urban Climate*, Lodz, Poland, 1–5 September, pp309–312

Sailor, D. J. (1998) 'Simulations of annual degree day impacts of urban vegetative augmentation', *Atmospheric Environment*, vol 32, no 1, pp43–52

Sasaki, K., Mochida, A., Yoshida, T., Yoshino, H. and Watanabe, H. (2006) 'A new method to select appropriate countermeasures against heat-island effects according to the regional characteristics of heat balance mechanism', in *Proceedings of the Fourth International Symposium on Computational Wind Engineering (CWE2006)*, Yokohama, Japan, 16–19 July, pp223–226

Sato, T., Murakami, S., Ooka, R. and Yoshida, S. (2006) 'Analysis of regional characteristics of the atmospheric heat balance in the Tokyo metropolitan area in summer', in *Proceedings of the Fourth International Symposium on Computational Wind Engineering (CWE2006)*, Yokohama, Japan, 16–19 July, pp231–234

Sera, T. (2006) 'Japan's policy instruments on urban heat island measures by MLIT', in *Proceedings of 1st International Conference on Countermeasures to Urban Heat Islands*, Tokyo, Japan, 3–4 August, pp233–240

Sham, S. (1972) 'Some aspects of urban micro-climate in Kuala Lumpur West Malaysia', *Akademika*, vol 1, pp85–94

Skamarock, WC, Klemp JB, Dudhia J, Gill DO, Barker DM, Wan W,. Powers JG, (2005). A description of the advanced research WRF version 2. *NCAR/TN-468 + STR, NCAR Technical Note. National Center for Atmospheric Sciences, Boulder, CO.*

Smith, K. (1997) *Do Skyscrapers Affect Weather in Cities? If So, How?* www.madsci.org/posts/archives/dec97/875205167.Es.r.html (accessed 21 October 2008)

Spalart, P. R. (1999) *Strategies for Turbulence Modelling and Simulations, Engineering Turbulence Modelling and Experiments – 4*, Elsevier Science, pp3–17

Squires, K. D. (2003) 'Perspective and challenges in simulation and modeling of unsteady flows', in *CD Proceedings of 17th Japan National Conference on Computational Fluid Dynamics*

Squires, K. D. (2006) 'Prediction of turbulent flows at high Reynolds numbers using detached-eddy simulation', in *Proceedings of the Fourth International Symposium on Computational Wind Engineering (CWE2006)*, Yokohama, Japan, 16–19 July, pp61–64

Stathopoulos, T. and Baskaran, A. (1996) 'Computer simulation of wind environmental conditions around buildings', *Engineering Structures*, vol 18, no 11, pp876–885

Steadman, R. G. (1979) 'The assessment of sultriness. Part 1: Temperature-humidity index based on human physiology and clothing science', *Journal of Applied Meteorology*, vol 18, no 7, pp861–873

Su, H.-B., Shaw, R. H., Paw, U. K. T., Moeng, C.-H. and Sullivan, P. P. (1998) 'Turbulent statistics of neutrally stratified flow within and above a sparse forest from large-eddy simulation and field observations', *Boundary-Layer Meteorology*, vol 88, no 3, pp363–97

Svensson, U. and Häggkvist, K. (1990) 'A two-equation turbulence model for canopy flows', *Journal of Wind Engineering and Industrial Aerodynamics*, vol 35, no 1–3, pp201–211

Swanson, B. (2007) *Human Effects on Weather*, www.usatoday.com/weather/resources/askjack/archives-human-effects.htm (accessed on 21 October 2008)

Taha, H. (1996) 'Modeling impacts of increased urban vegetation on ozone air quality in the south coast air basin', *Atmospheric Environment*, vol 30, no 20, pp3423–3430

Taha, H. (1997) 'Modeling impacts of large-scale albedo changes on ozone air quality in the south coast air basin', *Atmospheric Environment*, vol 31, no 11, pp1667–1676

Timofeyef, N. (1998) 'Numerical study of wind mode of a territory development', in *Proceedings of the Second East European Conference on Wind Engineering*, 7–11 September, Prague, Czech Republic

Tominaga, Y., Mochida, A., Shirasawa, T., Yoshie, R., Kataoka, H., Harimoto, K. and Nozu, T. (2004) 'Cross comparisons of CFD results of wind environment at Pedestrian level around a high-rise building and within a building complex', *Journal of Asian Architecture and Building Engineering*, vol 3, no 1, pp63–70

Tominaga, Y., Yoshie, R., Mochida, A., Kataoka, H., Harimoto, K. and Nozu, T. (2005) 'Cross comparison of CFD prediction for wind environment at pedestrian level around buildings (Part 2)', in *Proceedings of the Sixth Asia-Pacific Conference on Wind Engineering (APCWE-VI)*, Seoul, Korea, 12–14 September, pp2661–2670

Tominaga, Y., Mochida, A., Yoshie, R., Kataoka, H., Nozu, T., Yoshikawa, M. and Shirasawa, T. (2008a) 'AIJ guidelines for practical applications of CFD to pedestrian wind environment around buildings', *Journal of Wind Engineering and Industrial Aerodynamics*, vol 96, no 10–11, pp1749–1761

Tominaga, Y., Mochida, A., Murakami, S. and Sawaki, S. (2008b) 'Comparison of various k-ε models and LES applied to flow around a high-rise building model with 1:1:2 shape placed within the surface boundary layer', *Journal of Wind Engineering and Industrial Aerodynamics*, vol 96, no 4, pp389–411

Uehara, K., Murakami, S., Oikawa, S. and Wakamatsu, S. (2000) 'Wind tunnel experiments on how thermal stratification affects flow in and above urban street canyons', *Journal of Atmospheric Environment*, vol 34, no 10, pp1553–1562

UNFPA (United Nations Population Fund) (2007) *State of World Population 2007 – Unleashing the Potential of Urban Growth*, UNFPA report

Uno, I., Ueda, H. and Wakamatsu, S. (1989) 'Numerical modeling of the nocturnal urban boundary layer', *Boundary-Layer Meteorology*, vol 49, no 1–2, pp77–98

Vu, T. C., Ashie, Y. and Asaeda, T. (2002) 'A k-e turbulence closure model for the atmospheric boundary layer including urban canopy', *Boundary-Layer Meteorology*, vol 102, no 3, pp459–90

Wang, Y. P. and Leuning, R. (1998) 'A two-leaf model for canopy conductance, photosynthesis and partitioning of available energy: I. Model description and comparison with a multilayered model', *Agricultural and Forest Meteorology*, vol 91, no 1–2, pp89–111

Watanabe, H., Mochida, A., Sakaida, K., Yoshino, H., Jyu-nimura, Y., Iwata, T., Hataya, N. and Shibata, K. (2005) 'Field measurement of thermal environment and pollutant diffusion in street canyon to investigate the effects of its form and roadside trees', in *Proceedings of International Symposium on Sustainable Development of Asia City Environment (SDACE 2005)*, Xi'an, China, 23–25 November, pp509–515

Westbury, P. S., Miles, S. D. and Stathopoulos, T. (2002) 'CFD application on the evaluation of pedestrian-level winds', in *Workshop on Impact of Wind and Storm on City Life and Built Environment, Cost Action C14, CSTB*, Nantes, France, 3–4 June

Wilson, J. D., Finnigan, J. J. and Raupach, M. R. (1998) 'A first-order closure for disturbed plant canopy flows, and its application to winds in a canopy on a ridge', *Quarterly Journal of the Royal Meteorological Society*, vol 124, no 547, pp705–32

Wippermann, F. K. and Gross, G. (1986) The wind induced shaping and migration of an isolated dune: A numerical experiment. *Boundary-Layer Meteorology* **36**: 319–334

Yamada, T. (1982) 'A numerical model study of turbulent airflow in and above a forest canopy', *Journal of the Meteorological Society of Japan*, vol 60, no 1, pp439–454

Yamada, T. and Bunker, S. (1988) Development of a nested grid, second moment turbulence closure model and application to the 1982 ASCOT Brush Creek data simulation. *Journal of Applied Meteorology* **27**: 562–578

Yang, W., Jin, X., Jin, H., Quan, Y. and Gu, M. (2006) 'Research on the parameters of turbulence model and modeling of equilibrium atmosphere boundary layer in CWE', in *Proceedings of The Fourth International Symposium on Computational Wind Engineering (CWE2006)*, Yokohama, Japan, 16–19 July, pp901–904

Yinka, R. A. (1990) 'Aspects of the variation in some characteristics of radiation budget within the urban canopy of Ibandan', *Atmospheric Environment*, vol 24, Part B, no 1, pp9–17

Yee, E. and Biltoft, C. A. (2004) 'Concentration fluctuation measurements in a plume dispersing through a regular array of obstacles', *Boundary-Layer Meteorology*, vol 111, no 3, pp363–415

Yoshida, S., Murakami, S., Ooka, R., Mochida, A. and Tominaga, Y. (2000) 'CFD prediction of thermal comfort in microscale wind climate', in *Abstracts of Papers Presented at the 3rd International Symposium on Computational Wind Engineering*, Birmingham, UK, 4–7 September, pp27–30

Yoshida, S., Ooka, R., Mochida, A., Murakami, S. and Tominaga, Y. (2006) 'Development of three dimensional plant canopy model for numerical simulation of outdoor thermal environment', in *Proceedings of 6th International Conference on Urban Climate (ICUC 6)*, Gothenburg, Sweden, 12–16 June

Yoshie, R., Mochida, A., Tominaga, Y., Kataoka, H. and Yoshikawa, M. (2005) 'Cross comparison of CFD prediction for wind environment at pedestrian level around buildings (Part 1)', in *Proceedings of the Sixth Asia-Pacific Conference on Wind Engineering (APCWE-VI)*, Seoul, Korea, 12–14 September, pp2648–2660

Yoshie, R., Mochida, A. and Tominaga, Y. (2006) 'CFD prediction of wind environment around a high-rise building located in an urban area', in *Proceedings of The Fourth International Symposium on Computational Wind Engineering (CWE2006)*, Yokohama, Japan, 16–19 July, pp129–132

Yoshie, R., Mochida, A., Tominaga, Y., Kataoka, H., Harimoto, K., Nozu, T. and Shirasawa, T. (2007) 'Cooperative project for CFD prediction of pedestrian wind environment in the Architectural Institute of Japan', *Journal of Wind Engineering and Industrial Aerodynamics*, vol 95, no 9–11, pp1551–1578

publishing for a sustainable future

Post-Occupancy Evaluation: An Inevitable Step Toward Sustainability

Isaac A. Meir, Yaakov Garb, Dixin Jiao and
Alex Cicelsky

Abstract

Post-occupancy evaluation (POE) is a platform for the systematic study of buildings once occupied, so that lessons may be learned that will improve their current conditions and guide the design of future buildings. Various aspects of the occupied buildings' functioning and performance can be assessed in a POE, both chemo-physical (indoor environment quality (IEQ), indoor air quality (IAQ) and thermal performance) as well as more subjective and interactional (space use, user satisfaction, etc.). POE draws on an extensive quantitative and qualitative toolkit: measurements and monitoring, on the one hand, and methods such as walk-throughs, observations and user satisfaction questionnaires on the other. POE may seem a necessary, indeed, axiomatic phase of the design and construction process, and exactly the kind of integrated assessment essential for the design of more sustainable buildings. Yet POE researchers have often been regarded with suspicion and even hostility, since their work may cause friction between different stakeholders. This chapter reviews material published in recent years in an attempt to trace the emergence of POE, describe its conceptual and methodological backdrop, its interaction with other issues related to sustainable design, and its increasing 'canonization' as a method. We argue that POE offers the potential to integrate a range of fragmented aspects of the construction process and of the relations of buildings to their environment and users. We propose that the acceptance of POE as a mandatory step in the design and commissioning of buildings, whose results are habitually fed backward and forward to other stages of the design and construction processes, is an important and probably inevitable step toward making buildings more sustainable.

■ *Keywords* – appropriate design; indoor air; indoor environment; monitoring; post-occupancy evaluation; survey; sustainability; thermal performance; user satisfaction; walk-through

doi:10.3763/aber.2009.0307 ■ © 2009 Earthscan ■ ISSN 1751-2549 (Print), 1756-2201 (Online) ■ www.earthscanjournals.com

INTRODUCTION
BACKGROUND
Whereas designers expend considerable resources in examining the actual functioning of and user satisfaction with everyday commodities (especially successful ones), and in refining their design accordingly, this is not the case with buildings. Although they are disproportionately more expensive than cars, audio or electrical and electronic equipment, buildings are very rarely revisited and reassessed once they are handed over to their users. This lack of evaluation and study stems from numerous reasons and leads to a situation in which every single building remains a unique specimen, design mistakes are repeated, and when some re-evaluation of the building as an end product is undertaken, it is often based on non-systematic troubleshooting. In many cases it is hard to compare the results of such studies due to lack of uniform, standard procedures and protocols. It has been claimed that unless a systematic approach is taken for the benchmarking of buildings, improvement of current practices is left to a haphazard process that does not necessarily promote sustainability (Roaf et al, 2004).

The absence of regularized feedback from performance to planning and construction phases becomes ever more relevant under the current conditions which include (Meir, 2008):

- continuous rise in the consumption of energy, both per capita and in absolute terms;
- buildings in industrialized countries consume some 40–50 per cent of overall energy, from 'cradle-to-grave', but primarily during the operational part of their lives (heating, cooling, ventilation, lighting, etc.);
- the realization that fossil fuels are being depleted and that their use has adverse environmental, health, social, political and security implications;
- people in industrialized countries (but not only!) spending 80–90 per cent of their lives in buildings, living, studying, working, entertaining themselves, consuming and even exercising, which means that the indoor conditions can have a strong imprint on well-being, health and productivity (Pearson, 1989; Wargocki et al, 1999). The indoors is, in a very real sense, the human 'environment'.

At the same time as there are demands to decrease the use of energy, these coincide with increasing demand for comfort in buildings (Zeiler and Boxem, 2008; Zeiler et al, 2008). A pointed question that post-occupancy evaluation (POE) may answer is whether these increasingly stringent environmental constraints must come at the expense of occupant comfort and satisfaction. Or might achievements be made *simultaneously* in both dimensions, balancing energy consumption and occupants' demands for physical, physiological and psychological needs? It is difficult to answer these and similar questions without the kind of insights that POE offers into how buildings actually function and are perceived.

The tools employed in POE include plan analysis, monitoring of indoor environment quality (IEQ), indoor air quality (IAQ) and thermal performance, and surveys including walk-throughs, observations, user satisfaction questionnaires, and semi-structured and structured interviews. With some first attempts in the 1960s, POE was introduced in response to significant problems experienced in building performance with particular emphasis on the building occupant perspective (Preiser, 1995). POE serves as a way of providing subjective and objective feedback that can inform planning and practice throughout the building's life cycle from the initial design to occupation. The benefits from POE can be in the short, medium and long term:

- short-term benefits include obtaining users' feedback on problems in buildings and the identification of solutions;
- medium-term benefits include feed-forward of the positive and negative lessons learned into the next building cycle;
- long-term benefits aim at the creation of databases and the update, upgrade and generation of planning and design protocols and paradigms (Preiser and Vischer, 2005).

Despite these obvious benefits, POE researchers are often regarded with suspicion and even hostility, since their work may cause friction between different stakeholders (including architects, consultants, clients, owners, managers and users) and between these and the authorities (planning and health, for example), expose some of them to liability lawsuits, and others to potential demand for upgrade investments. This institutional and professional fragmentation of authorities, perspectives and liabilities has hampered the uptake of POE as a self-evident part of the design and construction professions and industry.

This chapter attempts to outline the various issues related to POE, draw a picture of the current state of affairs and suggest some possible steps for the canonization of POE within the planning, design and construction domains. Various sources were reviewed in order to assess the evolution and state of the art of this field, among them peer-reviewed journals, electronic databases (Science Direct, Scirus, Web of Science, Google Scholar) and conference proceedings (including *Windsor 2004 Conference – Closing the Loop*; *Indoor Air* 2002, 2005 and 2008; *Healthy Buildings* 2003 and 2006; *Passive Low Energy Architecture (PLEA)* 1988–2008; and *Passive Low Energy Cooling (PALENC)* 2005 and 2007). In addition, curricula of selected schools of architecture were searched online for POE-oriented courses and programmes with a POE emphasis.

More than 100 papers from these sources were selected for more in-depth examination and categorized according to types of projects (residential, educational, public and institutional buildings, research facilities, clusters), aims and targets (energy consumption, IAQ, IEQ, user satisfaction) and tools and methods employed (walk-through, monitoring, questionnaires, surveys). The results have been compiled in Table 7.1, which may be used as a synoptic overview of the paper's bibliographic sources.

It is important to note that despite the wish to gain a full understanding of projects, particularly buildings, and the interaction of these and their systems with the building users, the parameters involved are numerous, the interrelations complicated and often not straightforward, and the resources needed for conducting a full POE are often beyond the reach of entrepreneur, designer, owner or user.

POE may be divided into two broad types: lateral studies investigating a limited number of parameters in a large number of case studies; and in-depth studies providing

TABLE 7.1 Synoptic overview of POE studies in selected papers

REFERENCE	TITLE	BUILDING TYPE	DIMENSIONS EVALUATED	METHODOLOGIES	SCOPE
Abbaszadeh et al (2006)	Occupant satisfaction with indoor quality in green buildings	Office	Pathogens, allergens Indoor environmental quality: thermal comfort, air quality, lighting, acoustics	Web-based IEQ survey	181 office buildings
Baird and Jackson (2004)	Probe-style questionnaire surveys of building users – an international comparison of their application to large-scale passive and mixed-mode teaching and research facilities	Academic, educational, office	Users satisfaction, use of space, thermal control	Public access to POE/PROBE surveys	3–5 days, 5 buildings (complexes) 1241 respondents
Bordass and Leaman (2004)	Probe: how it happened, what it found and did it get us anywhere?	Clinics, hospitals, offices, residential	Occupant control of systems and windows, maintenance	Review of published research, questionnaires	
Buhagiar (2004)	A post-occupancy evaluation of manipulating historic built form to increase the potential of thermal mass in achieving thermal comfort in heavyweight buildings in a Mediterranean climate	Historic buildings	Thermal comfort, occupants' satisfaction	Questionnaires survey, structured interviews	4 historic buildings
CABE (2006)	Assessing secondary school design quality	Secondary schools	Building functionality (access, space and uses), built quality (performance, engineering and construction), and impact (sense of place and effect on community)	Photographic walk-through, database, evaluations written by design and construction professionals trained as CABE enablers, client interviews, follow-up, web surveys of enablers for overall recommendations	2000–2005, 52 schools

CABE (2007)	A sense of place – what residents think of their homes	Residential	User satisfaction	Survey into the views of 643 residents living in 33 developments. In addition, 704 residents took part in census surveys at six case study developments	2006
Coulter et al (2008)	Measured public benefits from energy efficient homes	Residential house	Energy efficiency, owner satisfaction	Monitoring and questionnaires	7141 residential houses
Daioumaru et al (2008)	Thermal performance evaluations of DSF with vertical blinds	Public building	Thermal comfort	monitoring	
Davara et al (2006)	Architectural design and IEQ in an office complex	Public building, offices, multifunctional	IEQ, space usability, air temperature, relative humidity, light intensity, CO_2	Walk-through, interviews, spot measurements, short-term monitoring	1 multifunctional facility
Donnell-Kay Foundation, Denver, CO (2005)	School facility assessments: State of Colorado	School	Assessments of physical condition, educational suitability, technology readiness, site condition and capacity/utilization	BASIS® School facility assessment system	2004, 7 Colorado districts, 22 schools
Etzion (1994)	A bioclimatic approach to desert architecture	Residential	Indoor temperatures, thermal performance	Monitoring	1 single family detached house
Etzion et al (1993)	Project monitoring in the Negev and the Arava, Israel	Residential, office, educational	Indoor temperatures, thermal performance	Monitoring	Student accommodation, 1 multifunctional educational building

TABLE 7.1 Synoptic overview of POE studies in selected papers (*Cont'd*)

REFERENCE	TITLE	BUILDING TYPE	DIMENSIONS EVALUATED	METHODOLOGIES	SCOPE
Etzion et al (2000a)	A GIS framework for studying post-occupancy climate-related changes in residential neighbourhoods	Residential, clusters	Climate-related building changes	Survey, walk-through, GIS	Residential neighbourhoods
Etzion et al (2000b)	Climate-related changes in residential neighbourhoods: analysis in a GIS framework	Residential, clusters	Climate-related building changes	Survey, walk-through, GIS	Residential neighbourhoods
Etzion et al (2001)	An open GIS framework for recording and analyzing post-occupancy changes in residential buildings – a climate-related case study	Residential, clusters	Climate-related building changes	Survey, walk-through, GIS	Residential neighbourhoods
Frenkel et al (2006)	POE of a scientists' village complex in the desert – towards a comprehensive methodology	Educational complex	Energy consumption, IEQ	Short-time monitoring, observations, questionnaire surveys	Educational complex: office building, dorms, classrooms, facilities
Genjo and Hasegawa (2006)	Questionnaire survey on indoor climate and energy consumption for residential buildings related with lifestyle in cold climate areas of Japan	Residential	Thermal comfort, energy consumption	Questionnaire	Not noted
Hydes et al (2004)	Understanding our green buildings: seven post-occupancy evaluations in British Columbia	Academic, educational, office, industrial	User satisfaction, use of space, thermal comfort	Not noted	Not noted

Reference	Title	Building type	Focus	Method	Sample
Ito et al (2008)	Field survey of visual comfort and energy efficiency in various office buildings utilizing daylight	Office	Daylighting	Questionnaire, measurements	9 office buildings, 2002–2007
Kenda (2006)	Pneumatology in architecture: the ideal villa	Residential, clinic	User satisfaction, ventilation		
Kosonen et al (2008)	Perceived IEQ conditions: why the actual percentage of dissatisfied persons is higher than standards indicate?	Office	User satisfaction	A web-based IEQ survey	29 office buildings
Kowaltowski et al (2004)	From post occupancy to design evaluation: site planning guidelines for low income housing	Residential, public space	User satisfaction, use of space thermal control	Questionnaires of selected representative public	107 questionnaires were applied in five housing areas during a period of 4 months at the end of 2003
Langstone et al (2008)	Perceived conditions of workers in different organizational settings	Educational, office, commercial	User satisfaction, use of workspace in addition to thermal, visual, acoustic comfort	Questionnaires	2 years, 14 case studies, 555–4500 respondents
Leaman and Bordass (1999)	Productivity in buildings: the "killer" variables	Office	User satisfaction, use of workspace		
Levin (2005)	Integrating indoor air and design for sustainability	All	Material use in green building thermal performance, POE		
Lighthall et al (2006)	Renovation impact on student success	School	Impact of large scale renovations of school buildings on facilities, student achievement, attendance and suspension rates, as well as	Data were collected and analyzed from end-of-grade and end-of-course exams,	2005, 18 schools

TABLE 7.1 Synoptic overview of POE studies in selected papers (*Cont'd*)

REFERENCE	TITLE	BUILDING TYPE	DIMENSIONS EVALUATED	METHODOLOGIES	SCOPE
			the impact on stakeholder satisfaction	SAT scores, average daily attendance, out-of-school suspensions and parent satisfaction surveys. Interviews were also conducted with school staff regarding their satisfaction during and following renovations	1995–2005
Loftness et al (2006)	Sustainability and health are integral goals for the built environment	Offices, schools, hospitals	User satisfaction, worker productivity as function of all aspects of health and well being in built environment, SBS, energy consumption/conservation, VOC, TVOC, visual comfort, thermal comfort, ventilation, pathogens, allergens	Review of published research, correlation of results	
Mahdavi et al (2008)	Occupants' evaluation of indoor climate and environment control systems in office buildings	Office	IEQ, occupants' control	Interviews, long-term measurements	5 buildings, 68 respondents
Marmont (2004)	City Hall London: evaluating an icon	Office	User satisfaction, use of space	Evaluation of data	
McMullen (2007)	Determining best practices for design, implementation and service	University library	User satisfaction	Photographic walk-through, client interviews, professional analysis	2007, 19 interviewees

Reference	Title	Building type	Parameters	Method	Notes
Meir (1990)	Monitoring two kibbutz houses in the Negev desert	Residential	Indoor/outdoor temperatures	Monitoring	Symmetrical building discrepancies evaluation
Meir (1998)	Bioclimatic desert house – a critical view	Residential	Indoor climate, thermal comfort, energy consumption, water consumption	Monitoring	1 single family detached house
Meir (2000a)	Integrative approach to the design of sustainable desert architecture: a case study	Residential	Temperature, relative humidity, thermal comfort, water consumption, landscaping	Monitoring	Single family detached house
Meir (2000b)	Courtyard microclimate: a hot arid region case study	Courtyard of health facility	Outdoor air temperature	Short-term monitoring	Courtyard microclimate variability
Meir and Hare (2004)	Where did we go wrong? POE of some bioclimatic projects, Israel	Schools, visitor centre, residential, landscape	Occupant control of systems and windows, maintenance, training of occupants	Walk-through, interviews	Single buildings
Meir et al (1995)	On the microclimatic behaviour of two semi-enclosed attached courtyards in a hot dry region	Residential courtyards	Temperatures, shading simulations	Monitoring, CAD shading simulation/ visualization	Comparative behaviour of courtyards with different orientation
Meir et al (2007)	Towards a comprehensive methodology for post-occupancy evaluation (POE): a hot dry climate case study	Dormitory	User satisfaction, thermal control	Walk-through, survey, questionnaires	1–2 weeks for each building during winter and summer, 2 dormitories, 31 tenants

TABLE 7.1 Synoptic overview of POE studies in selected papers (*Cont'd*)

REFERENCE	TITLE	BUILDING TYPE	DIMENSIONS EVALUATED	METHODOLOGIES	SCOPE
Menzies and Wherrette (2005)	Windows in the workplace: examining issues of environmental sustainability and occupant comfort in the selection of multi-glazed windows	Office	Window controllability, lighting user satisfaction, environmental sustainability, productivity	Monitoring, questionnaires	4 office buildings
Mochizuki et al (2006)	Field measurement of visual environment in office building daylight from lightwell in Japan	Office	Daylighting, energy consumption, visual comfort	Questionnaires, measurements	
Morhayim and Meir (2008)	Survey of an office and laboratory university building – an unhealthy building case study	Multifunctional, educational, university, office	IEQ, usability, user satisfaction	Walk-through, surveys, questionnaires, interviews, measurements	Comprehensive analysis of one university building
Nakamura et al (2008)	The evaluation of productivity and energy consumption in 28 degrees office with several cooling methods for workers	Office	Energy consumption, thermal environment, performance, productivity	Simulation, evaluation	
Nordberg (2008)	Thermal comfort and indoor air quality when building low-energy houses	Residential house	Thermal comfort	Short- and long-term measurements	1 house, includes 3 units, without conventional heating systems
Patricio et al (2006)	Double-skin facades: acoustic, visual and thermal comfort indoors	Commercial and services building employed the DSF technology	Energy consumption, visual, acoustic and thermal comfort	Monitoring	
Pearlmutter and Meir (1995)	Assessing the climatic implications of lightweight housing in a peripheral arid region	Residential	Indoor temperatures, relative humidity, MRT	Monitoring	Heavy vs light construction

Pearlmutter and Meir (1998)	Lightweight housing in the arid periphery: implications for thermal comfort and energy use	Residential	Temperature, relative humidity, MRT, energy consumption	Summer/winter-monitoring, various operation modes, thermal simulation	1 heavy-, 2 lightweight housing units
Pearlmutter et al (1996)	Refining the use of evaporation in an experimental down-draft cool tower	Multifunctional, educational	Evaporative cooling potential, indoor temperature and relative humidity	Monitoring	Cooling potential, alternative evaporative cooling technologies
Pfafferott et al (2004)	Comparison of low-energy office buildings in summer using different thermal comfort criteria	Office	Thermal comfort	Monitoring	12 office buildings with passive cooling systems
Pitts and Douvlou-Beggiora (2004)	Post-occupancy analysis of comfort in glazed atrium spaces	Educational building	Thermal comfort	Measurements	Summer and winter, 300 respondents
Preiser (2004)	Evaluating Peter Eisenman's Aronoff Center: De-Bunked De-Constructivism	Academic, office, studio, hall	User satisfaction, use of space	Evaluation of data	
Roaf (2004)	Cave Canem: will the EU Building Directive bite?		Use of space		
Sanoff (2004)	Schools designed with community participation	Schools	User satisfaction	Walk-through and surveys by clients (teachers) POE in conjunction with design process	2000, 50 teachers participated

TABLE 7.1 Synoptic overview of POE studies in selected papers (Cont'd)

REFERENCE	TITLE	BUILDING TYPE	DIMENSIONS EVALUATED	METHODOLOGIES	SCOPE
Silva et al (2006)	Monitoring of a double skin façade building: methodology and office thermal and energy performance	Commercial and services building employing the DSF technology	Energy consumption, visual, acoustic and thermal comfort	Short-term monitoring	2 weeks, evaluations of all 14 houses in project; surveys and interviews, 2002 and follow up in 2004
Stevenson (2004)	Post occupancy – squaring the circle: a case study on innovative social housing in Aberdeenshire, Scotland	Residential	User satisfaction, use of space, thermal control, energy consumption	Tenant interviews, energy use data	79 single-family detached houses, solar neighbourhood
Vainer and Meir (2005)	Architects, clients and bioclimatic design: a solar neighbourhood POE	Residential neighbourhood, houses	User satisfaction, building performance, energy consumption	Plan analysis, questionnaires, structured interviews, walk-through and visual analysis	2004, 1300 questionnaires in 16 low energy German office buildings, summer and winter
Wagner et al (2007)	Thermal comfort and workplace occupant satisfaction – results of field studies in German low energy office buildings	Office	User satisfaction, use of workspace	Field study and questionnaires	3 projects
Watson (2003)	Review of building quality using post-occupancy evaluation	School, dormitory	User satisfaction	Walk-through and surveys by POE interviewers	

Author	Title	Type	Focus	Method	Details
Watson (2005)	Post-occupancy evaluation – Braes high school, Falkirk	School	User satisfaction	Walk-through and surveys by clients	2000, 55 stakeholders incl. pupils, staff and other school users, as well as council officials and technical staff involved in the design, construction and maintenance of the building
Woollett and Ford (2004)	How happy are we? Our experience of conducting an occupancy survey	Office	User satisfaction, use of space, thermal control	Written survey	4 hour, one time sample, 48 respondents
Xiong (2007)	The impact of exterior environmental comfort on residential behaviour from the insight of building energy conservation: a case study on Lower Ngau Tau Kok estate in Hong Kong	Residential	User satisfaction, thermal control	Interviews, measurements, simulations	One month (Aug), 40 sampling sites
Zagreus et al (2004)	Listening to the occupants: a web-based indoor environmental quality survey	Office	To introduce a-web based survey and accompanying online reporting tool	A web-base IEQ survey	3 case studies of office buildings

a detailed analysis of all possible parameters in a single case study. Several lateral studies were reviewed for this chapter, among them the European research project HOPE which surveyed 97 apartment buildings and 67 office buildings (Roulet et al, 2005, 2006), the Probe project (Bordass and Leaman, 2004) covering more than 20 office and public buildings in the UK, a study of office and institutional buildings in the US (Zagreus et al, 2004) and smaller studies such as those by Mahdavi and Proeglhoef (2008) on user control actions in office buildings in Austria. An example of an in-depth case study is recorded by Morhayim and Meir (2008) in which a university building including offices, laboratories and assorted facilities was investigated using observations and walk-throughs, monitoring, questionnaires and plan analysis.

Our goal in this chapter is not an exhaustive review of all material published on POE, but, rather, to trace current practices and methods, lacunae and problems, and point to potential modifications and protocols.

AIMS AND TARGETS OF POE

The nature and goals of POE depend on who is asked, as the prospects and hazards of this tool and approach are seen differently from the standpoint of each stakeholder.

The *entrepreneur* should have a vested interest in POE as a way to assess the design quality and potential gains – value for money invested – enabled by a better end product, i.e. the building. Against such potential gains, however, entrepreneurs do not always want too probing a light to be thrown on the performance of their buildings and, in extreme cases, they will be wary of their legal liability for malfunctioning or hazardous buildings.

The *building manager* should be interested in lowering energy consumption and maintenance costs, and an understanding of the actual operation of the building by the users is an essential step towards this. It has been demonstrated that often there is an acute discrepancy between objective comfort (such as thermal comfort defined by ASHRAE (1992)) and subjective comfort (such as defined by the adaptive thermal model (Nicol and Humphreys, 2002)). Studies have demonstrated energy waste alongside thermally uncomfortable interiors (overheated or overcooled), as well as increased energy consumption in buildings in which there is no control over one's personal space (air temperature, light intensity, etc.).

The *building user*. Here we can distinguish between the emphasis on well-being and health (in the case of the building's occupants, workers, tenants, students, etc.) and an emphasis on productivity (in the case of the company owner/manager of the building or the institutional entity responsible for it (the education system, etc.). These two emphases are clearly intertwined, although in reality clashes of interest exist (Davara et al, 2006).

The *architect* and *consultants* should be aiming at producing the best possible building within the existing economic, statutory, technical and other constraints. The responsibility of design professionals for the well-being of the people that occupy their buildings is an obvious but sometimes overlooked basic principle, sometimes inscribed in professional ethics codes and legislation. As an example, the second paragraph of the *Israeli Bylaws of Engineers and Architects* states that the first and foremost task of the architect and the engineer is to ensure public health and safety (IAUA, 1994), issues definitely associated with IAQ and IEQ. This often causes raised eyebrows among architecture students, as these mundane duties do not square with their initial glamorous image of the design professions!

Institutional stakeholders, i.e. the various governmental bodies concerned, on the national and political levels should be interested in the promotion of better design and building practices, such as would be enabled by a continuous process of assessment and upgrade that can be facilitated by POE. In severe cases, faulty buildings characterized by sick building syndrome (SBS) cause absenteeism, hospitalization and may create demands for potential compensation for long-term health and other damages. Institutional stakeholders will also be motivated to achieve the added longevity of better buildings and systems, minimizing the need for changes, refurbishment or demolition and reconstruction.

While each of these stakeholders approaches POE from differing and at times conflicting viewpoints, it is clear that all have much to gain from a thoroughgoing institutionalization of POE practices and from the extensive use of these methods for understanding flaws in current practices and producing solutions for the correction of these.

TOOLS AND METHODS USED IN POE

Since buildings are very complex systems, and their interaction with occupants further compounds the complexity of possible interrelations and potential malfunctions, it is imperative that the study of building post occupancy be based on a multi-level, multi-faceted system of checks and tests. These should involve thermal comfort alongside heating, ventilation and air-conditioning; illumination and visual comfort; occupants' satisfaction and behaviour; and, not least, physiological and psychological comfort, since all of these issues together will affect energy consumption and human well-being.

The methods and tools employed are both quantitative and qualitative, and may be classified in three rough categories on the basis of the information analysed and assessed.

Measurements, monitoring, sampling

Some key parameters that may be measured include temperature, relative humidity, air movement, light intensity, noise levels, pollutants, allergens and pathogens, volatile organic compounds (VOC) of various compositions and forms (e.g. formaldehyde) and their overall combination expressed as total volatile organic compounds (TVOC), gases of different types (CO, CO_2), electromagnetic fields and radiation (including radon), etc. While these may seem the most concrete and unequivocal aspects of a building – simple to measure and straightforwardly comparable with established objective standards – POE has shown that the picture is somewhat more complex than may be suggested by a traditional building physics approach (Pati and Augenbroe, 2006).

For example, while national and international standards exist for some of these parameters, recent research has questioned the validity of some of these, for example, the thermal comfort standards defined by ASHRAE *Standard 55* (2004). This is rigid in its upper and lower thresholds, yet its opening statement defining thermal comfort as 'the state of mind that expresses satisfaction with the surrounding environment' suggests that things may not be so clear cut. An alternative adaptive model advocated by Nicol and Humphreys (2002) assumes behavioural and cultural differences, as well as varying degrees of readiness to accept environmental conditions beyond these rigid thresholds. Thus, it may not be enough to measure the physical factors in a given environment, nor, indeed, to measure thermal parameters in isolation, since accumulating evidence indicates a significant degree of influence of psychological factors on the physiology of subjects (Mallick, 1996; Faruqui Ali et al, 1998) and the interaction of other attributes (such as noise) with thermal comfort levels (Pati and Augenbroe, 2006). These effects bring thermal comfort out of what has been assumed in recent decades to be the task of HVAC engineers.

Similarly, while visual comfort is addressed by different standards based on light intensity measurements in isolation, individuals relate differently to both the quantitative properties and non-quantitative qualities of light in different settings. Thus, whereas CIBSE *Code for Interior Lighting* (CIBSE, 1994) provides quantitative standards for different tasks, it says little on qualitative issues such as material properties like texture and colour which may influence visual requirements and comfort. Such complex interactions come into play when combining natural and artificial lighting, which has the potential to promote energy conservation, but also poses special challenges and problems for measurement and standard setting. This, together with the limited research on daylighting utilization possibilities in specific climatic regions (e.g. with high solar radiation throughout the year), limits the flexibility and options of designers and creates a dependence on electric lighting for performing visual tasks (Ochoa and Capeluto, 2006). With these limitations on the uses of theory and standards for designing for visual comfort, the emphasis on the kind of empirical measurements and feedback offered by POE becomes all the more important. POE, thus, becomes a bridge not only between pre- and post-occupation phases, but also between objective and subjective in the responses of people to buildings and between various domains of experience which interact in shaping overall satisfaction.

In addition to the complexities inherent in the subjective and interactional nature of temperature or lighting parameters, additional complexities may arise in the sampling procedures and the level of the standard themselves. Sampling and monitoring of various compounds differ from one country to another and even where such standards do exist, they are often relative and not absolute. For example, the maximum acceptable concentration of CO_2 is defined by ASHRAE *Standard 62* (1992) as up to 700 ppm *above* the outdoor levels, and radiation is often assessed in relation to *background levels*. Other compounds of biological significance are often not considered, and rarely sampled.

Compared with the above, energy consumption would seem a relatively easily quantifiable parameter, since it is already measured continuously at the electricity supply. However, more refined analysis of this parameter can provide useful insights into a building's qualities, properties and problems, such as the comparison of such basic measurements with a base case or a standard such as the PassivHaus (2008), which defines targets for energy consumption per floor area.

Finally, additional discrepancies or inaccuracies may arise related to the minimum accuracy desired, calibration procedures, minimum monitoring period and/or number of samples, which must be specified to avoid undue variance in the data or misleading results. Attempts to standardize such procedures and protocols do exist (see, for example, Spengler et al, 2000), but these can be compromised by different considerations and limitations, not least by actual on-site capacity, local and national differences, etc.

Surveys, questionnaires, cohort studies, observations, task performance tests

Such tools may be used by themselves or, preferably, in combination with the more quantitative measurements described above. While some may consider the type of tools drawn from the psychological or social sciences to be supplementary, and of use primarily to gauge user satisfaction, there are researchers who consider them no less accurate and representative than physical measurements and monitoring, so that a cleverly prepared questionnaire may provide as much as 80 per cent of all needed indicators for the assessment of building performance.

The main purpose of these tools is to help understand the intricate interrelations between a building, its users and the various systems that are part of the building's operation. Whereas measuring air temperature within a space seems to be rather straightforward, how this temperature is perceived by the individual is a totally different issue, often affected by parameters other than physiology or temperature *per se*. Such questionnaires are used to quantify the subjective perception of indoor parameters by asking interviewees to rank temperature, light, noise, ventilation, overall satisfaction and other parameters on five- or seven-degree scales. Questionnaires may be in hard copy or online, filled in by interviewee or surveyors.

Task performance tests are used in order to understand the influence of indoor parameters on the ability of the user to perform satisfactorily over a given period of time whether short or long term. This may be of importance both in terms of subjective well-being and objectively measured productivity. Such tests usually involve a repetition of a series of tasks such as word identification, form matching, typing, simple or complex mathematical calculations or other activities similar to those that users are typically expected to perform. Non-optimal indoor conditions – hotter or colder than neutral, flickering light or light levels that are too high or too low, lack of fresh air supply, noisy environments, smells, etc. – will eventually affect the outcome of the test, showing a decline in performance capabilities (Amai et al, 2007).

While theoretically dealing with the subjective perception of indoor conditions, and individual interventions, taken collectively, such assessments also offer a good overall indication of the indoor environment's condition and properties.

Document analysis, on-site observations

Document analysis can be divided into two main groups. In the pre-construction stage, drawings, briefs and specifications can be critically analysed to allow correction of potential mistakes. These may range from architectural details such as oblique angles of walls and structural elements causing not only usable space loss (Marmont, 2004) but also being the cause of behavioural problems within the spaces (Preiser, 2004).

Surveys and stationary and walk-through observations are also used to identify various building or building system problems, among them the actual use of spaces and details (such as user-devised *ad hoc* shading solutions common in fully glazed facade buildings, open windows in conditioned buildings indicating indoor conditions outside the comfort zone or lack of sufficient outdoor air supply) or other indicators such as mould and stains on HVAC outlets, walls and ceiling indicating potential health hazards.

DISCUSSION

Having briefly reviewed POE from the standpoints of various stakeholders in buildings, and some of its methods and goals, we can reflect on how POE can play a role in mediating and bridging some key tensions in contemporary building design.

POE IN THE BALANCE OF CREATIVITY AND UTILITY IN BUILDING DESIGN

For example, POE can inform the debate regarding the trade-offs of creativity and utility in buildings. In the modern age, the former has found more widespread expression as monumental buildings are designed not only to house religious or public institutions: spectacular museums, libraries and universities are joined today by apartment blocks and offices as landmarks. Everyman and company can commission a building that says 'art', not just architecture. Today's new tools allow architectural forms to be constructed within a spectacularly wide design scale. Buildings with non-repeating unique structural components are now commonly engineered and constructed. How do these capabilities interact with the familiar constraints of cost, and the forward-looking constraints of energy efficiency?

The contention is that the market is full of spectacular and unique buildings that may be jewels to view but are unethical in the use of energy, land and budget. Even those that gesture to sustainability may fall short: their rugs may meet the highest standards of recycled content, the paints free of VOCs and the high-performance windows the most insulated available; at the same time, however, they may have unwarranted use of some of these components in the building, in particular if the building has a curious shape. In other words, even though the building may use more efficient materials, it may use more of them, or, more of another kind of less efficient material. For example, in the case of wrap-around glazing, true energy savings could be attained with fewer windows altogether.

POE can play a role in attempts to determine an acceptable balance between creativity and utility. It does this by bringing in the element of user satisfaction as well as the actual functioning of the building, which together constitute its utility, and can help assess if and how the more imaginative or artistic elements interact with these. Currently, POE of 'green' buildings assumes that they are utilitarian in design and, therefore, measuring satisfaction of the air quality, lighting, thermal conditions, energy use and perhaps workspace comfort is sufficient. Roaf (2004) is critical of the current definitions of 'green' and 'sustainable' for all projects (that is, the idea that all projects today need to be sustainable) and in particular when confronted with 'signature buildings' that may be green in material selection and thermal standards of each element, but lack sense when it comes to total material use, use (read 'waste') of space in work areas and in public access zones of these buildings (Wilson and Austin, 2004). Roaf (2004) contends that POE needs to be expanded to contend with design and layout parameters in addition to the ones covered in research and surveys to date.

POE IN EDUCATIONAL BUILDINGS

Perhaps one of the areas in which POE has a most compelling role, and is also most likely to make inroads in institutional terms, is in the design and construction of schools. As opposed to private and corporate construction processes, schools are in the public domain and need to balance utility and innovation and, in many districts, must respond to serious public accountability.

The stakes are large. The magnitude of the education building business both in the US and the world increases annually.[1,2] More than $20 billion was spent in the US on new elementary, middle and high school construction alone in 2007 (Abramson, 2008). Trade journals, architectural websites, research foundation reports and government documentation show that innovative design is a prime component in the new construction of schools. Architectural firms that specialize in school design publicize that they lead community inclusion in the design process, usually limited to pre-design/charrette stages (AAF, 2008). The school building business includes using POE as a marketing tool for architect and construction firms[3] where (almost exclusively) successes are showcased and awarded. An exception is the US National Clearinghouse for Educational Facilities (NCEF) which makes some public sector evaluations available[4] (Sanoff, 2002). Historically these evaluations consider the facility's physical condition, usage (as a function of area appropriated for each type of use, e.g. classroom, music room, cafeteria, student lounge) and energy use by using POE and evaluation database programmes usually facilitated by professional assessors. The commissioning of buildings with stated budgets for both cost and energy use is increasing. Space utility is a dynamic issue as education styles and populations change quickly (Lighthall et al, 2006).

The issue of signature/innovative designs and their association with educational theory coalesce with stakeholder participation during the design stages but analysis of the utility of these architectural features has not been addressed systematically in POE.

The UK has put an emphasis on determining better design practices based on POE for educational buildings and community involvement in the design process. This process (based on work by the Commission for Architecture and the Built Environment – UK (CABE) and furthered by the Design Quality Indicator – UK (DQI) evaluation process) promotes ways to design more usable educational facilities (CABE, 2006). New and colourful publications[5] show new buildings, innovative interior and exterior spaces and describe their intended use. It does not include POE results that would determine whether the goals could be realized. Analysing the use of space in these new facilities should be a priority in light of the CABE 2006 research report *Assessing Secondary School Design Quality* which includes insightful sections such as:

Weaknesses: A large number of schools surveyed failed to function spatially. The survey identifies that teaching, key ancillary spaces and circulation are often inappropriate for their function. (CABE, 2006, section 7.2)

A school may be designed in accordance with all the conventional and green criteria but in practice may not lend itself to allowing the occupants to use it to its potential. These errors can only be corrected if POE addresses these issues and the results are honestly and openly publicized.

There are school districts worldwide that are committed to evaluating the use of space and user satisfaction (Watson, 2003, 2005). These POE and feasibility studies do not delve deeper than giving a numerical grade to user satisfaction (DKF, 2005). There is no place in the assessments to ponder whether wasted space was produced by the design features.

SUSTAINABILITY: FROM DECLARATION TO PERFORMANCE THROUGH POE

Architect Alexi Marmont took the designers of the new London City Hall to task by questioning the cost, spatial and energy efficiency of this landmark project (Marmont, 2004). Her review of published data suggested that the unique interior and exterior shapes were part of the reasons that led to a good, but not excellent, usability rating. The building cost more than others in its class and user thermal comfort was mixed. Of most concern is the fact that no data pertaining to energy use had been published and maintenance of the building will be relatively costly because of its idiosyncrasies. Touted as a sustainable building, life cycle analysis (LCA) of materials used and energy use data needs to be added to the evaluations already performed. Perhaps sustainability cannot be determined by a single mechanism such as the proposed 'next generation LEED', which incorporates performance and a wide range of LCA metrics, but needs to be professionally evaluated by a group of professionals using a variety of measurements, POE and other post-construction data. The issue of sustainability, holistic by definition, may be too complex to determine by measurements alone.

Obviously user sensibility and satisfaction must play a pre-eminent role in evaluating all types of facilities and therefore they must play an active part in building performance evaluations of all types (Leaman and Bordass, 1999; Wagner et al, 2007; Langston et al, 2008). The questions at hand demand that the occupant can express his/her satisfaction with the immediate workspace as well as give an opinion on how much the built environment is beneficial, neutral or negatively affects the satisfaction rating. Increasingly, buildings are not simply a workplace or a classroom but where people spend the best part of their lives (Baird and Jackson, 2004), with potentially profound effects on health and productivity (Wargocki et al, 1999). Nevertheless it may be unreasonable to include questions about material choices at the post-occupancy stage for all types of buildings. The POE is a tool that must relate to the job at hand. Gordon and Stubbs (2004) touch on the different goals of POE in five case studies of buildings selected for continued long-term review by the AIA's Building Performance Committee. They note that architectural practice Fox and Fowle's selection of a curtain wall of high-performing glass was part of the specification for the LEED Silver residential high-rise development in Manhattan, The Epic (also known as St Francis of Assisi). Even if the glazing was of superior quality and energy efficient relative to other windows, it is less energy efficient and contains higher embodied energy than many opaque wall options. Does the entire wall of the building need to be glazed for aesthetic reasons? In order to maintain sale value? Is it possible to structure questions in a POE to elicit constructive responses from a building's tenants with reference to their views of glazed curtain walls? There are developers of commercial urban projects that ask such questions such as at the Solaire residential tower in New York City's Battery Park City, but they are few and have not been externally evaluated after construction (DOE, 2004).

Achieving comfort in conventional buildings, especially the glass-and-steel blocks characteristic of the past 30–40 years, results in substantial energy consumption and adverse environmental impacts. A survey of 200 detached houses conducted in the cold climatic area of Japan to clarify the characteristics of indoor climate and energy consumption (Genjo and Hasegawa, 2006), showed that the indoor climate in buildings constructed in recent years was better than that in older buildings, and the energy consumption in the former was higher than in the latter. There is a persistent discrepancy between the increasing demand for comfort in buildings and the need to decrease the use of energy (Zeiler and Boxem, 2008). However, under the mounting pressure of energy shortage, one approach to minimize the contradiction is to design sustainable buildings in an informed and responsive way (Zeiler et al, 2008). Therefore, integration between end-user needs and building performance is of significant importance.

POE studies have the potential to clarify discrepancies, loopholes and problems in different ways. They can indicate problems in the design process (for the architect and related disciplines), the operation (for the occupant, user and building manager) or in the building as a system.

Even (or, perhaps, especially) in initiatives that are declared to be green, and for architects who plan for them, POE has an important role to play in providing feedback. A POE study of a solar neighbourhood in Israel (Vainer and Meir, 2005) showed discrepancies between the planners' environmentally friendly intentions and the final outcome. The neighbourhood was constructed in three stages over time. Analysis of quantitative data, such as south-facing fenestration as a percentage of the overall area of each unit, showed a gradual decline over the course of the three-stage construction, substantially restricting the potential for passive heating and natural ventilation, and deviating from initial intentions. Whereas two monitored houses in stage 1 showed minimal or no auxiliary energy use for backup heating in winter and none for cooling in the summer (Etzion, 1994; Meir, 2000a), monitoring of houses in the later stages revealed indoor winter temperatures to be significantly below thermal comfort even with backup heating, and incorporation and operation of air conditioners in the summer, owing to wrong design. The study revealed that whereas the houses in stage 1 were mainly designed by local architects who were acquainted both with the local climate and with the bioclimatic strategies, the latter were designed primarily by non-local architects with little or no acquaintance with the potential provided by the plan, misunderstanding the principles of sustainable design and bringing about the unintended outcome.

The inputs available from POE can also identify where the behaviour of building users undermines their functioning and where the education of users is critical in order to prevent this and increase their capacity to operate features optimally (Hydes et al, 2004). A POE survey of seven energy conserving projects in Israel showed that, to some degree, each of them malfunctioned after several years of occupancy, mainly due to the lack of communication between the architect and users over time, often caused by the introduction of new users (Meir and Hare, 2004). As a result, systems and features were operated inappropriately, leading to poor indoor conditions and eventually to building changes. An evaluation of the refurbishment of four historic buildings dating from the 16th century showed that inherent traditional physical features were abandoned for new technological solutions such as air-conditioning, since users were largely unaware of the potential of the original building features and their potential to modify the indoor environment (Buhagiar, 2004).

A study of office buildings in Austria indicated considerable levels of dissatisfaction with certain aspects of the indoor climate and environmental control systems. Occupants interviewed considered their knowledge of their offices' environmental systems as insufficient and would welcome clarification on the operation of such systems (Mahdavi et al, 2008).

Besides instructions on use and operation, it is necessary to explain to the user the rationale and potential of sustainable design. A survey of a bioclimatic complex in a desert climate showed that although there were detailed instructions on the appropriate operation of buildings and their systems and details (shutters, wind chimney, etc.) in every unit, tenants were often confused and doubtful about the actual effects of such measures (Meir et al, 2007). Actually, not all tenants who are living in green buildings feel committed to the concept of passive heating or cooling. Energy efficient building is hardly just a technology – it truly is a way of life and a tool to achieve a bigger goal. Thus this kind of building should be nurtured by education and not left to self-explanatory tools. POE can identify the where and illuminate the how.

THE RELATION OF POE TO LCA AND GREEN STANDARDS

POEs can supplement life cycle analysis, the set of mechanistic and analogical determinations based on energy use and quantifying the types and amounts of construction materials, increasingly used to compare the impact of buildings (Boecker, 2005; BFRL, 2007). These, it is hoped, can help quantify the carbon emissions from materials used in building and the potential emission savings inherent in using specific construction elements (Huberman and Pearlmutter, 2008). LCA has not been used widely for comparing buildings to date because of the multitude of variables that make up embodied energy calculations and the problems involved in attempting to attribute energy savings to elements that are dependent on how they are operated (Meir and Hare, 2004). Also, some work and public spaces have qualities that may not have an agreed use and are challenging to monitor, such as multistorey atriums (Atif and Galasiu, 2003). User intuition and feedback may play a guiding role in correlating the materials used and the value of the space created. Evaluation of a building's success in supplying a healthy and usable environment by its occupants and users as well as professionally produced appraisals are a necessary component of LCA (Gale, 2008), and this is where POE comes in.

As green buildings and low-energy houses began to catch on, the design community recognized the need for a rating system to assess how green a building is (Hydes et al, 2004). In 2000, the United States Green Building Council (USGBC) launched the first formal framework for rating green buildings in the US – Leadership in Energy and Environmental Design (LEED) (Abbaszadeh et al, 2006). The LEED system, with its

69-point scheme and third-party verification, offered a set of assessments of green buildings, which consisted of sustainable sites, water efficiency, energy and atmosphere, materials and resources, and indoor environmental quality (USGBC, 2002). However, it was not developed to be used as a POE tool, and has been criticized for its relative flexibility in mutually compensating items.

In 2002, the California-based Center for the Built Environment developed a web-based survey and accompanying online reporting tools to assess the performance of workspace, identify areas needing improvement and provide useful feedback to designers and operators about specific aspects of building design features and operation strategies from the occupants' perspective. The survey includes the following modules related to IEQ: office layout, office furnishing, thermal comfort, air quality, lighting, acoustics and building cleanliness and maintenance (Zagreus et al, 2004; Abbaszadeh et al, 2006; Kosonen et al, 2008). A seven-point semantic differential scale with endpoints 'very dissatisfied' and 'very satisfied' was used to evaluate occupant satisfaction quantitatively. If respondents indicate dissatisfaction with a survey topic, they branch to a follow-up page where they can specify the source of dissatisfaction. Thus, a web-based IEQ survey has been utilized as a diagnostic tool to identify specific problems and their sources (Kosonen et al, 2008). Furthermore, the survey implementation process is convenient and inexpensive, since the survey is delivered through a website, where occupants are given the ability to evaluate their workplace online. Responses are collected and added to a benchmarking database comprising the records of buildings investigated, enabling the comparison of occupant satisfaction of different buildings transversely.

POE IN DISENTANGLING THE NEXUS OF ENERGETIC, THERMAL AND VISUAL DIMENSIONS

POE becomes important, even essential, given the often unexpected interrelations between various aspects of building function. For example, buildings with sustainable features are not only expected to save energy, especially for heating and cooling, but also to provide their tenants with a better indoor environment. The following case studies focus on the evaluation of building performance relevant to energy consumption, referring to thermal comfort, visual comfort, occupant satisfaction, energy efficiency, etc. The majority of these buildings were designed with a variety of energy saving characteristics regarding the local climatic conditions, for instance, double-skin facade, thermal mass, natural ventilation and passive heating or cooling devices in the light of sustainability principles. The methodology involved in these case studies basically covered both physical measurement and monitoring, and occupants' subjective assessments. Furthermore, a few studies emphasized the comparison between green buildings and ordinary buildings, using an IEQ tool.

Three POE studies (Gossauer, 2005; Pfafferott et al, 2007; Wagner et al, 2007) conducted in Germany investigated green buildings referring to thermal comfort and occupant satisfaction. Although their research targets differed, the findings were similar and interrelated. It is noteworthy that a positive perception of thermal comfort reported by respondents who are working in these green buildings may occur beyond the temperature limits normally set for air-conditioned buildings. Actually, several research projects have demonstrated that occupants in naturally ventilated buildings perceive higher room temperatures as comfortable. Since the current regulations, standards and recommendations of design temperature only refer to air-conditioned buildings, being forced to apply them in the evaluation of green building performance would probably result in an unsatisfactory outcome and restrictions in passive cooling design. The limits of tenants' perception of comfort are important for naturally ventilated and passively cooled buildings.

Additionally, the occupants' ability to control the indoor environment influences their satisfaction. Questionnaires suggested that in the case of no control over one's environment, occupant satisfaction with indoor temperature was relatively lower, although it met the standards of ISO 7730. Obviously, perceived control is different during the different seasons. On the other hand, two studies concerning environmental control systems (Menzies and Wherrette, 2005; Mahdavi et al, 2008) showed that buildings with fully operable windows were marked with higher satisfaction than those without such windows, the former allowing occupants to enjoy natural ventilation and daylight.

Daylighting and visual comfort are fundamental aspects of the indoor environment and energy efficiency in buildings. For tenants, many studies have demonstrated that if daylight is the primary source of lighting, there is a great improvement in productivity, performance and well-being for occupants in general (De Carli et al, 2008). The most used parameter for quantification of daylighting in buildings is the daylight factor (DF) which can be defined as the illuminance received at a point indoor from a sky of known or assumed luminance distribution, expressed as a percentage of the horizontal illuminance outdoors from an unobstructed hemisphere of the same sky (Patricio et al, 2006). Other relevant parameters for the evaluation of daylighting and visual comfort performance in buildings are the uniformity ratio and glare. By introducing daylight into office space through windows and using it as the source of lighting, energy consumption for artificial illumination can be reduced substantially.

A field study in Japan investigated performance and daylighting during the operation phase by surveying visual comfort and energy efficiency in office buildings. Questionnaires and monitoring surveys were conducted in nine office buildings between 2002 and 2007. Installing a large window has the advantage that energy for lighting is saved and visual comfort is improved by introducing natural light and allowing views out (Ito et al, 2008). However, thermal and visual comfort near windows are not always improved – occupants may suffer from excessive heat and glare discomfort (Mochizuki et al, 2006; Ito et al, 2008). South-facing windows are especially problematic due to the penetration of direct radiation and their blinds are often kept shut. In that case, a large window may prove to be largely counterproductive (Ito et al, 2008).

Driven by the need to increase comfort and energy efficiency, double-skin facades (DSF) were developed as an architectural engineering solution that has been adopted mainly in office buildings. Since DSF were initially developed for colder climates to help reduce energy for heating, they can cause overheating problems and increased energy needs for cooling during the summer, as well as in temperate and hot climates, especially if appropriate shading and ventilation devices are not designed or properly operated. Three POE studies showed that blinds and mechanical ventilation systems in DSF can solve glare-related problems or excess solar radiation.

CONCLUSIONS: POE – BETWEEN THE EMBRYONIC AND THE INEVITABLE

In this chapter we have reviewed some of the key aspects of POE and presented a somewhat eclectic and opinionated account of how POE interacts with important debates and issues in making the construction and life cycle of buildings more sustainable. Now is the time to take a step back from the details to look at the big picture regarding the development and institutionalization of POE and its emerging role as a facilitator of sustainable building practices.

Our review of published, conference and web sources shows a rich and increasingly sophisticated set of practices associated with POE – the overall sense is of a field that is at the threshold of maturity, but not quite there yet. Indications of this are that despite a growing awareness of POE principles, procedures and their importance, their use and the way in which they relate to key debates remains erratic. Thus, while the number of studies has grown enormously, they lack agreed-upon protocols, measures and procedures, making comparison difficult. Another indicator of pre-maturity is the distinct but still relatively low level of use of POE in the education of architects and the other professionals involved in the building process.

What does strike us in our review, however, is the extent to which the path toward greater maturity, acceptance, consistency and formalization of POE is inevitable. This is because of the remarkable and increasingly demonstrable potential of POE to serve as an integrator of various realms. It is in this integration role where POE contributes to sustainability in the deeper senses. By integration we mean the following (among others):

- integration between the pre- and post-handover phases in the building life cycle;
- integration of the various stakeholders in the building process – particularly the designer, owner, operator and occupant;
- integration of the various building disciplines with one another;
- the merging of practice with research;
- integration of various tools and, indeed, with the suites of qualitative and quantitative research traditions;
- integration of subjective and objective dimensions of building use and experience, and their measurement;
- the ability to bridge the static performance conceived for the building versus the dynamic functioning when real users interact with and modify these static features;
- bringing conceptions and aspirations closer to actual practices and performances.

New buildings are required to meet increasingly demanding standards with respect to comfort, safety, cost-effectiveness and sustainability, while still allowing creative expression. And they must do so across a time horizon now stretched backwards and forwards in new ways by perspectives such as life cycle analysis. In this milieu, the kinds of integration sketched above are no longer a luxury, but an imperative for survival (Roaf et al, 2005). And this is why we argue that it is inevitable that POE – which can facilitate so many of these forms of integration – will take on an increasing and, ultimately, indispensable role in the building process.

NOTES

1 Foundations and Firms American Architectural Foundation http://www.archfoundation.org/aaf/aaf/Publications.htm
2 Organisation for Economic Co-operation and Development, Directorate for Education, Programme on Educational Building PEB Exchange No 63 – June 2008 and previous journals
3 www.designshare.com, www.learningbydesign.biz/
4 NCEF www.edfacilities.org/
5 www.dqi.org.uk/DQI/Common/PSD-book.pdf

AUTHOR CONTACT DETAILS

Isaac A. Meir: Desert Architecture and Urban Planning Unit, Department of Man in the Desert (MID), Blaustein Institutes for Desert Research (BIDR), Ben Gurion University of the Negev (BGU), Sede Boqer Campus, Israel; sakis@bgu.ac.il
Yaakov Garb: Social Studies Unit, MID, BIDR, BGU, Israel
Dixin Jiao and **Alex Cicelsky**: Albert Katz International School for Desert Studies, BIDR, BGU, Israel

REFERENCES

AAF (American Architectural Foundation) (2008) *Winter 2008 School Design Institute*, AAF, www.archfoundation.org/aaf/aaf/pdf/sdi.winter08.pdf (accessed 15 July 2008)

Abbaszadeh, S., Zagreus, L., Lehrer, D. and Huizenga, C. (2006) 'Occupant satisfaction with indoor environmental quality in green buildings', in E. de Oliveira Fernandes et al (eds), *Healthy Buildings: Creating a Healthy Environment for People, Proceedings of HB2006 International Conference*, Lisbon, vol III, pp365–370

Abramson, P. (2008) *The 2008 Annual School Construction Report*, School Planning and Management, www.peterli.com/spm/resources/rptsspm.shtm (accessed 2 August 2008)

Amai, H., Tanabe, S. I., Akimoto, T. and Genma, T. (2007) 'Thermal sensation and comfort with different task conditioning systems', *Building and Environment*, vol 42, no 12, pp3955–3964

ASHRAE (1992) *Standard 62 – Ventilation for Acceptable Indoor Air Quality*, American Society of Heating, Refrigerating and Air-Conditioning Engineers, Atlanta, GA

ASHRAE (2004) *Standard 55 – Thermal Environmental Conditions for Human Occupancy (ANSI Approved)*, American Society of Heating, Refrigerating and Air-Conditioning Engineers, Atlanta, GA

Atif, M. R. and Galasiu, A. D. (2003) 'Energy performance of daylight-linked automatic lighting control systems in large atrium spaces: Report on two field-monitored case studies', *Energy and Buildings*, vol 35, no 5, pp441–461

Baird, G. and Jackson, Q. (2004) 'Probe-style questionnaire surveys of building users – an international comparison of their application to large-scale passive and mixed-mode teaching and research facilities', in *Proceedings of SBSE Conference Closing The Loop: Post Occupancy Evaluation: The Next Steps*, Windsor, UK, Society of Building Science Educators, 29 April–2 May, CD-Rom

BFRL (Building and Fire Research Laboratory) (2007) *BEES – Building for Environmental and Economic Sustainability*, National Institute of Standards and Technology, Office of Applied Economics, www.bfrl.nist.gov/oae/software/bees/ (accessed 2 August 2008)

Boecker, J. (2005) *Going Green: Where to Find Green Product Information*, AIA Best Practices, http://soloso.aia.org/stellent/idcplg?IdcService=GET_FILE&dID=25505&Rendition=Web (accessed 2 August 2008)

Bordass, W. and Leaman, A. (2004) 'Probe: How it happened, what it found and did it get us anywhere?', in *Proceedings of SBSE Conference Closing The Loop: Post Occupancy Evaluation: The Next Steps*, Windsor, UK, Society of Building Science Educators, 29 April–2 May, CD-Rom

Buhagiar, V. M. (2004) 'A post-occupancy evaluation of manipulating historic built form to increase the potential of thermal mass in achieving thermal comfort in heavyweight buildings in a Mediterranean climate', in *Proceedings of SBSE Conference Closing The Loop: Post Occupancy Evaluation: The Next Steps*, Windsor, UK, Society of Building Science Educators, 29 April–2 May, CD-Rom

CABE (Commission for Architecture and the Built Environment) (2006) *Assessing Secondary School Design Quality*, research report, www.cabe.org.uk/AssetLibrary/8736.pdf (accessed 15 July 2008)

CABE (2007) *A Sense of Place – What Residents Think of Their New Homes*, research report, www.cabe.org.uk/AssetLibrary/10948.pdf (accessed 15 July 2008)

CIBSE (Chartered Institution of Building Services Engineers) (1994) *Code for Interior Lighting*, CIBSE, London

Coulter, J., Hannas, B., Swanson, C., Blasnik, M. and Calhoum, E. (2008) 'Measured public benefits from energy efficient homes', in *Proceedings of Indoor Air 2008 Conference*, Copenhagen, Denmark, 17–22 August

Daioumaru, K., Tanabe, S., Kitahara, T. and Yamamoto, Y. (2008) 'Thermal performance evaluation to double skin facade with vertical blinds', in *Proceedings of Indoor Air 2008 Conference*, Copenhagen, Denmark, 17–22 August

Davara, Y., Meir, I. A. and Schwartz, M. (2006) 'Architectural design and IEQ in an office complex', in E. de Oliveira Fernandes et al (eds), *Healthy Buildings: Creating a Healthy Environment for People, Proceedings of Healthy Building 2006 International Conference*, Lisbon, III, pp77–81

De Carli, M., De Giuli, V. and Zecchin, R. (2008) 'Review on visual comfort in office buildings and influence of daylight in productivity', in *Proceedings of Indoor Air 2008 Conference*, Copenhagen, Denmark, 17–22 August, Paper ID: 112

DKF (Donnell-Kay Foundation) (2005) *School Facility Assessments: State of Colorado*, www.dkfoundation.org/PDF/COSchoolFacilityAssessments-2005April.pdf (accessed 11 July 2008)

DOE (US Department of Energy) (2004) *20 River Terrace – The Solaire*, Buildings Database, http://eere.buildinggreen.com/overview.cfm?ProjectID=273 (accessed 2 August 2008)

Etzion, Y. (1994) 'A bio-climatic approach to desert architecture', *Arid Lands Newsletter*, vol 36, pp12–19

Etzion, Y., Meir, I. A., Pearlmutter, D. and Tene, M. (1993) 'Project monitoring in the Negev and the Arava, Israel', in *Solar Energy in Architecture and Urban Planning, Proceedings of the 3rd European Conference on Architecture*, Florence, Italy, May, Commission of the European Communities, H.S. Stephens and Associates, pp568–571

Etzion, Y., Portnov, B. A. and Erell, E. (2000a) 'A GIS framework for studying post-occupancy climate-related changes in residential neighbourhoods', in K. Steemers and S. Yannas (eds), *Architecture, City, Environment, Proceedings of the 17th PLEA International Conference*, James & James, London, pp678–683

Etzion, Y., Portnov, B. A., Erell, E., Meir, I. A. and Pearlmutter, D. (2000b) 'Climate-related changes in residential neighbourhoods: Analysis in a GIS framework', in K. Steemers and S. Yannas (eds), *Architecture, City, Environment, Proceedings of the 17th PLEA International Conference*, James & James, London, pp781–782

Etzion, Y., Portnov, B. A., Erell, E., Meir, I. A. and Pearlmutter, D. (2001) 'An open GIS framework for recording and analyzing post-occupancy changes in residential buildings: A climate related case study', *Building and Environment*, vol 36, no 10, pp1075–1090

Faruqui Ali, Z., Mallick, F. H., Ford, B. and Diaz, C. (1998) 'Climate, comfort and devotion. A study at the Krsna temple complex, Mayapur, India', in E. Maldonado and S. Yannas (eds), *Environmentally Friendly Cities, Proceedings of the PLEA98 International Conference*, Lisbon, Portugal, June, pp187–190

Frenkel, L., Fundaminsky, S., Meir, I. and Morhayim, L. (2006) 'Post-occupancy evaluation of a scientists village complex in the desert: Towards a comprehensive methodology', in R. Compagnon et al (eds), *Clever Design, Affordable Comfort:*

A Challenge for Low Energy Architecture and Urban Planning, Proceedings of the 23rd PLEA International Conference, Geneva, Switzerland, 6–8 September, pp907–912

Gale, S. F. (2008) *Built to Last: Measuring the Life Cycle of a Facility*, GreenerBuildings, www.greenerbuildings.com/feature/2008/04/10/built-last-measuring-life-cycle-a-facility (accessed 5 September 2008)

Genjo, K. S. M. and Hasegawa, K. (2006) 'Questionnaire survey on indoor climate and energy consumption for residential buildings related with lifestyle in cold climate area of Japan', in E. de Oliveira Fernandes et al (eds), *Healthy Buildings: Creating a Healthy Environment for People, Proceedings of Healthy Building 2006 International Conference*, Lisbon, vol III, pp355–360

Gordon, D. E. and Stubbs, S. (2004) *Building Performance: Where Do We Stand, and Where Are We Going?*, AIA Building Performance Committee Searches for Answers, The American Institute of Architects, www.aiahouston.org/cote/Building%20Performance%20-%20article.htm (accessed 5 September 2008)

Gossauer, E. A. W. (2005) *User Satisfaction at Workspaces: A Study in 12 Office Buildings in Germany*, CISBAT, Lausanne, Switzerland

Huberman, N. and Pearlmutter, D. (2008) 'A life-cycle energy analysis of building materials in the Negev desert', *Energy and Buildings*, vol 40, no 5, pp837–848

Hydes, K. P., McCarry, B., Mueller, T. and Hyde, R. (2004) 'Understanding our green buildings: Seven post-occupancy evaluations in British Columbia', in *Proceedings of SBSE Conference Closing The Loop: Post Occupancy Evaluation: The Next Steps*, Windsor, UK, Society of Building Science Educators, 29 April–2 May, CD-Rom

IAUA (Israel Association of United Architects) (1994) *Bylaws of Engineers and Architects*, IAUA, www.isra-arch.org.il/ (accessed 30 October 2008)

Ito, H., Yuming, W., Watanabe, S. and Tanabe, S. (2008) 'Field survey of visual comfort and energy efficiency in various office buildings utilizing daylight', in *Proceedings of Indoor Air 2008 Conference*, Copenhagen, Denmark, 17–22 August, Paper ID: 309

Kenda, B. (2006) 'Pneumatology in architecture: The ideal villa', in E. de Oliveira Fernandes et al (eds), *Healthy Buildings: Creating a Healthy Environment for People, Proceedings of Healthy Building 2006 International Conference*, Lisbon, vol III, pp71–75

Kosonen, R., Kajaala, M. and Takki, T. (2008) 'Perceived IEQ conditions: Why the actual percentage of dissatisfied persons is higher than standards indicate?', in *Proceedings of Indoor Air 2008 Conference*, Copenhagen, Denmark, 17–22 August, Paper ID: 861

Kowaltowski, D. C. C. K., Pina, S. A. M. G., Wilva, V. G. D., Labaki, L. C., Ruschel, R. C. and Moreira, D. C. (2004) 'From post-occupancy to design evaluation: Site-planning guidelines for low income housing in the state of São Paulo,' in *Proceedings of SBSE Conference Closing The Loop: Post Occupancy Evaluation: The Next Steps*, Windsor, UK, Society of Building Science Educators, 29 April–2 May, CD-Rom

Langston, C., Song, Y. and Purdy, B. (2008) 'Perceived conditions of workers in different organizational settings', *Facilities*, vol 26, no 1/2, pp54–67

Leaman, A. and Bordass, B. (1999) 'Productivity in buildings: The "killer" variables', *Building Research and Information*, vol 27, no 1, pp4–19

LEED Policy Manual (2006) *Leadership in Energy and Environmental Design*, www.usgbc.org/ShowFile.aspx?DocumentID =2039 (accessed 18 March 2008)

Levin, H. (2005) 'Integrating indoor air and design for sustainability', in *Proceedings of the 10th International Conference on Indoor Air Quality and Climate*, Beijing, China, 4–9 September, item 1.7-34

Lighthall, C., Carruthers, W. and Zulli, R. A. (2006) *Renovation Impact on Student Success*, Wake County Public School System, North Carolina

Loftness, V., Hartkopf, V., Poh, L. K., Snyder, M., Hua, Y., Gu, Y., Choi, J. and Yang, X. (2006) 'Sustainability and health are integral goals for the built environment', in E. de Oliveira Fernandes et al (eds), *Healthy Buildings: Creating a Healthy Environment for People, Proceedings of HB2006 International Conference*, Lisbon, Portugal, vol I, Plenary lectures, pp1–10

Mahdavi, A. and Proeglhoef, C. (2008) 'Observation-based models of user control actions in buildings', in *Proceedings of PLEA – Passive and Low Energy Architecture 2008 Conference,* University College Dublin, Ireland, 22–24 October, (electronic version only) paper 169

Mahdavi, A., Kabir, E., Mohammadi, A. and Lambeva, L. (2008) 'Occupants' evaluation of indoor climate and environmental control systems in office buildings', in *Proceedings of Indoor Air 2008 Conference*, Copenhagen, Denmark, 17–22 August

Mallick, F. H. (1996) 'Thermal comfort and building design in the tropical climates', *Energy and Buildings*, vol 23, no 3, pp161–167

Marmont, A. (2004) 'City Hall, London: Evaluating an icon', in *Proceedings of SBSE Conference Closing The Loop: Post Occupancy Evaluation: The Next Steps*, Windsor, UK, Society of Building Science Educators, 29 April–2 May, CD-Rom

McMullen, S. (2007) 'Libraries in transition: Evolving the information ecology of the Learning Commons: A sabbatical report'. *Librarian Publications,* Paper 10, http://docs.rwu.edu/librarypub/10 (accessed 2 August 2008)

Meir, I. A. (1990) 'Monitoring two kibbutz houses in the Negev Desert', *Building and Environment*, vol 25, no 2, pp189–194

Meir, I. A. (1998) 'Bioclimatic desert house: A critical view', in *Environmentally Friendly Cities, Proceedings of the 15th PLEA International Conference*, Lisbon, Portugal, 1–3 June, James & James, London, pp245–248

Meir, I. A. (2000a) 'Integrative approach to the design of sustainable desert architecture: A case study', *Open House International*, vol 25, no 3, pp47–57

Meir, I. A. (2000b) 'Courtyard microclimate: A hot arid region case study', in K. Steemers and S. Yannas (eds), *Architecture City Environment, Proceedings of the 17th PLEA International Conference*, Cambridge, James & James, London, pp218–223

Meir, I. A. (2008) 'Apology for architecture', in S. Roaf and A. Bairstow (eds), *The Oxford Conference: A Re-evaluation of Education in Architecture*, WIT Press, Southampton, Boston, pp33–36 (invited presentation)

Meir, I. A. and Hare, S. (2004) 'Where did we go wrong? POE of some bioclimatic projects, Israel 2004', in *Proceedings of SBSE Conference Closing The Loop: Post Occupancy Evaluation: The Next Steps*, Windsor, UK, Society of Building Science Educators, 29 April–2 May, CD-Rom

Meir, I. A., Pearlmutter, D. and Etzion, Y. (1995) 'On the microclimatic behavior of two semi-enclosed attached courtyards in a hot dry region', *Building and Environment*, vol 30, no 4, pp563–572

Meir, I. A., Motzafi-Haller, W., Krüger, E. L., Morhayim, L., Fundaminsky, S. and Oshry-Frenkel, L. (2007) 'Towards a comprehensive methodology for Post Occupancy Evaluation (POE): A hot dry climate case study', (keynote presentation) in M. Santamouris and P. Wouters (eds), *Building Low Energy Cooling and Advanced Ventilation in the 21st Century, Proceedings of the 2nd PALENC and 28th AIVC Conference*, Crete, II, 27–29 September 2007, pp644–653

Menzies, G. F. and Wherrette, J. R. (2005) 'Windows in the workplace: Examining issues of environmental sustainability and occupant comfort in the selection of multi-glazed windows', *Energy and Buildings*, vol 37, no 11, pp623–630

Mochizuki, E., Watanabe, S., Kobayashi, K., Wei, Y., Tanabe, S., Takai, H. and Shiratori. Y. (2006) 'Field measurement on visual environment in office building daylight from light-well in Japan', in E. de Oliveira Fernandes et al (eds), *Healthy Buildings: Creating a Healthy Environment for People, Proceedings of HB2006 International Conference*, Lisbon, vol II, pp201–206

Morhayim, L. and Meir, I. (2008) 'Survey of an office and laboratory university building: An unhealthy building case study', in *Proceedings of Indoor Air 2008 Conference*, Copenhagen, Denmark, 17–22 August, Paper ID: 933 (electronic proceedings version only)

Nakamura, S., Tanabe, S. I., Nishihara, N. and Haneda, M. (2008) 'The evaluation of productivity and energy consumption in 28 degrees office with several cooling methods for workers', in *Proceedings of Indoor Air 2008 Conference*, Copenhagen, Denmark, 17–22 August, Paper ID: 129 (electronic proceedings version only)

Nicol, J. F. and Humphreys, M. A. (2002) 'Adaptive thermal comfort and sustainable thermal standards for buildings', *Energy and Buildings*, vol 34, no 6, pp563–572

Nordberg, M. (2008) 'Thermal comfort and indoor air quality when building low-energy houses', *Proceedings of Indoor Air 2008 Conference*, Copenhagen, Denmark, 17–22 August

Ochoa, C. E. and Capeluto, I. G. (2006) 'Evaluating visual comfort and performance of three natural lighting systems for deep office buildings in highly luminous climates', *Building and Environment*, vol 41, no 8, pp1128–1135

PassivHaus (2008) www.passivhaustagung.de/Passive_House_E/passivehouse.html

Pati, D. and Augenbroe, G. (2006) 'Modeling relative influence of environmental and sociocultural factors on context-specific functions,' *International Journal of Physical Sciences*, vol 1, no 3, pp154–162.

Patricio, J., Santos, A. and Matias, L. (2006) 'Double-skin facades: Acoustic, visual and thermal comfort indoors', in E. de Oliveira Fernandes et al (eds), *Healthy Buildings: Creating a Healthy Environment for People, Proceedings of Healthy Building 2006 International Conference*, Lisbon, vol II, pp37–42

Pearlmutter, D. and Meir, I. A. (1995) 'Assessing the climatic implications of lightweight housing in a peripheral arid region', *Building and Environment*, vol 30, no 3, pp441–451

Pearlmutter, D. and Meir, I. A. (1998) 'Lightweight housing in the arid periphery: Implications for thermal comfort and energy use', in H. J. Bruins et al (eds), *The Arid Frontier: Interactive Management of Environment and Development*, Kluwer Academic Publishers, Dordrecht, pp365–381

Pearlmutter, D., Etzion, Y., Erell, E., Meir, I. A. and Di, H. (1996) 'Refining the use of evaporation in an experimental down-draft cool tower', *Energy and Buildings*, vol 23, no 3, pp191–197

Pearson, D. (1989) *The Natural House Book*, Conran Octopus, London, UK

Pfafferott, J., Herkel, S., Kalz, D. and Zeuschner, A. (2007) 'Comparison of low-energy office buildings in summer using different thermal comfort criteria', *Energy and Buildings*, vol 39, no 7, pp750–757

Pitts, A. E. and Douvlou-Beggiora, E. (2004) 'Post-Occupancy Analysis of Comfort in Glazed Atrium Spaces', in *Proceedings of SBSE Conference Closing The Loop: Post Occupancy Evaluation: The Next Steps*, Windsor, UK, Society of Building Science Educators, 29 April–2 May, CD-Rom

Preiser, W. F. E. (1995) 'Post-occupancy evaluation: How to make buildings work better', *Facilities*, vol 13, no 11, pp19–28

Preiser, W. F. E. (2004) 'Evaluating Peter Eisenman's Aronoff Center: De-bunked de-constructivism', in *Proceedings of SBSE Conference Closing The Loop: Post Occupancy Evaluation: The Next Steps*, Windsor, UK, Society of Building Science Educators, 29 April–2 May, CD-Rom

Preiser, W. F. E. and Vischer, J. (eds) (2005) *Assessing Building Performance*, Butterworth-Heinemann, Oxford

Roaf, S. (2004) 'Cave Canem: Will the EU Building Directive bite?', in *Proceedings of SBSE Conference Closing The Loop: Post Occupancy Evaluation: The Next Steps*, Windsor, UK, Society of Building Science Educators, 29 April–2 May, CD-Rom

Roaf, S. with Horsley, A. and Gupta, R. (2004) *Closing the Loop. Benchmarks for Sustainable Buildings*, RIBA Enterprises, London

Roaf, S., Crichton, D. and Nicol, F. (2005) *Adapting Buildings and Cities for Climate Change: A 21st Century Survival Guide*, Elsevier Architectural Press, Oxford

Roulet, C. A., Foradini, F., Cox, C., Maroni, M. and de Oliveira Fernandez, E. (2005) 'Creating healthy and energy-efficient buildings: Lessons learned from the HOPE project', in *Proceedings of the 10th International Conference on Indoor Air Quality and Climate*, Beijing, China, 4–9 September

Roulet, C. A., Bluyssen, P. M., Cox, C. and Foradin, F. (2006) 'Relations between perceived indoor environment characteristics and well-being of occupants at individual level', in E. de Oliveira Fernandes et al (eds), *Healthy Buildings: Creating a Healthy Environment for People, Proceedings of Healthy Building 2006 International Conference*, Lisbon, vol III, pp163–168

Sanoff, H. (2002) *Schools Designed with Community Participation*, National Clearinghouse for Educational Facilities, www.edfacilities.org/pubs/sanoffschools.pdf (accessed 25 August 2008)

Silva, F. M., Duarte, R. and Cunha, L. (2006) 'Monitoring of a double skin facade building: Methodology and office thermal and energy performance', in E. de Oliveira Fernandes et al (eds), *Healthy Buildings: Creating a Healthy Environment for People, Proceedings of Healthy Building 2006 International Conference*, Lisbon

Spengler, J. D., McCarthy, J. F. and Samet, J. M. (2000) *Indoor Air Quality Handbook*, McGraw-Hill, New York

Stevenson, F. (2004) 'Post-occupancy – squaring the circle: A case study on innovative social housing in Aberdeenshire, Scotland', in *Proceedings of SBSE Conference Closing The Loop: Post Occupancy Evaluation: The Next Steps*, Windsor, UK, Society of Building Science Educators, 29 April–2 May, CD-Rom

USGBC (United States Green Buildings Council) (2002) *Leadership in Energy and Environmental Design (LEED) Green Building Rating System for New Construction and Major Renovation (LEED-NC)*

Vainer, S. and Meir, I. A. (2005) 'Architects, clients and bioclimatic design: A first POE of a solar neighborhood', in M. Santamouris (ed), *Passive and Low Energy Cooling for the Built Environment, Proceedings of the PALENC 2005 International Conference*, May, Santorini, II, pp1059–1064

Wagner, A., Gossauer, E., Moosmann, C., Gropp, T. and Leonhart, R. (2007) 'Thermal comfort and workplace occupant satisfaction: Results of field studies in German low energy office buildings', *Energy and Buildings*, vol 39, no 7, pp758–769

Wargocki, P., Wyon, D. P., Baik, Y. K., Clausen, G. and Fanger, P. O. (1999) 'Perceived air quality, sick building syndrome symptoms and productivity in an office with two different pollution loads', *Indoor Air*, vol 99, no 3, pp165–179

Watson, C. (2003) 'Review of building quality using post occupancy evaluation', *Journal of the Programme on Educational Building*, OECD, www.postoccupancyevaluation.com/publications/pdfs/POE%20OECD%20V4.pdf (accessed 15 July 2008)

Watson, C. (2005) *Post Occupancy Evaluation – Braes High School, Falkirk*, Scottish Executive, www.scotland.gov.uk/Publications/2006/01/23112827/11 (accessed 15 July 2008)

Wilson, A. and Austin, B. (2004) 'Post occupancy evaluation case study advanced naturally ventilated office', in *Proceedings of SBSE Conference Closing The Loop: Post Occupancy Evaluation: The Next Steps*, Windsor, UK, Society of Building Science Educators, 29 April–2 May, CD-Rom

Woollett, S. and Ford, A. (2004) 'How happy are we? Our Experience of conducting an occupancy survey', in *Proceedings of SBSE Conference Closing The Loop: Post Occupancy Evaluation: The Next Steps*, Windsor, UK, Society of Building Science Educators, 29 April–2 May, CD-Rom

Xiong, Y. (2007) 'The impact of exterior environmental comfort on residential behaviour from the insight of building energy conservation: A case study on Lower Ngau Tau Kok estate in Hong Kong', in *Building Low Energy Cooling and Advanced Ventilation in the 21st Century, Proceedings of the 2nd PALENC and 28th AIVC Conference*, Crete, 27–29 September, vol 2, pp1141–1145

Zagreus, L., Huizenga, C., Arens, E. and Lehrer, D. (2004) 'Listening to the occupants: A web-based indoor environmental quality survey', *Indoor Air*, vol 14 (Supplement 8), pp65–74

Zeiler, W. and Boxem, G. (2008) 'Sustainable schools: Better than traditional schools?', in *Proceedings of Indoor Air 2008 Conference*, Copenhagen, Denmark, 17–22 August, Paper ID: 10 (electronic proceedings version only)

Zeiler, W., Savanovic, P. and Boxem, G. (2008) 'Design decision support for the conceptual phase of sustainable building design', in B. W. Olesen et al (eds), *The 11th International Conference on Indoor Air Quality and Climate*, Copenhagen, Technical University of Denmark, vol CD, pp1–8

8

Guidelines to Avoid Mould Growth in Buildings

H. Altamirano-Medina, M. Davies, I. Ridley, D. Mumovic and T. Oreszczyn

Abstract

There is now widespread acceptance that mould growth in buildings should be avoided as it may lead to adverse health effects. Consequently, it is critically important to have appropriate guidelines that address this issue. As well as reviewing the existing literature with regard to the state of the art of relevant mould-related research, this chapter reports on work aimed at developing mould-related Building Regulation guidance for dwellings in England and Wales. The major findings are, first, that although the factors that influence mould growth are well known, in buildings the variation and interrelationships of and between those are complex and, second, to deal with this complexity there is a fundamental choice between setting specific moisture performance criteria or using a 'mould model' to demonstrate compliance with regulations. At present, for England and Wales, the setting of moisture criteria is preferable and this chapter makes relevant suggestions for such criteria.

■ *Keywords* – mould growth; relative humidity; dwellings; guidelines

INTRODUCTION

A recent report from the World Health Organization (WHO, 2007) states that mould growth on interior surfaces and in building structures should be prevented as it may lead to adverse health effects. This chapter addresses the state of the art with respect to this issue and suggests relevant practical guidelines.

The initial sections of this report examine the body of literature that investigates the relationships between mould growth and environmental conditions in dwellings. While not intended to be exhaustive, the review demonstrates the extensive work that has been undertaken in this area and describes the main findings to date. From a methodological point of view, the studies can be divided into three major groups as follows:

ADVANCES IN BUILDING ENERGY RESEARCH ■ 2009 ■ VOLUME 3 ■ PAGES 221–236

doi:10.3763/aber.2009.0308 ■ © 2009 Earthscan ■ ISSN 1751-2549 (Print), 1756-2201 (Online) ■ www.earthscanjournals.com

- *Studies based on monitored data from real dwellings.* This section contains studies based on physical surveys in real dwellings that provide a valuable insight into the parameters such as moisture generation, ventilation and internal temperature that may lead to mould growth.
- *Studies based on laboratory experiments.* This section focuses on laboratory experiments that aim to develop a better understanding of the factors that control mould development. This section highlights how those factors determine germination times and growth rates under steady-state and transient conditions.
- *Studies based on numerical modelling.* This section presents studies related to the implementation of modelling tools for the prediction of mould growth in buildings.

Finally, in the last section, a study tasked with forming specific regulatory guidance for dwellings in England and Wales is described and its recommendations noted.

STUDIES BASED ON MONITORED DATA IN REAL DWELLINGS

Designers of new dwellings and those refurbishing existing properties require some mechanism for assessing the impact of design changes on the risk of mould growth. However, the problems of mould growth in dwellings are not purely technical but are also related to occupant behaviour. Research into a number of parameters affecting mould growth in dwellings over the past 30 years has led to a substantial improvement in the understanding of the problem. A selection of the relevant work is described in this section.

A study carried out in Israel (Becker, 1984) involved the use of surveys and interviews in 200 dwellings. It was noted that out of 28 cases of severe mould contamination, 76 per cent occurred on walls in 'dry' rooms (living rooms and bedrooms) and 100 per cent on thermal bridges. It was found that 100 per cent of severe cases occurred in dry rooms, whereas only 65 per cent occurred in wet rooms. In addition, the study concluded that the major factors affecting the extent and severity of mould growth were location and orientation of dwellings, occupancy density, cooking habits and type of wall covering.

Woolliscroft (1997) stated that the high level of condensation and mould in the UK is the consequence of the small size of the dwellings, low temperatures, high humidity of the incoming air, and high occupancy of dwellings. 35 per cent of dwellings were affected by condensation and 17 per cent by mould growth.

An analysis of the 2001 *English House Condition Survey* database (ODPM, 2003) was undertaken to investigate the sensitivity of mould growth to occupancy levels (Ridley et al, 2005). The analysis suggests that mould growth is related to the number of occupants in the dwelling. As household size and occupant density increase, the prevalence of mould increases.

The influence of house characteristics on mould growth was addressed by testing 59 houses in Canada (Lawton et al, 1998). All the houses were subject to a detailed building survey including air infiltration testing, carbon dioxide and volatile organic compound (VOC) measurements, temperature and relative humidity (RH) monitoring. The air change rates were determined using the tracer gas decay method. It was found that low air leakage and natural ventilation were not associated with higher levels of mould growth as measured by ergosterol concentrations. Furthermore, the analysis showed that specific moisture sources such as condensation on external walls and ceilings of bathrooms, floor wicking, driving rain and damp because of plumbing accidents were a more significant factor for mould growth than the relative humidity.

A sample of 33 Belgium dwellings built between 1972 and 2002 was re-examined to evaluate the importance of nine different factors on mould growth:

1 external temperature and relative humidity;
2 internal temperature;
3 moisture generation;
4 ventilation;
5 thermal performance of building envelope;
6 architectural layout of building;
7 moisture buffer capacity of walls;
8 wall finishing;
9 surfaces with high condensation risk (Hens, 2005).

Although it was not possible to assess all the factors systematically, the low internal temperature, moisture generation underpinned by rain penetration and built-in moisture, poor ventilation, the layout of the room, and the thermal performance of the envelope (including thermal bridges) were noted to be significantly important.

The occurrence of mould and a number of housing characteristics in 186 dwellings in The Netherlands (Ginkel and Hasselaar, 2005) was investigated. The number of showers (more than 14 per week) taken by the occupants and the age of the ventilation box (more than 6 years) were found to be the most important predictors of mould growth.

Analysis based on detailed measurements of indoor temperatures, relative humidity and mould in low-income households undertaken for England's Home Energy Efficiency Scheme (known as 'Warm Front') showed that the risk of mould growth increases above values of standardized RH of around 45 per cent (standardized for external reference conditions of temperature (5°C) and RH (80 per cent)) (Oreszczyn et al, 2006). The study included 1604 dwellings undergoing Warm Front improvements over the winters of 2001–2002 and 2002–2003 in five urban areas of England: Birmingham, Liverpool, Manchester, Newcastle and Southampton. Figure 8.1 shows a very small risk of mould even in dwellings with standardized RH below 40 per cent, but above this there was a clear gradient of increasing risk, reaching, at 80 per cent standardized RH, around 40 per cent risk of having an MSI (mould severity index) greater than 1 (i.e. mould is reported in at least one room). The fact that a small proportion of homes appear to have mould even at very low standardized RH is most likely to be attributable to mould occurring in localized areas of microclimate.

Recently the author assessed vapour pressure excess as an indicator of mould growth. 15 apartments (20 rooms) were monitored during the heating season of 2006 and 2007. Mould was found in 78 per cent of flats. A strong correlation was found between mould occurrence and standardized vapour pressure excess (SVPX) (standardized for external reference conditions of temperature (5°C) and RH (80 per cent)). In addition, the

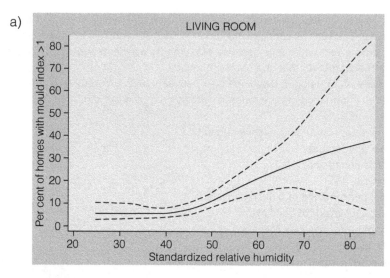

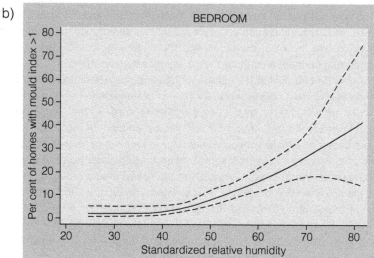

FIGURE 8.1 Risk of mould – standardized relative humidity

data were analysed according to methods suggested by Janssens and Hens (2003), the British Standard BS5250 and an alternative humidity classification based on UK dwellings (Ridley et al, 2007). It was found that almost 100 per cent of rooms affected with mould were on the higher levels of humidity regarding each categorization system.

The studies described above provide useful information regarding the conditions required for mould growth. However, such work needs to be combined with information obtained from the rather more controlled environment of the laboratory. The next section deals with a series of relevant studies.

STUDIES BASED ON LABORATORY EXPERIMENTS

Relevant work in this area is not new. For example, Tomkins (1929) focused his studies on the time taken for spores to germinate under specific conditions of humidity, temperature and access to nutrients. Similarly, other researchers centred their attention on the effect of temperature and moisture on the growth of individual species (Tyler, 1948; Ayerst 1969). From such work, isopleth diagrams (to describe either growth rates or the germination time of specific species) were determined under steady-state conditions, i.e. constant relative humidity and temperature (Figure 8.2).

The moisture requirement for mould growth was studied by Grant et al (1989). He found that the lowest water activity level recorded for growth on malt extra agar was 0.76 (76 per cent RH), while for building materials such as painted woodchip wallpaper it was 0.79. He also observed that increasing the temperature and the amount of nutrients led to a reduction in the water activity required for mould growth but, what is more, that if mould growth is to be avoided, water activity in building materials should be maintained below 0.80.

Pasanen et al (1991) carried out several experiments in order to investigate the relationship between fungal growth and temperature and humidity. Two common fungi were analysed and exposed to a range of temperatures (4–30°C) and relative humidities (11–96 per cent). It was reported that fungal growth started shortly after exposure to favourable conditions. In addition, it was found that water on surfaces has a more direct effect on mould growth than air relative humidity, since mould will grow if there is enough water on surfaces even under very low levels of air humidity.

The effect of transient boundary conditions on mould growth was studied by Adan (1994). Various experiments were performed in order to establish the fungal defacement of interior finishes such as plain and coated gypsum. Adan introduced the term 'time of

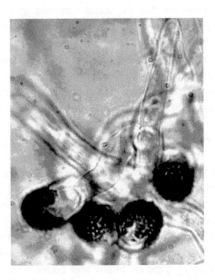

FIGURE 8.2 Spore germination of *Aspergillus niger*

wetness' (TOW), which represented the water availability to fungal growth under transient conditions. TOW was defined as the ratio between the wet period (RH >80 per cent) and the total period. It was concluded that on these substrates, the growth rate increases as the TOW increases; and for values above TOW = 0.5, mould should grow faster and more abundantly. In addition, the effect of the frequency of RH changes was studied, concluding that exposure to a sequence of short, high RH periods with the same TOW have no effect on the resulting mould growth unless a very high frequency of RH is applied, which normally does not occur indoors.

Zöld (1990) investigated the number of daily hours when mould starts to grow at temperatures below 20°C and relative humidities of more than 75 per cent. A range is regarded as safe, if, over a long period, the relative humidity of 75 per cent is not exceeded more than 8 to 12 hours per day, and if the limit of 75 per cent relative humidity is not exceeded more than 12 hours on three successive days. A state is described as critical when this limit is exceeded over a period of more than 12 hours on five successive days. As part of the findings from a large study, Rowan et al (1999) recommend maintaining surface RH below 75 per cent to limit fungal growth in buildings.

Deacon (1997) listed all parameters of significant importance for mould growth. Apart from factors such as humidity, temperature and the substrate (nutrients), he characterized time as an important factor. Light and oxygen were found to be less important.

The effect of transient climatic conditions on the growth of mould fungi was studied by varying air humidity, air temperature, air speed, surface humidity and temperature (Gertis et al, 1999). Building materials such as plasters, wallpapers and dispersion paints were tested. The surface temperatures used were 14°C and 18.5°C. The growth intensity of the mould fungus was evaluated according to a five-stage classification. Here the areas infested with mould fungus on the material surface were assessed with regard to the growth intensity by means of a microscope, without determining the respective individual fungus species. It was observed that on some materials after a period of 6 weeks at 18.5°C and a humidity load of 95 per cent for 6 hours a day, mould fungus infestation occurs with and without heavy contamination. With the daily humidity load being reduced to 1 or 3 hours, fungus growth takes place only with existing contamination.

Mould growth under fluctuating moisture and temperature conditions was investigated using gypsum board, particleboard and wood board (Pasanen et al, 2000). The materials were taken from buildings under repair and no additional inoculation of fungi was undertaken. The materials were exposed to four different environmental conditions each 4–8 weeks long: capillary absorption of water, drying at 30 per cent RH, condensation and drying at 50 per cent RH. The study showed that capillary absorption of water in selected building materials resulted in rapid mould growth.

The influence of relative humidity and temperature on mould growth was investigated using eight different mould species on 21 different types of building material (Nielsen et al, 2004). Fungi were incubated at four different temperatures (5, 10, 20 and 25°C) and three humidity levels in the range 69–95 per cent over a period of 4 months. It was shown that wood and wood-based materials exposed to 78 per cent RH at 20–25°C could trigger mould growth. In the case of gypsum board and ceramic material, the values of RH required for growth exceeded 86 per cent.

The main findings of the experiments conducted as a part of the Danish 'Mould in Buildings' research programme (Rode, 2005) showed that wood, wallpapers and materials containing starch have the lowest resistance to mould growth. Plaster and concrete, which have a very modest content of organic material, can be infested with mould growth only if RH is in the region of 95 per cent. It was shown that higher temperatures, up to 28°C, and dust on rough surfaces facilitate mould growth. It was noted that mould growth rates change under different transient conditions.

The value information obtained from the field and laboratory studies described in the previous two sections is enhanced by its incorporation into models of mould germination and growth. The following section describes relevant work in this area.

STUDIES BASED ON NUMERICAL MODELLING

A mathematical model of mould growth on wooden materials was developed in 1996 in the Technical Research Centre of Finland (VTT). The model describes the material response and the critical conditions for mould growth on surfaces of pine and spruce sapwood, taking into account variable humidity conditions. In addition, the development of mould growth expressed via a mould index (MI) was proposed (Hukka and Viitanen, 1999). This model is being improved in an ongoing project carried out by the Tampere University of Technology, Finland (TUT) in collaboration with VTT. The project involves the modelling of moisture behaviour and mould growth in building envelopes, which include both laboratory and field work. It considers a wider range of climatic conditions and a greater variety of construction materials such as: edge glued spruce board, polyurethane (paper-coated and grounded), glass wood, expanded polystyrene, polyester wool, concrete, autoclaved aerated concrete and expanded light aggregated concrete. Initial results from this project, which is expected to be finished during 2009, have shown that mould was found growing in every material tested, but in smaller quantities than those found in the original model, which was based on wood materials only (Lähdesmäki et al, 2008).

A mould growth model based on six generic mould categories was formulated and incorporated within the ESP-r system for building energy and environmental simulation (Clarke et al, 1999). The principal mould species affecting UK dwellings were identified and their minimum growth requirements established based on a literature review. The moulds were assigned to one of six categories ranging from xerophilic to hydrophilic. The model is, however, unable to indicate the effects of temperature and relative humidity *fluctuations* on mould growth.

The Condensation Targeter II model allows the impact of key variables associated with mould growth, such as fabric type, ventilation, heating system, occupant fuel affordability and occupant density, to be assessed in order that the impact of design and policy decisions can be tested (Oreszczyn and Pretlove, 1999a). Condensation Targeter II is a monthly steady-state model, based on the Bredem 8 algorithm, that predicts the monthly mean indoor air temperature and relative humidity in two zones, zone 1 being the living room area and zone 2 being the rest of the dwelling (including the bedrooms). The model had been tested and comparison between modelled and measured data for 36 dwellings showed good agreement (within 10 per cent). Further analysis highlighted the variation of vapour pressure excess with external temperatures. Further studies are needed to investigate the cause of this variation including the extent to which occupants change ventilation and moisture production as a function of external temperature, and the possible role of moisture buffering of surface finishes on this variation. Another version of the tool included an MI that measures the risk of mould occurring on the coldest surfaces within the dwelling each month of the year (Oreszczyn and Pretlove, 1999b).

The Biohygrothermal model 'WUFI bio' (Sedlbauer, 2001), predicts the formation of mould based on the comparison of 'favourable' relative humidities, temperatures and substrates for mould growth and transient environmental conditions that occur in buildings. In the model, the spore itself is treated as an additional layer on the wall. As this assumption leads to a non-realistic diffusion resistance for the wall, the calculations for the 'biological' layer are done separately. The inner surface temperature and humidity of the building wall determined by WUFI serve as boundary conditions on both sides of the 'biological layer', i.e. spore. Note that the biological processes, which affect the moisture balance of the spore, were not taken into account in this model. It has been assumed that the spores begin to germinate only if the moisture content exceeds a certain minimum level. In the model, four categories of substrate are considered and the length of time in days after which germination occurs and the growth rate in mm per day is predicted.

WUFI bio was applied in order to assess mould growth at corners, middle of walls and behind furniture (Sedlbauer, 2002). It was found that the locally prevailing conditions (lower temperature and higher humidity values) result in fast germination of spores. Similarly, the impact of furniture layout on the risk of condensation and mould growth was investigated in another study by varying the convective heat transfer coefficient (Oreszczyn et al, 2005a). It was shown that the predicted effect of furniture placed close to an external wall could increase the surface RH by the order of 10 per cent.

Another modelling study investigated the importance of fluctuating environmental conditions on the assessment of condensation risk and mould growth in dwellings (Mumovic et al, 2006). Numerical results of the condensation risk calculation under both steady-state and transient conditions were compared and evaluated. In addition, the effect of different moisture generation, ventilation rates and the thermal mass of the wall on the predicted surface relative humidity was studied for a number of RCDs (robust construction details) (DTLR, 2002). Significant differences were apparent between the predictions of the simple (steady-state) and complex (transient) methods for all construction details modelled.

In an attempt to capture some of the complexity involved with the prediction of mould, a nomogram was developed (Oreszczyn et al, 2005b) using Condensation Targeter to quantify the effect of workmanship on thermal performance of both 'standard' and 'as built' RCDs. The fitted curves, linking the surface RH with the air change rate, moisture generation and temperature factors were derived using the modelling result from 54 simulations (Figure 8.3).

An attempt to explain unexpected mould growth in houses that would typically not show up in a standard deterministic simulation was based on a mixed simulation approach (Moon and Augenbroe, 2003). By introducing uncertainties in relevant input parameters (air infiltration, temperature factor, etc.), this approach generates a statistical

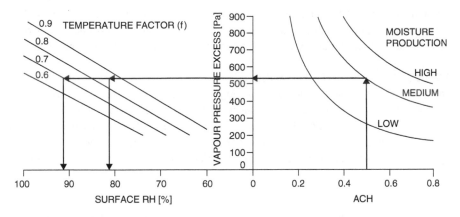

FIGURE 8.3 Condensation risk nomogram; it shows the complex relationship between surface RH and the air change rate, moisture generation and temperature factors. Note that this diagram is derived using the mean January temperature for the London region (3.7°C)

distribution of time-aggregated mould growth conditions at a number of 'trouble spots' in a specific building case. This distribution was then translated into an overall mould risk indicator. Note that accuracy of the mould risk indicator results depends on the accuracy of a large number of models – whole building energy and moisture models, hygrothermal building envelope models, thermal bridge assessment models and computational fluid dynamics (CFD). The authors selected the following as the top five dominant parameters: air mass flow coefficient, moisture source, convective heat transfer coefficient, wind exponent and insulation conductivity.

It is clear then that much work has been undertaken to date with regard to the conditions that are required for mould germination and growth to occur in buildings. The challenge is to incorporate the findings of this extensive body of work in the formation of meaningful and usable guidance to prevent mould growth. The next section deals with this issue.

DETERMINATION OF MOISTURE PERFORMANCE STANDARDS IN DWELLINGS – AN INTEGRATED APPROACH

Building on the extensive body of existing work that has been outlined in the previous sections, the authors of this chapter have investigated the extent to which the existing performance criteria for the control of mould introduced as part of the ventilation regulation for England and Wales (HMSO, 2006) is the most appropriate criteria for UK dwellings. The work undertaken adopted an integrated approach where laboratory work, modelling and field work were used in both the testing of the existing guidelines and developmental work for a suggested revision.

The current criterion noted in Approved Document F (ADF) (HMSO, 2006) is that there should be no visible mould growth on external walls. The document states that this criterion will be met if 'the relative humidity (RH) in a room does not exceed 70 per cent for more than 2 hours in any 12 hour period, and does not exceed 90 per cent for more than 1 hour in any 12 hour period during the heating season'. In the study, a literal interpretation of the guideline was first assumed – if the criterion is exceeded once in a heating season, then the system is deemed to have failed to demonstrate compliance.

TESTING THE CURRENT GUIDANCE

Initially an experiment was carried out involving the use of environmental chambers where repeating transient profiles of temperature and RH were constructed. A variety of representative mould species (Rowan et al, 1996) – *Aspergillus repens* and *versicolor, Penicillium chrysogenum, Cladosporium sphaerospermum* and *Ulocladium consortiale* (Figure 8.4) – were then exposed to these profiles. In order to test the current ADF

Ulocladium consortiale

Penicillium chrysogenum

Aspergillus repens

Aspergillus versicolor

Cladosporium sphaerospermum

Aspergillus niger

FIGURE 8.4 Representative moulds

guidelines, an attempt was made to grow the mould species in the environmental chambers under typical conditions which an external wall may be exposed to if the air conditions specified by ADF for mould growth are met. Surface relative humidities were calculated with standard psychrometrics using both the surface temperature calculated by a multidimensional steady-state computer thermal model (TRISCO) at a wall/ceiling junction and 12 hour air RH profiles with more than 2 hours of RH above 70 per cent.

Subsequently, the proposed guideline (ADF) was correlated with field data in order to determine the number of dwellings complying with it. Field data from the Warm Front project data set (1498 dwellings), which represents the current housing stock regarding comparison with the 2001 *English House Condition Survey* (EHCS), and the Milton Keynes Energy Park data set (18 dwellings) (MKEP), which have energy efficiency features incorporated in their design and constructions, were used in the analysis.

The results of the test did not support the current guidelines. In the experiments neither mould germination was observed on agar samples nor on building material (plaster board) exposed to the RH profile created over a period of almost a month, even though the surface RH results from the analysis were higher than the 10 per cent elevation suggested by BS5250 (BSI, 2005) and hence more likely to result in mould growth. In the case of correlation with field data, it was found that a significant proportion of buildings fail the ADF guidance, albeit many do not have mould growth. In Warm Front, 35 per cent of dwellings do not meet the ADF criterion for an RH of 70 per cent (more than 2 hours in a 12 hour period), while 3 per cent of them do not meet the ADF criterion for an RH of 90 per cent (more than 1 hour in a 12 hour period). For MKEP, 78 per cent of dwellings do not meet the ADF criterion for an RH of 70 per cent (more than 2 hours in a 12 hour period) and 11 per cent of dwellings do not meet the ADF criterion for an RH of 90 per cent (more than 1 hour in a 12 hour period).

SUGGESTED REVISION

The authors concluded that the current guidance requires amendment. In order to deal with the complexity demonstrated in the previous sections of this chapter, there is a fundamental choice between setting specific moisture performance criteria or using a 'mould model' (of the type described in a previous section in this chapter) to demonstrate compliance with regulations. For the 2010 revision of ADF, the introduction of a modelling based approach was not feasible (it may be for later revisions) and so the work focused on establishing the most appropriate performance criteria.

It was suggested that the surface RH guideline suggested by IEA Annex 14 (IEA Annex 14, 1990) was the most suitable basis for a potential change, in combination with a minimum thermal quality (i.e. minimum temperature factors) of the building envelope as set down in other parts of the regulations (HMSO, 2006).

The Annex 14 criteria (Table 8.1) acknowledge the water activity of any substrate as being of fundamental importance in determining whether mould growth will occur.

The Annex 14 work, however, was published in 1990 and thus the University College London (UCL) team compared the Annex 14 recommendations with that of more recent work. The Annex 14 criteria are not as strict as the isopleths reported by Sedlbauer (2001) where, for example, on substrate class 1 (i.e. most building surfaces) and a surface

TABLE 8.1 IEA Annex 14 criteria – surface water activity required for mould growth

PERIOD	F	SURFACE WATER ACTIVITY
1 month	1	0.8
1 week	1.11	0.89
1 day	1.25	1.00

Note: F is a correction factor to acknowledge that different maximum average surface water activities are appropriate for different timescales)

temperature of 15°C, mould will grow if the surface RH is kept at 93 per cent RH for a period of a day. In contrast, the IEA criterion is 100 per cent RH for a day. Likewise, Sedlbauer suggests mould will grow if the surface RH is kept at 83 per cent RH for a period of a week. In contrast, the IEA criterion is 89 per cent RH for a week.

However, to create the substrate specific isopleth systems, Sedlbauer used the least demanding conditions for mould to grow that could be found in literature. For example, if there were different spore germination times for the same material with the boundary conditions being the same, the shorter times were taken. Furthermore, species were taken into account that only rarely occur in buildings. Classification of the different building materials was also done in a way that they were always assigned to the more unfavourable substrate group. The combined effect of all these assumptions results in the worst case scenario.

We suggest that the Annex 14 criteria as detailed in Table 8.1 be adopted as the initial basis for the ADF 2010 guidance. However, in light of the more recent work of Sedlbauer and others, consideration should also be given to applying an extra margin of safety to Table 8.1. Table 8.2 thus provides a suggested final set of surface criteria with an additional safety margin applied to the IEA Annex 14 criteria.

These figures move towards those noted by Sedlbauer. They are still not the *absolute* worst case for weekly and daily but are judged to be a *reasonable* worst case. The monthly value is that noted by Sedlbauer as this value has also been reported – post IEA Annex 14 – by other workers (Rowan et al, 1999).

Given that surface water activity is not generally measured or modelled by building practitioners, we recommend that, where possible, guidance on *air* RH values should also form part of the ADF criteria (based on the IEA Annex 14 *surface* values).

In order to facilitate this approach, the relationship between surface water activity and surface RH can be utilized. An assumption is that, over the daily, weekly and monthly averaging periods, the *average* surface RH (divided by a factor of 100) and the *average*

TABLE 8.2 Surface water activity

PERIOD	SURFACE WATER ACTIVITY
1 month	0.75
1 week	0.85
1 day	0.95

TABLE 8.3 Room air relative humidity

PERIOD	ROOM AIR RELATIVE HUMIDITY
1 month	65%
1 week	75%
1 day	85%

surface water activity are equal. This supposes that the water activity is a result only of interactions with the moisture in the room air. If instead water is introduced via rain penetration, flooding or as a result of accumulated condensation, for example, this assumption is less likely to hold. Given this assumption, the IEA Annex 14 criteria thus state that average surface RH *below* values corresponding to the critical water activities (for each averaging period) should result in no mould growth in dwellings.

A suitable method of converting *air* RH values to *surface* RH values is then required. BS5250 (BSI, 2005) notes that the addition of 10 per cent to the air RH is an appropriate value for conversion to the internal surface RH of an external wall. UCL undertook some work to better understand how the air and surface RHs related under transient situations. The conclusion from this work was that, for new dwellings, 10 per cent was a reasonable value to apply over the whole season.

For regulatory guidance purposes for *new* dwellings, the moisture criteria in Table 8.2 are likely to be met if the average relative humidity in a room is less than the values noted in Table 8.3 during the heating season.

For *existing* dwellings, the much broader range of thermal quality (i.e. temperature factors) precludes similar guidance for the room air relative humidity values. Instead, the room air relative humidity should be such as to result in average surface water activity values less than those noted in Table 8.2.

CONCLUSIONS

This chapter has demonstrated the complexity involved with attempting to set appropriate guidelines with regard to the prevention of mould growth in dwellings. To deal with this complexity there is a fundamental choice between setting specific moisture performance criteria or using a 'mould model' to demonstrate compliance with regulations. At present, for England and Wales, the setting of moisture criteria is preferable and this chapter makes relevant suggestions for such criteria based on modifications to the Annex 14 criteria. The suggested guidance is able to capture enough of the complexity outlined in this chapter to be physically meaningful while at the same time being not too complex as to preclude its use by practitioners.

AUTHOR CONTACT DETAILS
H. Altamirano-Medina, M. Davies, I. Ridley, D. Mumovic (corresponding author) and T. Oreszczyn: The Bartlett, University College London, Gower Street, WC1E 6BT London, UK; d.mumovic@ucl.ac.uk

REFERENCES

Adan, O. C. G (1994) *On the Fungal Defacement of Interior Finishes*, PhD dissertation, Technical University of Eindhoven

Ayerst, G. (1969) 'The effects of moisture and temperature on growth and spore germination in some fungi', *Journal of Stored Products Research*, vol 5, pp127–141

Becker, R. (1984) 'Condensation and mould growth in dwellings: Parametric and field study', *Building and Environment*, vol 19, no 4, pp243–250

BSI (British Standards Institution) (2005) BS5250: 2002 (as amended 2005), *Code of Practice for the Control of Condensation in Buildings*, BSI, London

Clarke, J. A., Johnstone, C. M., Kelly, N. J., McLean, R. C., Anderson, J. A, Rowan, N. J. and Smith, J. E. (1999) 'A technique for the prediction of the conditions leading to mould growth in buildings', *Building and Environment*, vol 34, no 4, pp515–521

Deacon, J. W. (1997) *Modern Mycology*, 3rd edn, Blackwell Science, Oxford

DTLR (Department for Transport, Local Government and the Regions) (2002) *Limiting Thermal Bridging and Air Leakage: Robust Construction Details for Dwellings and Similar Buildings*, TSO, Norwich

Gertis, K., Erhorn, H. and Reiß, J. (1999) 'Klimawirkungen und Schimmelpilzbildung bei sanierten Gebäuden', in *Proceedings Bauphysik Kongreß*, Berlin, pp241–253

Ginkel, J. T. V. and Hasselaar, E. (2005) 'Housing characteristics predicting mould growth in bathrooms', in *Proceedings of Indoor Air 2005*, Beijing, China, 4–9 September, pp2425–2429

Grant, C., Hunter, C. A., Flannigan, B. and Bravery, A. F. (1989) 'The moisture requirements of moulds isolated from domestic dwellings', *International Biodeterioration*, vol 25, no 4, pp259–284

Hens, H. (2005) 'Mould in dwellings: Field studies in a moderate climate', IEA-EXCO Energy Conservation in Buildings and Community Systems, *Annex 41 'Moist-Eng'* Kyoto meeting, April 2006

HMSO (2006) *The Building Regulations, Approved Document F – Ventilation (2006 edition)*, The Stationery Office, London

Hukka, A. and Viitanen, H. (1999) 'A mathematical model of mould growth on wooden material', *Wood Science and Technology*, vol 33, no 6, pp475–485

IEA Annex 14 (1990) *Condensation and Energy: Guidelines and Practice*, Belgium, KU Leuven, Laboratory for Building Physics

Janssens, A. and Hens, H. (2003) 'Development of indoor climate classes to assess humidity in dwellings', in *Proceedings of AIVC 24th Conference & BETEC Conference – Ventilation, Humidity Control and Energy*, Washington DC, pp41–46

Lähdesmäki, K., Vinha, J., Viitanen, H., Salminen, K., Peuhkuri, R., Ojanen, T., Paajanen, L., Iitii, H. and Strander, T. (2008) 'Development of an improved model for mould growth: Laboratory and field experiments', in *Proceedings of Building Physics 2008 – 8th Nordic Symposium*, Copenhagen, Denmark, 16–18 June, pp935–942

Lawton, M. D., Dales, R. E. and White, J. (1998) 'The influence of house characteristics in a Canadian community on microbiological contamination', *Indoor Air*, vol 8, pp2–11

Moon, H. J. and Augenbroe, G. (2003) 'Evaluation of hygrothermal models for mold growth avoidance prediction', in *Proceedings of Eighth International IBPSA Conference*, Eindhoven, Netherlands, pp895–902

Mumovic, D., Ridley, I., Oreszczyn, T. and Davies, M. (2006) 'Condensation risk: Steady-state and transient hygrothermal modelling methods', *Building Services Engineering and Technology*, vol 27, no 3, pp219–235

Nielsen, K. F., Holm, G., Uttrup, L. P. and Nielsen, P. A. (2004) 'Mould growth on building materials under low water activities: Influence of humidity and temperature on fungal growth and secondary metabolism', *International Biodeterioration and Biodegradation*, vol 54, no 4, pp325–336

ODPM (Office of the Deputy Prime Minister) (2003) *English House Condition Survey 2001*, ODPM, London

Oreszczyn, T. and Pretlove, S. (1999a) 'Condensation Targeter II: Modelling surface relative humidity to predict mould growth in dwellings', *Building Services Engineering Research and Technology*, vol 20, no 3, pp143–153

Oreszczyn, T. and Pretlove, S.E.C (1999b) 'Mould index', in F. Nicol and J. Rudge (eds), *Cutting the Cost of Cold: Affordable Warmth for Healthier Homes*, E & FN Spon, London

Oreszczyn, T., Mumovic, D., Ridley, I. and Davies, M. (2005a) 'The reduction in air infiltration due to window replacement in UK dwellings: Results of a field study and telephone survey', *International Journal of Ventilation*, vol 4, no 1, pp71–78

Oreszczyn, T., Mumovic, D., Davies, M., Ridley, I., Bell, M., Smith, M. and Miles-Shenton, D. (2005b) *Condensation Risk – Impact of Improvements to Part L and Robust Details on Part C*, Office of Deputy Prime Minister, Building Regulations Division Project Number: CI 71/6/1 BD2414, London

Oreszczyn, T., Ridley, I., Hong, S. and Wilkinson, P. (2006) 'Mould and winter indoor relative humidity in low income households in England', *Indoor and Built Environment*, vol 15, no 2, pp125–135

Pasanen, A.-L., Kalliokoski, P., Pasanen, P., Jantunen, M. J. and Nevalainen, A. (1991) 'Laboratory studies on the relationship between fungal growth and atmospheric temperature and humidity', *Environmental International*, vol 17, no 4, pp225–228

Pasanen, A.-L., Kasanen, J.-P., Rautiala, S., Ikaheimo, M., Rantamaki, J., Kaariainen, H. and Kalliokoski, P. (2000) 'Fungal growth and survival in building materials under fluctuating moisture and temperature conditions', *International Biodeterioration and Biodegradation*, vol 46, no 2, pp117–127

Ridley, I., Pretlove, S., Ucci, M., Mumovic, D., Davies, M. and Oreszczyn, T. (2005) *Sensitivity of Humidity and Mould Growth to Occupier Behaviour in Dwellings Designed to the New Air Tightness Requirements*, Office of Deputy Prime Minister, Building Regulations Division Project Number: CI 71/6/11 (cc 2424), London

Ridley, I., Davies, M., Hong, S. H. and Oreszczyn, T. (Warm Front Study Group) (2007) 'Vapour pressure excess in living rooms and bedrooms of English dwellings: Analysis of the Warm Front Dataset', *Annex 41 Moist-Eng*, Florianapolis, Brazil, 16–18 April

Rode, C. (2005) *Danish Mould Research Programme*, Department of Civil Engineering, Technical University of Denmark, October

Rowan, N. J., Anderson, J. G., Smith, J. E., Clarke, J. A., Johnstone, C., Kelly, N. J. and McLean, R. C. (1996) 'Development of a technique for the prediction/alleviation of conditions leading to mould growth in houses', *Final Report No 68017*, Scottish Homes, Edinburgh

Rowan, N. J., Johnstone, C. M., McLean, R. C., Anderson, J. G. and Clarke, J. A. (1999) 'Prediction of toxigenic fungal growth in buildings by using a novel modelling system', *Applied and Environmental Microbiology*, vol 65, no 11, pp4814–4821

Sedlbauer, K. (2001) *Prediction of Mould Manifestation on and in Buildings* (in German), PhD dissertation, University of Stuttgart, Germany

Sedlbauer, K. (2002) 'Unwanted biological growth in and around buildings', *Rosenheimer Fenstertage 2002*, Germany

Tomkins, R. G. (1929) 'Studies of the growth of Mould-I', in *Proceedings of the Royal Society of London*, Series B, vol 105, no 738, pp375–401

Tyler, J. (1948) 'A study of the temperature and humidity requirements of *Aspergillus niger*', *Mycologia*, vol 40, no 6, pp728–738

World Health Organization (2007) *Development of WHO Guidelines for Indoor Air Quality: Dampness and Mould*, report on a Working group meeting, WHO Regional Office for Europe, October

Woolliscroft, M. (1997) 'PH-97-8-3. Residential ventilation in the United Kingdom: An overview', in *ASHRAE Transactions: Symposia*, vol 103, no 1, Paper Number: PH-97-8-3, pp706–716

Zöld, A. (1990) 'Mindestluftwechsel im praktischen Test', in *HLH – Heizung Lüftung, Haustechnik*, vol 41, no 7, pp620–622

9

Thermal Impact of Strategic Landscaping in Cities: A Review

Chen Yu and Wong Nyuk Hien

Abstract

Greenery in cities has long been recognized as an ecological measure to mitigate some environmental issues. The chapter discusses three major types of urban greenery: public green areas, rooftop gardens and vertical landscaping from both a systematic and thermal benefit point of view. Much related research carried out worldwide is reviewed. In order to achieve strategic landscaping in cities, some general guidelines are given.

■ *Keywords* – greenery; public green areas; rooftop gardens; vertical landscaping

INTRODUCTION

With rapid urbanization, there has been tremendous growth in populations and buildings in cities. As a result, the high concentration of hard surfaces in a built environment can trigger many environmental issues, such as global warming, urban heat island (UHI) effect, air/water pollution, etc. Greenery in cities is essentially an ecological measure to mitigate some of the environmental issues. Plants can create an 'oasis effect' and reduce the warming issue at both the macro- and micro-level. When greenery is arranged throughout a city in the form of nature reserves, urban parks, neighbourhood parks, rooftop gardens or even vertical landscaping, the energy balance of the whole city can be modified through the addition of more evaporating surfaces. More absorbed radiation can be dissipated to become latent heat rather than sensible heat. Hence the urban temperature can be reduced.

In a built environment, thermal issues can be described as conflicts between buildings and the urban climate. In consideration of the positive impact of plants on the conflict, a conceptual model was proposed by Chen and Wong (2005) to further reveal the interactions among the three critical components in the built environment (Figure 9.1). More accurately, the model consists not only of the three components, but also their mutual impacts (PB, PC, BC and CB). Two hypotheses are subsequently generated from the model (Figure 9.2). When climate and buildings are influenced to a greater extent by plants, the overlapping shaded area between them is decreased which means there is less conflict, i.e. less active energy is consumed to mitigate the conflicts (see hypothesis I).

doi:10.3763/aber.2009.0309 ■ © 2009 Earthscan ■ ISSN 1751-2549 (Print), 1756-2201 (Online) ■ www.earthscanjournals.com

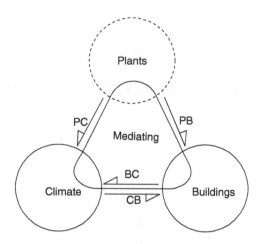

FIGURE 9.1 Model of environment (plants are considered to be the major component of environmental control)

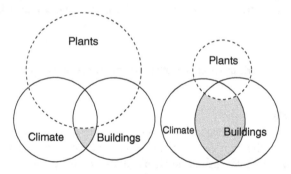

FIGURE 9.2 Graphical interpretation of hypothesis I (left) and hypothesis II (right)

On the other hand, the overlapping shaded area is greater when the influence of plants on climate and buildings is less. It indicates that there are more negative conflicts, i.e. more active energy is used to mitigate the negative effects (see hypothesis II).

To prove the above conceptual model, it is necessary to obtain some quantitative data. Much research related to the thermal benefits of urban greenery has been carried out worldwide. In these studies, urban greenery has been simply classified into three major groups: public green areas, rooftop gardens and vertical landscaping.

Public green areas are the areas of rural greenery in open spaces within cities. The size of them is not important, but they often have crucial roles in establishing the image of a city and improving the quality of the built environment. With increasing populations and rapid urbanization, public green areas are turning into environmental luxuries in cities.

Rooftop gardens are artificial green spaces on the rooftops of buildings. The attractive benefit of a green roof is its visual beauty which can soften the ugly roof surfaces and provide a recreation space for people living below.

Vertical landscaping is man-made space where plants are vertically introduced into the facades of buildings. Compared with urban parks and rooftop gardens, vertical landscaping is not common at the moment. However, there is great potential for vertical landscaping to be implemented because of the vast surface area of walls; more research has been conducted on this area recently.

REVIEW OF STUDIES RELATED TO PUBLIC GREEN AREAS

According to their dimensions, public green areas in cities can be simply classified as national/city parks, neighbourhood parks and green belts. National/city parks are usually large and owned and maintained by a national or local government (Figure 9.3). The purpose of creating such parks is to introduce variety, preserve a unique landscape, provide active or passive recreation or shape the image of a city. Smaller green areas are classified as neighbourhood parks and generally take the form of more flexible green areas in between developments (Figure 9.4). As community assets, they are normally

FIGURE 9.3 National Parks near Gold Coast, Australia (left) and Vancouver, Canada (right)

FIGURE 9.4 A neighbourhood park in Singapore

created to beautify the harsh built environment and provide immediate access to nature for residents. Green belts (Figure 9.5) are defined as small, irregular green patches (such as plants in playgrounds and trees beside roads) in cities. They are often introduced to protect locations such as playgrounds and roads, etc. from excessive solar exposure, noise, wind and so on. It is believed that all kinds of public green areas, no matter how small, can benefit the urban climatic conditions, but their impacts vary a lot.

Bernatzky (1982) observed that a small green area in Frankfurt (around 500 m^2) could reduce the air temperature by up to 3.5°C. Meanwhile, relative humidity was increased by 5–10 per cent which is welcomed as a further effect of air cooling. According to Bernatzky's understanding, trees in towns can reduce pollutant and noise levels even in winter when the trees have no leaves, as they retain 60 per cent of their 'cleaning' efficiency in this state. In the 1980s, Bernatzky proposed that grassed areas and trees should be planted more systematically in towns.

Jauregui (1990/1991) measured the urban air temperature of Chapultepec Park (around 500 ha) in Mexico City over a period of 4 years. The author observed that the park was about 2–3°C cooler on average compared with its surroundings. The cooling range extended about 2 km from the park (which measured about 2 km in width). Within the range, the maximum temperature difference obtained at the end of the dry season was

FIGURE 9.5 Green belts in Singapore

up to 4°C, although during the wet season it was only 1°C. It was also observed that the park heated up more slowly than the nearby built-up area at Tacubaya on sunny mornings.

Kawashima (1990/1991) explored the effects of vegetation density on surface temperatures in urban and suburban areas of Tokyo on a clear winter day by using satellite images. It was found that in urban areas the impact of plants on surface temperatures was marginal during the daytime. However, in suburban areas, lower surface temperatures could be easily detected in the more vegetated areas during the daytime, whereas the reverse situation was observed at night. In terms of surface temperature variations, fluctuations could be easily observed in the heavily planted areas in the urban environment whereas the fluctuations were not so obvious in the suburbs.

Honjo and Takakura (1990/1991) developed some numerical models to estimate the cooling impact of green areas on their surroundings by calculating the distribution of temperatures and humidity in Japan. In their calculations, the width of green areas ranged from 100 to 400 m, while their intervals (distances between the green areas) varied from 100 to 200 m. It was found that the range of the influenced area near a green area is a function of the width of the green area and the interval between the green areas. The authors' conclusion was that effective cooling of surrounding areas by green areas can be maximized by locating small green areas at carefully calculated intervals.

Saito et al (1990/1991) investigated the meteorological data and greenery distribution in Kumamoto City, Japan. In their study, air temperatures were measured by moving vehicles which can cover a relatively large area. Other parameters, such as dry/wet bulb temperatures, globe temperature, wind direction and velocity, were measured in smaller areas. Meanwhile, the land cover and surface temperature were also examined by using remote-sensing data. It was found that even a fairly small green area (60 x 40 m) can have a cooling effect on the surroundings. The maximum air temperature difference could be up to 3°C between the park and its surroundings. The range of the cooling area caused by the greenery varied with the wind direction. Saito et al concluded that air temperature distribution in an urban area was closely related to the distribution of greenery.

Ca et al (1998) carried out field observations to explore the cooling impact of Tama Central Park (0.6 km²) on the surroundings during summertime in Tama New Town which is west of Tokyo Metropolitan Area. It was found that, at noon, the surface temperatures measured inside the park were 19 and 15°C lower than those measured on the asphalt and the concrete surfaces, respectively. Similarly, the air temperature measured at noon at a height of 1.2 m inside the park was more than 2°C lower than that measured at the same height in the surrounding built environment. The park acted as a cool source immediately after sunset since the surface temperatures of the planted area were lower than the air temperatures. In conclusion, with the help of the park, the air temperature in a busy commercial area 1 km downwind could be reduced by up to 1.5°C on a hot day. From 1 to 2 pm, a maximum 4000 kWh of cooling energy for that area could be saved due to the presence of the park. Based on energy prices of 1998, the overall cooling energy savings could amount to US$650.

Shashua-Bar and Hoffman (2000) explored the cooling effect of small urban greens in Tel-Aviv, Israel. An empirical model was specially developed to predict the cooling effect of vegetation. The model was based on the statistical analysis carried out previously on

714 experimental observations at 11 different wooded sites. Two significant parameters were found to explain the air temperature variance within the green areas – the partially shaded area under the tree canopy and the air temperature of the adjacent built environment. The study found that the cooling impact of the planted area due to its geometry and tree characteristics, apart from the shading, was limited (about 0.5 K (degrees Kelvin)). However, the average cooling impact could be up to 3 K at noon when shading was considered. Meanwhile, it was found that the cooling range caused by small planted areas could extend up to about 100 m in the adjacent streets outside the green areas. Shashua-Bar and Hoffman (2002) also extended the general analytical model (Green CTTC (cluster thermal time constant) model) to predict air temperature variations in urban wooded sites. It was concluded that the model is an appropriate tool to assess the climatic effect of greenery in a built environment. In further work, Shashua-Bar and Hoffman (2004) predicted the impact of trees within streets and attached courtyards with the Green CTTC model. It was found that tree density has a significant impact on the cooling effect. The cooling effect could be up to 2.5 K and 3.1 K in densely planted streets and courtyards, respectively. Other factors, such as the deepening of tree clusters, albedo modification, and orientation, may also influence the cooling impact of trees. A cooling effect of up to 4.5 K could be achieved when all the factors are appropriately considered.

Picot (2004) investigated the impact of trees in a newly built square in Milan, Italy. Field measurements, a simplified thermal comfort evaluation method and an energy budget method were employed to study the thermal comfort level in the built environment. The study highlighted the clear difference between the conditions with young trees and mature trees. The shading effect provided by an aged tree canopy clearly shows a reduction of the absorbed radiation by users and generates an energy budget very close to a comfort energy budget under 50 W/m² even with a high ambient air temperature. It was also found that the trees could help to reduce the overall radiation absorbed by people by reducing the diffused solar radiation absorption.

Chen and Wong (2006) investigated two urban parks (12 and 36 ha, respectively) in Singapore using field measurements and simulations. In general, the average temperatures recorded in the two parks were lower than those in the surrounding environment. The surrounding built environment benefited from the parks but the cooling impact varied with the distance from the parks. A maximum air temperature difference of 1.3°C was observed among the locations outside the parks. The temperature difference was translated into cooling energy savings by performing building energy simulations. A reduction of up to 10 per cent of the cooling load was achievable in buildings near the parks. Moreover, the simulations also illustrated the fact that the loss of greenery may result in poor thermal conditions, not only in the original park site but also in the surroundings, especially when greenery was replaced by buildings or hard surfaces.

Wong and Chen (2005) explored the severity of the UHI effect and the cooling impact of green areas in Singapore through a mobile survey. A maximum air temperature difference of 4.01°C was observed between the highly vegetated area (Lim Chu Kang) and the Central Business District area. Wong et al (2006) also explored the cooling impact of roadside trees planted in three streets of an industrial area in Singapore. It was found that the variation of temperature within the streets was closely related to the density of

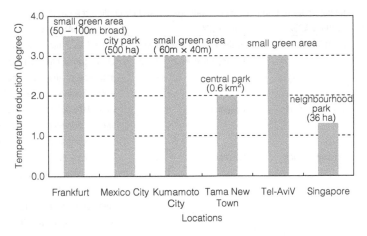

FIGURE 9.6 The temperature reduction caused by public green areas worldwide

roadside trees planted. The lowest mean air temperature obtained over a 4-week period in a street with very mature and dense roadside trees was 0.6°C lower than that of a street with very sparse and young trees. In addition, the air temperatures measured at night in the street with dense trees were 1.0–1.5°C lower than those obtained from the street with only young trees. Energy simulations were carried out with the data derived from the above measurements. By using an environment with young roadside trees as the benchmark, it was observed that up to 23 per cent and 38 per cent of cooling energy savings can be achieved for a factory in an environment with extremely dense roadside trees during a clear day and a cloudy day, respectively.

In conclusion, regardless of the size and location, public green areas have been shown to provide thermal benefits and energy savings for buildings located nearby. The most direct impact is the reduction in ambient air temperature. Figure 9.6 shows the comparison of temperature reductions caused by public green areas worldwide. A 2–3°C reduction of ambient air temperature as a result of public greenery is common in many cities around the world. Without doubt, the reduction of ambient temperatures can be translated into cooling energy savings in buildings near the green areas (Table 9.1). From the planning point of view, small green areas that are strategically arranged around buildings should be promoted. However, this does not mean that large urban parks are not effective in terms of improving the urban climate. But they may be considered a luxury in a heavily built-up environment where land is so valuable and scarce.

REVIEW OF STUDIES RELATED TO ROOFTOP GARDENS
Two major types of rooftop garden exist at the moment: intensive and extensive. Intensive green roofs, on which various species of plants can grow, require a relatively thick growing media and a lot of maintenance (Figure 9.7). They are normally accessible as a recreation space for residents and so incorporate areas of paving, seating, children's playgrounds and other architectural features. In contrast, extensive green roofs feature a lightweight growing media and self-generative plants such as turf/grass and require only

TABLE 9.1 The summary of potential energy savings caused by public green areas

TYPE OF PUBLIC GREEN AREA	LOCATION	ENERGY SAVINGS
Tama Central Park (0.6 km²)	at the west of the Tokyo Metropolitan Area	From 1 to 2 pm, maximum 4000 kWh of electricity can be saved due to the presence of the park and the cooling energy saving is about US$650.
Bukit Batok Nature Park (36 ha)	Singapore	Up to 10% reduction of the cooling load was achievable in buildings in the surrounding areas near to the parks.
A street with extensive mature trees	Singapore	Up to 23% and 38% of energy saving can be achieved in an environment with extremely dense road trees in a common stand-alone factory on a clear day and a cloudy day respectively.

FIGURE 9.7 Intensive rooftop garden in Singapore (atrium in International Business Park, Jurong East)

low maintenance (Figure 9.8). They are mainly designed for aesthetic and environmental benefits rather than for use by residents. Although they are commonly used on flat roofs, in some European countries, extensive systems can also be found on pitched roofs.

Hoyano (1988) conducted a series of experiments to measure the cooling effect of plants in Japan. One of the experiments was conducted on a rooftop using mock-ups of two turf-planting layers and a bare soil layer. The study shows that the solar radiation reflectance of turf was about 22–26 per cent of the incident radiation. The study also

FIGURE 9.8 Extensive rooftop gardens in Singapore (HDB multistorey car park, Puggol Road)

showed that the turf surface–air temperature difference was very much governed by the incident solar radiation. However, a negative correlation was found between the turf surface–air temperature difference and the wind speed. The daily amount of evaporation was found to be about 50 and 20 L/m² per day in summer and winter, respectively.

Harazono (1990/1991) developed a vegetation system to investigate the thermal effects of plants on rooftops. The measurements were conducted over 1 year on the rooftop of a building at a height of 16 m above ground level at the University of Osaka, Japan. In summer, the absolute humidity notably increased above the vegetation section. Air temperatures, measured at a height of 1 m above the rooftop, were also different between the vegetation and control (bare concrete) sections. The air temperature above the vegetation section was lower than that above the control section by approximately 2°C. However, in winter, there was no clear difference in terms of humidity and air temperature between the planted and control sections.

In summer, the surface temperature of the rooftop surface without vegetation rose to around 60°C on clear days, while the peak surface temperature beneath the plants was around 35°C. The difference of indoor air temperatures under planted and exposed roofs was around 2–3°C. In winter, the surface temperature of the control section rose to more than 30°C, but the surface temperature under the vegetation remained at around 11°C. The indoor air temperature under the planted roof was 0.5–1°C lower than that of the control section. The same data were also used to calculate the heat flux through the structure. The heat flux through the structure with vegetation was fairly small, while it was large (up to 200 W/m²) in the control section throughout the day. Harazono believed that the whole vegetation system (including the vegetation and the substrate) would be a good insulation layer for the rooftop.

Barrio (1998) simulated the dynamic thermal behaviour of green roofs. The cooling impacts of green roofs during the summer were analysed through the results obtained from the mathematical model. It was believed that green roofs work as insulation layers rather than as cooling devices. Some critical parameters that could influence the performance of green roofs were identified as the leaf area index (LAI), foliage geometrical characteristics, soil density, soil thickness and soil moisture. For example, increasing the LAI from 2 to 5 could decrease the solar transmittance dramatically and increase the foliage temperature during the daytime. However, LAI was found to have little effect on the air temperature in the plant canopy.

To explore the cooling effect of greenery covering a building, four concrete roof models with different coverings (bare concrete, soil layer, soil layer with turf, and soil layer with ivy) were investigated by Takakura et al (2000) at the University of Tokyo. It was observed that the greenery cover had an effective cooling impact on the internal air temperature. The inside air temperatures were stable for the concrete model with the soil and ivy layer. It ranged from 24–31°C throughout the day. In contrast, the inside air temperature under the bare concrete layer reached almost 40°C during the daytime and dropped to 20°C at night. Takakura et al also measured the heat flow through the unit surface areas of the four concrete models. For the bare concrete surface, more heat was gained during the daytime. With soil or greenery cover, heat loss could occur in the models even during the daytime. Furthermore, due to the evapotranspiration process of greenery, the concrete roof with plants lost more heat than the model with only the soil layer during the daytime. It was found that, with the increase of LAI, the plants brought about greater heat loss. Takakura et al believed that plants with a larger LAI can effectively cool the building and result in a lower cooling load due to the evaporation effect of the plants.

Onmura et al (2001) explored the cooling effect of a roof lawn garden through field measurements and a wind tunnel test. The surface temperature of the roof slab was reduced by 30°C due to the lawn garden during the daytime. The solar absorptivity of the lawn was around 0.78 which is consistent with Hoyano's (1988) study in which the radiation reflectance of turf was measured at 22–26 per cent. It was calculated that the thermal protection provided by the lawn was equivalent to a 50 per cent reduction of heat flux into the building. The evaporation rates of lawn samples generated from the wind tunnel test ranged from 0.1059 to 0.3130 kg/m²h. A simple transport model of heat and moisture was developed from the results generated from the wind tunnel test and the calculated results were in fairly good agreement with the data derived from the field measurements.

Niachou et al (2001) conducted field measurements on a green roof in order to examine its thermal performance and energy saving in Greece. Surface temperatures observed on the green roof varied according to the species of plants. They ranged from 28 to 40°C. On the other hand, surface temperatures measured on the bare roof fluctuated between 42 and 48°C. Meanwhile, it seemed that the green roof also benefited the room below. With the green roof, the percentage of maximum indoor air temperature exceeding 30°C was 15 per cent whereas it was 68 per cent without the green roof during a 24-hour period. The possible energy savings during a whole year in non-insulated buildings with green roofs could be 37–48 per cent. However, the possible energy saving caused by green roofs on moderate or well-insulated buildings was not impressive (only 4–7 per cent on moderately insulated buildings and less than 2 per cent on a well-insulated one).

Therefore, the authors believed that the impact of green roofs was beneficial on some old buildings with moderate or no insulation.

The direct and indirect thermal impacts of an intensive rooftop garden in a tropical climate were investigated by Wong et al (2003a) through field measurements carried out in Singapore. The study shows that the surface temperatures measured under plants were much lower than those measured on hard surfaces. The maximum surface temperature reduction caused by plants was around 30°C. The temperature measured under the vegetation varied according to the density (leaf area index) of the plants. Lower surface temperatures were measured under vegetation with dense foliage. Meanwhile, heat flux through the bare roof was greater than that through the planted roof. Compared with the bare roof, up to 78 per cent less heat gain was observed through a planted roof. The cooling effect of plants was also confirmed by the ambient air temperature reduction observed at different heights. The maximum temperature difference of 4.2°C was obtained between the locations with and without plants. Less long-wave heat emission from the planted surface was also confirmed by the measured global temperatures/mean radiant temperatures (MRT) on site. The maximum difference of the globe temperature and the MRT were 4.05 and 4.5°C, respectively, just after sunset between the planted roof and the bare concrete.

Wong et al (2003b) also carried out energy simulations to estimate the energy savings for a five-storey hypothetical commercial building in Singapore with an intensive rooftop garden. Essential inputs for the simulation were derived from the field measurements. It was found that 1–15 per cent of annual energy consumption could be saved due to the presence of a green roof. The best performance was achieved by dense shrubs on the green roof. Extra savings could also be achieved when the soil thickness of the green roof was increased. Wong et al (2007) also did some before and after measurements on extensive rooftop gardens in Singapore. It was found that the extensive green roof tended to experience lower surface temperatures compared with a bare roof. A maximum surface temperature difference of 18°C was observed between the well-planted roof and the bare one. However, the substrate surface temperature could be very high, up to 73.4°C, when it was exposed and dry. The heat flux through the roof structure was greatly reduced by the installation of the extensive green roof system. More than 60 per cent of heat gain could be intercepted by the system.

Theodosiou (2003) did a parametric study to evaluate the key characteristics that may affect the performance of a planted roof as a passive cooling technique. His results were validated by the real data taken from an existing construction in the Mediterranean area. It was found that the influence of the soil layer thickness was not obvious from the total heat flux during the whole summer period. The most important parameter was still the foliage density (LAI values). The foliage height did not influence the performance much unless it was coupled with the density of the vegetation layer. The role of the insulation layer thickness was also evaluated. It was found that in the absence of the insulation layer, heat flux from the relatively hot interior air to the cooler vegetated layers was enhanced. On the other hand, a high insulation layer could easily blur the cooling capability of the planted roof. Two climatic factors, relative humidity and wind speed, were considered important. It was believed that a dry, windy environment enhances evapotranspiration and the cooling capabilities of planted roofs.

Lazzarin et al (2005) carried out some measurements on a green roof in the Vicenza Hospital, Italy. A predictive numerical model was developed by the authors to calculate the thermal performance and energy saving in a building with a green roof under the local climatic conditions. The authors believed that the role of the latent flux due to the evapotranspiration of the greenery was significant. In a summer period, when the soil was dry, the green roof could prevent nearly 40 per cent of heat gain entering the room underneath compared with a traditional roof with insulation. The higher solar reflection of the greenery played an important role in this respect. Furthermore, when the soil was wet, there was no heat gain but only a slight heat loss under the same climatic conditions. Therefore, the authors concluded that the green roof worked as a passive cooler due to the evapotranspiration of the green roof. During the winter, the evapotranspiration process was strongly driven by the air vapour pressure deficit. In this case, the green roof played a negative role and produced an outgoing thermal flux from the roof. The heat loss was nearly 40 per cent higher than that of a high solar absorbing and insulated roof.

Kumar and Kaushik (2005) developed a mathematical model for evaluating the potential cooling impacts of green roofs and solar protection in buildings in Yamuna Nagar, India. It was found that plants with larger LAI values could reduce the canopy air temperature and eventually reduce the penetrating heat flux. The authors believed that the potential cooling impact of green roofs was adequate – the average indoor air temperature could be reduced by 5.1°C compared with that in a room without a green roof. The green roof provided a cooling potential of around 3.02 kWh per day and the average room air temperature could be maintained at 25.7°C when the planting was dense (LAI = 5).

Alexandri and Jones (2007) investigated the effect of green roofs on mitigating urban temperatures through a dynamic, one-dimensional model. The model was validated by an experiment carried out in Cardiff, UK. Some significant factors, such as the convective heat transfer coefficient, stomatal resistance and thermal diffusion, were examined in order to generate greater accuracy in the prediction.

In conclusion, both intensive and extensive rooftop gardens were found to greatly reduce surface temperatures and create energy savings for buildings. However, due to the differences in their composition, the performances of the two types of green roof are not the same. The most immediate impact of a green roof is its ability to reduce roof surface temperatures. Figure 9.9 shows a comparison of the maximum reduction of roof surface temperature caused by green roofs. Regardless of whether it is an extensive or intensive system, a reduction of up to 30°C of roof surface temperature can be achieved in different locations. According to the review, the ability to reduce the surface temperature is mainly governed by the density or more accurately, leaf area index, of plants. Intensive systems are supposed to be better in terms of supporting denser plants, such as shrubs or even trees. However, the review indicates that properly arranged turf/lawn on extensive systems can also achieve good surface temperature reductions on roofs.

Ambient air temperature reduction is another significant thermal benefit brought by green roofs. Figure 9.10 shows a comparison of the maximum reduction of air temperature caused by green roofs. Significant differences (around 2°C) can be observed between the two systems in reducing both indoor and outdoor air temperatures.

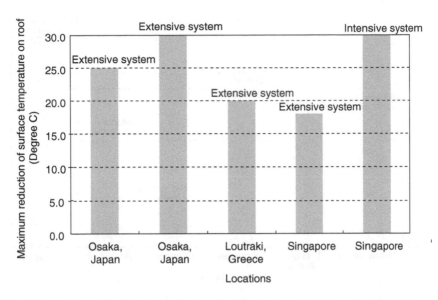

FIGURE 9.9 Comparison of maximum reduction of roof surface temperature caused by green roofs

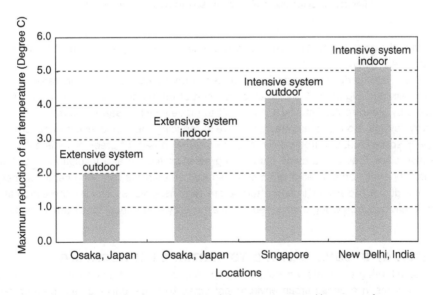

FIGURE 9.10 Comparison of maximum reduction of air temperature caused by green roofs

Intensive rooftop systems have denser plants, which can offer better evaporative cooling impact to the surroundings, and thick growing media, which works like extra insulation to the building roofs. However, the installation of intensive rooftop gardens usually requires extra structural loading and maintenance. Thus, instead of considering just the thermal

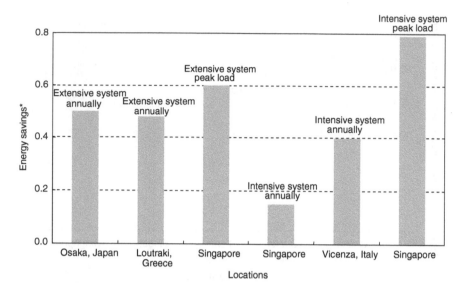

Note: The units on the y-axis are the proportion of the cooling energy consumption as compared to similar buildings without rooftop greenery

FIGURE 9.11 Comparison of potential cooling energy savings caused by green roofs

benefits, the potential cost of installation and maintenance of green roofs should also be carefully examined.

Finally, both surface temperature and air temperature reductions can be translated into energy savings. Figure 9.11 shows a comparison of the potential cooling energy savings caused by green roofs. Actually it is difficult to judge which system is better in terms of energy savings. Both of them are impressive in reducing the annual energy consumption or peak space load. In a tropical climate, the thermal impact of green roofs is potentially positive since they can provide outstanding solar protection and insulation to building roofs. However, some researchers mentioned that green roofs can bring negative energy savings during the winter in some temperate regions. This is because the evaporative effect may occur on wet growing media and eventually lead to heat loss from roofs during winter.

REVIEW OF STUDIES RELATED TO VERTICAL LANDSCAPING

Compared with green roofs, vertically arranged greenery can probably cover more exposed hard surfaces in a dense urban environment when high-rise buildings are predominant. Yeang (1998) believes that: 'This will significantly contribute towards the greening of the environment if a skyscraper has a plant ratio of one to seven, then the facade area is equivalent to almost three times the site area. So if you cover, let's say, even two-thirds of the facade you have already contributed towards doubling the extent of vegetation on the site. So in fact a skyscraper can become green. And if you green it, you're actually increasing the organic mass on the site.' The concept of introducing plants vertically into buildings is not new. However, incorporating plants strategically into building facades is still a challenge due to the lack of research and development in this area.

According to species of plants, types of growing media and construction methods, vertical landscaping (VL) or vertical greenery systems can be divided into four major categories – tree-against-wall type, wall-climbing type, hanging-down type and module type (Table 9.2). The tree-against-wall type (Figure 9.12) is not a genuine vertical landscaping method. However, similar to other vertical landscaping methods, trees placed strategically around buildings can bring thermal benefits directly to the walls of buildings. The wall-climbing type is a very common vertical landscaping method (Figure 9.13). Climbing plants can cover the walls of buildings naturally (time-consuming) or they can climb with the help of a trellis or other supporting system. The hanging-down type (Figure 9.14) is also a popular vertical landscaping method. The good point of this system is that it is not a time-consuming method for greening the whole facade of a building since plants can be planted at every storey to cover an entire facade. Finally, the module type (Figure 9.15) is a fairly new concept compared with the others. More complicated design considerations are required to set up and maintain such a system. In terms of cost, it is probably the most expensive method.

TABLE 9.2 Comparison of four types of vertical landscaping method

TYPE	PLANTS	GROWING MEDIA	CONSTRUCTION TYPE
Tree-against-wall type	Trees	Soil on the ground	No supporting structure
Wall-climbing type	Climbing plants	Soil on the ground or planted box	Supporting structure needed sometimes
Hanging-down type	Plants with long hanging-down stems	Planted box	Supporting structure needed sometimes
Module type	Short plants	Light-weight panel of growing media (such as compressed peat moss)	Supporting structure needed

FIGURE 9.12 Tree-against-wall type near a factory, Singapore

FIGURE 9.13 Wall-climbing type on a hotel (left, natural style) and on an office building (right, artificial style), Singapore

FIGURE 9.14 Hanging-down type on a car park (left) and on a university building (right), Singapore

McPherson et al (1988) used computer simulations to calculate the impact of trees on energy savings in similar houses (143 m²) in four US cities – Madison, Salt Lake City, Tucson and Miami. Irradiance reduction and wind reduction caused by trees are the two significant parameters considered in the simulations. With the reduced irradiance, annual heating costs in cold climates were increased (by $128 or 28 per cent in Madison),

FIGURE 9.15 Module-type outside hoarding (left) and on a university building (right), Singapore

whereas cooling costs in hot climates were reduced (by $249 or 61 per cent in Miami). Due to the dense shade on all surfaces of the houses, peak cooling loads could be reduced by 31–49 per cent (around 3108–4086 W). With 50 per cent wind reduction, on the other hand, annual heating costs were reduced by $63 (11 per cent) in Madison, but were increased by $68 (15 per cent) in Miami. In conclusion, plants placed strategically around buildings are necessary for different climatic conditions. Winter winds should be avoided but solar access to buildings should be encouraged in cold and temperate climates. In hot climates, however, high-branching shade trees and low ground cover should be planted to promote both shade and wind.

Hoyano (1988) conducted a series of experiments to measure the cooling effect of plants in Japan. A vine sunscreen designed for a south-west veranda was measured. The surface temperature difference of the veranda floor with and without the vine screen was 13–15°C, while the difference between air temperatures with and without the sunscreen was 1–3°C. In the late afternoon, both the air temperature and the globe temperature behind the sunscreen were higher than without sunscreen because the sunscreen stopped the outgoing radiation from the veranda and degraded the ventilation ability.

A test was also conducted using an ivy-covered west-facing concrete wall. The surface temperature of the exterior wall without ivy exceeded the outdoor temperature by around 10°C at 3 pm. For the wall with the ivy screen, the air temperature behind the ivy screen rose slightly. It was also found that the solar transmittance of the ivy sunscreen had a high inverse correlation with the distance between the wall and plant layer.

A row of evergreens forming a dense canopy next to a west-facing wall was investigated. More than 95 per cent of the solar radiation could be intercepted with the same density but a different arrangement of trees. With larger planting intervals, the ratio of solar radiation on the wall increased. However, the range of surface temperature distribution on the wall was not very wide. The experiment indicated that the temperature of surfaces exposed to sunlight was 3°C higher than that of the shaded parts. The leaf received increased solar radiation while the shaded parts remained almost the same temperature as the air. With a narrow distance between planting rows and the wall, the air layers tend to stagnate. As a result, the temperature of in-between air layers was lower than air temperatures in the morning but higher in the afternoon.

Wilmers (1988) measured some infrared temperatures in Hanover, Germany. His measurements indicated that some planted surfaces showed considerably different temperatures from other surfaces. During two sunny days, the air temperature ranged from 8 to 19°C. The surface temperatures in a garden court were measured on two concrete walls, which were facing south and west, respectively. Both walls were partly covered with climbing plants. In the morning, the temperatures of the two walls were similar and relatively higher than the air temperature. The temperature beneath the vegetation, however, was below the air temperature. In the early afternoon, the south wall with the plants was warmer than the air temperature, but 10°C below the temperature of the exposed south wall. The temperature of the shaded west wall was 20°C lower than the exposed south wall. In the evening, the temperature of the west wall was slightly higher than that of the south wall when the temperature of the vegetation was similar to the air temperature.

Holm (1989) adapted the DEROB (dynamic energy response of buildings) hourly computer programme to simulate the thermal effect of deciduous and evergreen vegetation on an exterior wall. A standard building in South Africa was chosen to observe the different thermal conditions with and without climbing plants. Without the plants, the indoor temperature was at least 2°C above the outdoor temperature during the daytime and dropped below 18°C at night. With the climbing plants, the indoor temperature was lowered by 5°C and was slightly raised at night. With the leaf cover, there was an improvement in the indoor thermal conditions. Also, the indoor temperature was within the standard comfort temperature range. This computer model was also applied to various building orientations, climates, seasons and building masses in South Africa.

Akbari et al (1997) measured the peak power and cooling energy savings resulting from shade trees at two houses in Sacramento, California. Seasonal cooling energy savings of 47 and 26 per cent were observed at the two houses. The corresponding daily savings are 3.6 and 4.8 kWh. The peak demand savings are 0.6 and 0.8 kW, respectively. The DOE-2.1E simulation programme was used to estimate the energy savings but some discrepancies were found between the simulation results and the field measurements.

Simpson and McPherson (1998) carried out measurements on 254 residential properties in Sacramento, California. There was an average of 3.1 trees for each property. The annual air-conditioning saving per tree was 7.7 and 5.6 per cent for the newer and older buildings, respectively. The peak air-conditioning saving per tree was 2.5 and 1.9 per cent for the newer and older buildings, respectively. Averaged over all properties, annual net savings of $10.00 and $4.00 per tree caused by the shading and the reduction of wind speed, respectively, were detected. The annual net saving caused by trees for a single building was around $43.00.

Simpson (2002) also provided a simplified method for estimating the impacts of trees on building cooling and heating loads by reducing incident solar radiation. Based on a limited number of discrete tree locations to represent all possible tree azimuths and tree–building distances, some lookup tables and tree distributions to quantify the impacts of trees were developed. A simple calculation of energy consumption can be done through the lookup tables. The level of agreement was less than 10 per cent from a comparison between the lookup table results and a detailed analysis of 178 properties in Sacramento, California. It was concluded that cooling reduction factors per tree were smaller than those for heating. Therefore, adding more trees for cooling in summer is less effective than for heating in winter. On the other hand, it is believed that the maximum impact on energy consumption can be achieved when larger trees are planted at west and east orientations to buildings at a distance ranging from 4.6 to 15 m.

Papadakis et al (2001) conducted measurements to explore the ability of trees to offer solar protection to building facades in Greece. The results showed that nearly 70–85 per cent of the incoming solar radiation can be intercepted by dense trees before reaching the facade. The ambient air temperature behind the shaded area and the surface temperature of the shaded facades were relatively lower than those without trees. Therefore, the radiative and thermal loads were significantly lower on the shaded walls. It was confirmed that plants can be used as an effective passive system for solar control of buildings.

Chen (2002) carried out some field measurements and simulations in some residential buildings in Singapore. Compared with sun shading devices, plants can 'consume', rather than simply reflect, solar radiation through the biological processes of photosynthesis and transpiration. Therefore, proper placement of vegetation vertically can cool facades of buildings through the interception and dissipation of heat before the solar radiation reaches the hard surfaces. It was found that solar radiation measured behind dense plants was kept constantly lower at around 50 W/m^2. A maximum surface temperature decrease of 12°C was also observed from the field measurements. Plants can also decrease the ambient air temperature by 1–2°C.

Stec et al (2005) conducted some laboratory tests on the thermal performance of a double-skin facade with plants. A numerical simulation was subsequently developed and validated with the data derived from the laboratory tests. It is believed that plants can provide more effective shading than blinds. For the same incident solar radiation, the temperature of plants was roughly half of that of blinds. Around 20 per cent cooling capacity can be saved by the introduction of plants. On the other hand, the introduction of plants may require more heating energy compared with a system of blinds since the plants cut down the solar heat gain during winter. During the warm period, the operation time of ventilation in a naturally ventilated building can be reduced, whereas it is increased during the cold period.

Wong et al (2006) also carried out a series of measurements and simulations to explore the cooling impact of trees planted near west- and east-facing facades of a factory in Singapore. As building facades are vulnerable to strong incident radiation in a tropical climate, high surface temperatures were observed and subsequently heat could enter the buildings and increase the cooling energy consumption. Vertical landscaping is a passive way to solve this problem. A maximum surface temperature difference of 13.6°C was

TABLE 9.3 Energy savings caused by plants on vertical facades at different locations

LOCATION	SAVINGS
Miami	Cooling costs in hot climate reduced by 61%; peak cooling loads reduced by 31–49%
Sacramento	Seasonal cooling energy reduced by 26–47%
Sacramento	The annual AC saving per tree by 5.6–7.7%; the peak AC saving per tree by 1.9–2.5%; annual net savings of $10.00 and $4.00 per tree; annual net savings caused by trees for single building at around $43.00.
Singapore	Cooling energy saving reduced by 10%

observed between the walls with and without tree shading. By using energy simulations, it was found that about 10 per cent of cooling energy can be saved in a stand-alone factory with tree shading on the east- and west-facing facades.

In conclusion, vertical landscaping can also bring thermal benefits to buildings through proper solar protection on the facades. The most direct impact is the decrease of surface temperatures of walls. In Singapore, the maximum temperature difference with and without vertical shading ranges from 12 to 13.6°C. Similar or smaller surface temperature reductions were also found in other countries. Compared with those experienced on green roofs (up to 30°C), the surface temperature reductions on the facades of buildings caused by plants are not so remarkable. This is understandable since the intensity of solar radiation incident on roofs and walls is different. Therefore, the exposed surface temperatures on roofs are usually much higher than those on facades. The impact of vertical landscaping on reducing ambient air temperatures was also observed but it is limited since vertical landscaping (except for tree-against-wall type) usually has relatively small leaf area index values compared with green roofs. The energy saving can be achieved through proper shading on the walls of buildings. However, currently most research has focused on the tree-against-wall type (Table 9.3). Other types of vertical landscaping are considered as fairly 'new' approaches and there is still a lack of in-depth studies being carried out on these approaches.

CONCLUSION

There is no doubt that public green areas, green roofs and vertical landscaping can bring thermal and environmental benefits to buildings and residents in a built environment. However, comprehensive understanding of these approaches is necessary before any of these strategies can be implemented (Figure 9.16). The conceptual model proposed by all the authors and the available quantitative data reviewed are useful references in generating some general guidelines:

● Large public green areas serve as the green heart of a city. Without them, the urban heat island effect will be aggravated. However, it is not easy to create more large parks due to land-use constraints in many cities. The current strategy should be to preserve as much of the existing public green spaces as possible.
● A network of public green areas should be formed to maximize the impact of greenery in cities. According to the review, scattered smaller green areas are

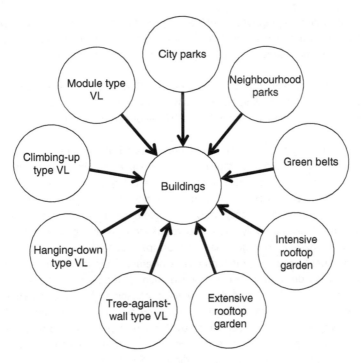

FIGURE 9.16 Greening possibilities in a built environment

favoured over a concentrated large green area. The aim is to segregate the harsh built environment into small sections with the use of plants.

- Trees and low-rise and even medium-rise buildings can be integrated. Lines of trees can be planted strategically near buildings in order to cast shadows on roofs and facades facing unfavourable orientations.
- Rooftop gardens and low-rise buildings can also be integrated. For a low-rise building, the rooftop garden can result in a considerable cooling energy saving since the roof plays a predominant role in determining the daytime heat gain during the hot season. On the other hand, rooftop greenery on low-rise buildings can also benefit surrounding buildings in terms of reducing the reflected radiation and providing aesthetic value.
- In terms of cooling energy savings, green roofs seem to be effective on roofs without an insulation layer. To prevent the wastage of materials, greenery and growing media should be integrated with roofs as both insulation and solar protection layers.
- Vertical planting and high-rise buildings can also be integrated. Compared with low-rise buildings, the area of vertical facades plays an important role in determining the daytime heat gain. For the facades beyond tree-top level, only vertical planting can extend the thermal benefits of plants. On the other hand, moisture concentration should be avoided by planting vegetation away from windows/openings and encouraging natural ventilation.

- Some vertical landscaping types, such as the module type, are still considered to be relatively new. More research and development should be carried out to better understand related issues such as safety, maintenance, impact on the facade integrity, etc.
- The density of greenery at macro-level and LAI values at micro-level are critical in reaping the significant cooling benefits caused by plants. In order to obtain better thermal protection, denser or large LAIs of greenery are recommended.

It is worth mentioning that the UHI effect encountered in big cities worldwide can be mitigated by strategically introducing plants into the built environment. In terms of reducing ambient air temperature, the priority should be to create large areas of greenery in the form of parks, roadside trees, landscaping within the vicinity of buildings, etc. Although a single rooftop garden or vertical landscaping may also have a cooling impact on its surroundings, it is usually very localized. Only rooftop gardens and vertical greenery, which can be implemented extensively on large areas of rooftops and facades, can contribute to the macro-environment by reducing the urban air temperature. On the other hand, the surface temperatures of facades and roofs could be significantly reduced by introducing plants on to individual buildings. Without the direct shading of plants, the surface temperatures of buildings cannot be effectively reduced even if they are located near large urban parks. Overall, improving urban thermal conditions by the use of plants should be carried out in a more strategic way. Figure 9.17 shows an example of current greening strategies and their possible application in different types of buildings in Singapore.

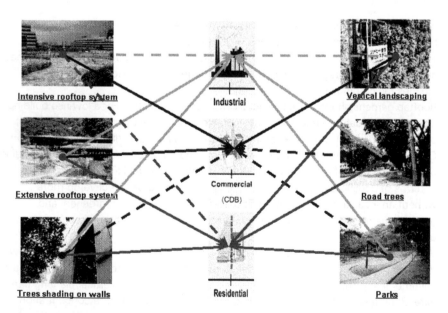

FIGURE 9.17 Potential greening strategies applied for different built environment in Singapore (solid line – very applicable; dotted line – not very applicable)

AUTHOR CONTACT DETAILS

Chen Yu and Wong Nyuk Hien: Department of Building, School of Design and Environment, National University of Singapore, 4 Architecture Drive, Singapore 117566; bdgwnh@nus.edu.sg

REFERENCES

Akbari, H., Kurn, D. M., Bretz, S. E. and Hanford, J. W. (1997) 'Peak power and cooling energy savings of shade trees', *Energy and Buildings*, vol 25, no 2, pp139–148

Alexandri, E. and Jones, P. (2007) 'Developing a one-dimensional heat and mass transfer algorithm for describing the effect of green roofs on the built environment: Comparison with experimental results', *Building and Environment*, vol 42, no 8, pp2835–2849

Barrio, E. P. D. (1998) 'Analysis of the green roofs cooling potential in buildings', *Energy and Buildings*, vol 27, no 2, pp179–193

Bernatzky, A. (1982) 'The contribution of trees and green spaces to a town climate', *Energy and Buildings*, vol 5, no 1, pp1–10

Ca, V. T., Asaeda, T. and Abua, E. M. (1998) 'Reductions in air conditioning energy caused by a nearby park', *Energy and Buildings*, vol 29, no 1, pp83–92

Chen, Y. (2002) 'An investigation of the effect of shading with vertical landscaping in Singapore', in R. A. Smith (ed), *Proceedings of IFPRA Asia-Pacific Congress 2002*, Singapore, 18–20 October, pp45–48

Chen, Y. and Wong, N. H. (2005) 'The intervention of plants in the conflict between building and climate in the tropical climate', in S. Murakami (ed), *Sustainable Building 2005*, Tokyo, Japan, 27–29 September, pp134–140

Chen, Y. and Wong, N. H. (2006) 'Thermal benefits of city parks', *Energy and Buildings*, vol 38, no 2, pp105–120

Harazono, Y. (1990/1991) 'Effects of rooftop vegetation using artificial substrates on the urban climate and the thermal load of buildings', *Energy and Buildings*, vol 15–16, no 3–4, pp435–442

Holm, D. (1989) 'Thermal improvement by means of leaf cover on external walls: A simulation model', *Energy and Buildings*, vol 14, no 1, pp19–30

Honjo, T. and Takakura, T. (1990/1991) 'Simulation of thermal effects of urban green areas on their surrounding areas', *Energy and Buildings*, vol 15, no 3–4, pp443–446

Hoyano, A. (1988) 'Climatological uses of plants for solar control and the effects on the thermal environment of a building', *Energy and Buildings*, vol 11, no 1–3, pp181–199

Jauregui, E. (1990/1991) 'Influence of a large urban park on temperature and convective precipitation in a tropical city', *Energy and Buildings*, vol 15, no 3–4, pp457–463

Kawashima, S. (1990/1991) 'Effect of vegetation on surface temperature in urban and suburban areas in winter', *Energy and Buildings*, vol 15, no 3–4, pp465–469

Kumar, R. and Kaushik, S. C. (2005) 'Performance evaluation of green roof and shading for thermal protection of buildings', *Building and Environment*, vol 40, no 11, pp1505–1511

Lazzarin, R. M., Castellotti, F. and Busato, F. (2005) 'Experimental measurements and numerical modelling of a green roof', *Energy and Buildings*, vol 37, no 12, pp1260–1267

McPherson, E. G., Herrington, L. P. and Heisler, G. M. (1988) 'Impacts of vegetation on residential heating and cooling', *Energy and Buildings*, vol 12, no 1, pp41–51

Niachou, A., Papakonstantinou, K., Santamouris, M., Tsangrassoulis, A. and Mihalakakou, G. (2001) 'Analysis of the green roof thermal properties and investigation of its energy performance', *Energy and Buildings*, vol 33, no 7, pp719–729

Onmura, S., Matsumoto, M. and Hokoi, S. (2001) 'Study on evaporative cooling effect of roof lawn gardens', *Energy and Buildings*, vol 33, no 7, pp653–666

Papadakis, G., Tsamis, P. and Kyritsis, S. (2001) 'An experimental investigation of the effect of shading with plants for solar control of buildings', *Energy and Buildings*, vol 33, no 8, pp831–836

Picot, X. (2004) 'Thermal comfort in urban spaces: Impact of vegetation growth. Case study: Piazza della Scienza, Milan, Italy', *Energy and Buildings*, vol 36, no 4, pp329–334

Saito, I., Ishihara, O. and Katayama, T. (1990/1991) 'Study of the effect of green areas on the thermal environment in an urban area', *Energy and Buildings*, vol 15, no 3–4, pp493–498

Shashua-Bar, L. and Hoffman, M. E. (2000) 'Vegetation as a climatic component in the design of an urban street: An empirical model for predicting the cooling effect of urban green areas with trees', *Energy and Buildings*, vol 31, no 3, pp221–235

Shashua-Bar, L. and Hoffman, M. E. (2002) 'The green CTTC model for predicting the air temperature in small urban wooded sites', *Building and Environment*, vol 37, no 12, pp1279–1288

Shashua-Bar, L. and Hoffman, M. E. (2004) 'Quantitative evaluation of passive cooling of the UCL microclimate in hot regions in summer, case study: Urban streets and courtyards with trees', *Building and Environment*, vol 39, no 9, pp1087–1099

Simpson, J. R. (2002) 'Improved estimates of tree-shading effects on residential energy use', *Energy and Buildings*, vol 34, no 10, pp1067–1076

Simpson, J. R. and McPherson, E. G. (1998) 'Simulation of tree shade impacts on residential energy use for space conditioning in Sacramento', *Atmospheric Environment*, vol 32, no 1, pp69–74

Stec, W. J., van Paassen, A. H. C. and Maziarz, A. (2005) 'Modelling the double skin facade with plants', *Energy and Buildings*, vol 37, no 5, pp419–427

Takakura, T., Kitade, S. and Goto, E. (2000) 'Cooling effect of greenery cover over a building', *Energy and Buildings*, vol 31, no 1, pp1–6

Theodosiou, T. G. (2003) 'Summer period analysis of the performance of a planted roof as a passive cooling technique', *Energy and Buildings*, vol 35, no 10, pp909–917

Wilmers, F. (1988) 'Green for melioration of urban climate', *Energy and Buildings*, vol 11, no 1–3, pp289–299

Wong, N. H. and Chen, Y. (2005) 'Study of green areas and urban heat island in a tropical city', *Habitat International*, vol 29, no 3, pp547–558

Wong, N. H., Chen, Y., Ong, C. L. and Sia, A. (2003a) 'Investigation of thermal benefits of rooftop garden in the tropical environment', *Energy and Buildings*, vol 38, no 3, pp261–270

Wong, N. H., Cheong, K. W., Yan, H., Soh, J., Ong, C. L. and Sia, A. (2003b) 'The effects of rooftop garden on energy consumption of a commercial building in Singapore', *Energy and Buildings*, vol 35, no 4, pp353–364

Wong, N. H., Chen, Y., Wong, S. T. and Chung, C. (2006) 'Exploring the thermal benefits of plants in industrial areas with respect to the tropical climate', in R. Compagnon et al (eds), *23rd International Conference on Passive and Low Energy Architecture (PLEA 2006)*, Geneva, Switzerland, 6–8 September, pp299–304

Wong, N. H., Tan, P. Y. and Chen, Y. (2007) 'Study of thermal performance of extensive rooftop greenery systems in the tropical climate', *Building and Environment*, vol 42, no 1, pp25–54

Yeang, K. (1998) 'The skyscraper bioclimatically considered: A design primer', in A. Scott (ed), *Dimensions of Sustainability: Architecture, Form, Technology, Environment, Culture*, E & FN Spon, London, pp109–116

publishing for a sustainable future

10

Urban Heat Island and its Impact on Building Energy Consumption

Rajagopalan Priyadarsini

Abstract

Urban areas tend to have higher air temperatures than their surroundings as a result of man-made alterations. This phenomenon is known as the urban heat island (UHI) effect. UHI is considered to be one of the major problems encountered by the human race this century. Solar radiation that is absorbed during the day by buildings is re-emitted after sunset creating high temperatures in urban areas. Also, anthropogenic heat sources such as air conditioners and road traffic add to the rise in temperatures. A number of studies have indicated that UHI has a significant effect on the energy use of buildings. In mid- and low-latitude cities, heat islands contribute to urban dwellers' summer discomfort and significantly higher air-conditioning loads. This chapter summarizes and reviews the latest research methodologies and findings about the effect of increased temperatures on the energy consumption of buildings. The latest developments in the heat island mitigation strategies are remarkable. However, more attention needs to be given to the implementation and testing of these strategies in full-scale buildings.

■ *Keywords* – urban heat island; temperature; energy; cool surfaces; vegetation

INTRODUCTION

Increasing urbanization and industrialization has caused the urban environment to deteriorate. The urban climate and the environmental efficiency of buildings are influenced by the deficiencies in proper development control (Santamouris, 2001). As a consequence of changes in the heat balance, air temperatures in densely built urban areas are higher than the temperatures of the surrounding country. This phenomenon, known as the urban heat island (UHI) effect, is a reflection of the totality of microclimatic changes brought about by man-made alterations of the urban surface (Landsberg, 1981). UHI was first identified by Luke Howard in 1820. He found out that in London, nights were 3.7°F warmer and days were 0.34°F cooler in the city than in the country. Heat island intensity differs in different parts of the city – the greatest intensity usually being in the most densely built areas. In high-latitude cities with cooler weather, heat islands can be an asset in reducing heating loads, but in mid- and low-latitude cities, heat islands contribute to urban dwellers' summer discomfort and significantly higher air-conditioning loads.

CAUSES OF URBAN HEAT ISLAND

Buildings are considered to be one of the main reasons for the urban heat island effect. Building masses increase the thermal capacity, which has a direct bearing on the city temperature. They reduce wind speed and give off heat either directly or indirectly. The heat that is absorbed during the day by the buildings, roads and other constructions in an urban area is re-emitted after sunset, creating high temperature differences between urban and rural areas. The urban heat island phenomenon is due to many factors, the most important of which are summarized as follows (Oke et al, 1991):

- the canyon radiative geometry that contributes to the decrease in long-wave radiation loss from within the street canyon due to the complex exchange between buildings and the screening of the skyline;
- the thermal properties of materials, which increase storage of sensible heat in the fabric of the city;
- the anthropogenic heat released from the combustion of fuels and animal metabolism;
- the urban 'greenhouse', which contributes to the increase in the incoming long-wave radiation from the polluted and warmer urban atmosphere;
- the canyon radiative geometry, which decreases the effective albedo of the system because of the multiple reflection of short-wave radiation between the canyon surfaces;
- the reduction of evaporating surfaces in the city, which means that more energy is put into sensible heat and less into latent heat; and
- the reduced turbulent transfer of heat from within streets.

SURFACE ALBEDO

The albedo of an object is the extent to which it diffusely reflects light from the sun. It is therefore a more specific form of the term reflectivity. Typically, urban albedos are in the range of 0.10 to 0.20 but in some cities these values can be exceeded. Using high albedo materials reduces the amount of solar radiation absorbed through building envelopes and urban structures and keeps their surfaces cooler. Increasing the reflectance or emittance lowers the surface temperature of any material which in turn reduces the heat transmitted into the building and lowers the heat transmitted to the ambient air. This in turn reduces the building cooling energy.

An investigation of the significant factors causing the UHI in Singapore using numerical simulation found that at very low wind speeds, the effect of facade materials and their colours was very significant and the temperature at the middle of a narrow canyon increased up to 2.5°C with the facade material having low reflectance (Rajagopalan et al, 2008).

GEOMETRY AND ORIENTATION

Urban geometry plays an important role in the transport and removal of pollutants. One of the main reasons for heat build-up in the heat island is poor ventilation. If pollutants land in sheltered areas like street canyons, they may stay there longer than they would in a rural environment. Detailed study of the air flow can help to eliminate problems of channelling and turbulence at the base of buildings and help to explore how the urban canyon effect can be utilized beneficially by designing the canyons in such a way that more parts of the city can be ventilated, thus optimizing heat extraction. Moreover, the orientation of the streets determines the amount of solar radiation received by the canyon surfaces. The degree to which this influences the air temperature in a street is a very interesting problem.

EVAPOTRANSPIRATION

Evapotranspiration from vegetation systems is another effective moderator of near-surface climates, particularly in the warm and dry, mid and low latitudes. Given the right conditions, evapotranspiration can create 'oases' that are cooler than their surroundings. In extreme oasis conditions, the latent heat flux can be so large that the sensible heat flux becomes negative, meaning that the air above vegetation and over the dry surroundings must supply sensible heat to the vegetated area and the Bowen ratio (ratio of sensible to latent heat fluxes) becomes negative (Taha, 1997). Urban areas, with extensive impervious surfaces have generally more runoff than their rural counterparts. The runoff water drains quickly and in the long run, less surface water remains available for evapotranspiration, thus affecting the urban surface energy balance. The lower evapotranspiration rate in urban areas is a major factor in increasing daytime temperatures.

ANTHROPOGENIC HEAT

The main source of anthropogenic heat is vehicle transport and heat discharge from air conditioners. The energy consumption of air conditioners is closely related to the climate of the city. The heat emitted by air conditioners in summer will induce a rise in the ambient air temperature. Also, the rise in ambient air temperature will increase the energy consumption of air conditioners. Studies have shown that condensers placed at narrow re-entrants of apartment buildings can cause significant heat build-up due to inadequate air flow (Chow et al, 2000). Very high on-coil temperature not only leads to energy wastage but also affects the equipment operation.

URBAN GREENHOUSE EFFECT

The greenhouse effect on the Earth can be identified in terms of the difference between the energy emitted by the Earth's surface and the energy emitted back into space by the upper atmosphere. Thus it is the long-wave energy, which is trapped in the atmosphere by feedback mechanisms, that causes climate changes. This effect is attributed to the property of greenhouse gases that absorb strongly in the infrared region of the electromagnetic spectrum. It is not only CO_2 (carbon dioxide) but also the other greenhouse gases such as CH_4 (methane), N_2 (nitrogen), O_3 (ozone) and CFCs (chlorofluorocarbons) that contribute to this phenomenon. These gases absorb infrared energy at wavelengths corresponding to the water-vapour spectral window. Thus, an increase in atmospheric CO_2 decreases the transmissivity of the atmosphere in this spectral window and reinforces the greenhouse effect.

IMPACT ON ENERGY CONSUMPTION

Many researchers have investigated the impact of UHI on building energy consumption. With the help of measured weather data as input parameters, most of the studies used numerical simulations to calculate the effect of increased temperatures on the cooling energy and peak demand. Some of the previous studies are discussed below.

INCREASED AIR TEMPERATURE AND COOLING LOAD

In Athens, annual cooling energy and peak demand were investigated to estimate the effect of high temperatures on cooling energy and peak demand. Both were found to be significantly increased as a result of the urban heat island effect, highlighting the need to reduce cooling energy by natural means (Hassid et al, 2000). Kolokotroni et al (2007) used measured air temperature data as input to a building energy simulation computer programme to assess the heating and cooling load of a typical air-conditioned office building positioned at 24 different locations within the London heat island. It was found that the effect of the London heat island on energy used for heating and cooling depends on the degree of urbanization in a particular location, radial distance from the city centre and the relative contribution of solar gain to total gains in a building.

Studies have also been conducted in city blocks to explore the interaction between summertime outdoor thermal conditions and cooling energy demand in urban areas. Kikegawa et al (2003) developed a numerical simulation system adopting a new one-dimensional urban canopy meteorological model coupled with a simple sub-model for the building energy analysis. This system was applied to the Ootemachi area, a central business district in Tokyo. The simulated temperature sensitivity of the peak-time cooling electric demand in the Ootemachi area was found to be almost consistent with the actual regional average sensitivity over all business districts in the central part of Tokyo, which was estimated using actual electricity demand data provided by the Tokyo electric power company.

ANTHROPOGENIC HEAT AND ENERGY

The main source of building-related anthropogenic heat is air conditioners. Although air-conditioning can improve the indoor thermal environment of a building, waste heat is dumped into the atmosphere making the urban thermal environment worse. Most of the commercial buildings in the central business districts of major cities have central air-conditioning where heat dissipation occurs through cooling towers. However, high-density residential buildings have window and split air-conditioning systems, the condensers of which dissipate heat to the atmosphere. Very high on-coil temperature not only leads to energy wastage but also affects the equipment operation and consumes more energy to cool the buildings to the specified temperature setting. A series of studies have indicated that summer cooling energy can be reduced through proper ventilation and also by cutting off the anthropogenic heat (Kikegawa et al, 2003; Kolokotroni et al, 2006).

Santamouris et al (2001) used data from Athens' urban climate to evaluate the increase in cooling load of urban buildings. It was found that for a representative building, the cooling load almost doubled in the central Athens area, while peak electricity load may be tripled for higher setpoint temperatures. Also the minimum coefficient of performance (COP) value for air conditioners in the central Athens area was reduced by 20 per cent. Hsieh et al (2007) evaluated the heat rejection, building energy use and air temperature distributions of residential apartments using window-type air conditioners in Taipei using both building energy programme and computational fluid dynamics (CFD) software. The ambient air temperature, the air temperature next to the building envelope and the air temperatures around the air conditioner were evaluated to clarify the thermal environment around buildings and the total amount of additional electricity consumption for air conditioners was found to be 10.7 per cent in the time period of 7.01 pm to 2 am. Wen and Lian (2009) established a model to calculate the rise in atmospheric temperature caused by the waste heat of urban air conditioners. The temperature rise in the city of Wuhan – which is known as the 'hot stove' in China, where the actual air temperature in the centre of the city may be above 40°C due to the heat island effect – was analysed using this model. The results demonstrated that the rise in atmospheric temperature is 2.56°C under inversion conditions and 0.2°C under normal conditions which indicates that thermal pollution is serious when the atmosphere is stable.

EFFECT OF HEAT ISLAND MITIGATION STRATEGIES

There are many attempts to modify the urban design elements to ameliorate the heat island effects. These include strategies limited to building scale (e.g. geometry and albedo modification, and modification of anthropogenic heat sources). City-scale strategies were also evaluated by looking into the dynamic interaction between buildings and the urban environment surrounding them. Some of the proposed strategies could be implemented during the design and planning stage and some others could be implemented after the building is constructed.

It is well recognized that urban areas have darker surfaces and less green areas than their surroundings. Although concentrated vegetation in the form of parks can lower the temperature in their immediate surrounding zones (Yu and Wong, 2006), they are unable to manipulate the thermal environment of concentrated high-density built spaces where people live. A number of mitigation strategies have been proposed by many researchers. Specifically, they seek to reduce the solar radiation absorbed by the surface or increase the latent heat flux away from the surface. Hence one of the most accepted heat island mitigation strategies is to incorporate vegetation on roofs and walls. Another effective tactic is to replace facades with high reflectance material. Researchers have also looked into using water as a cooling element.

GREEN WALLS AND ROOF

Most urban buildings have large areas of walls and roofs on which vegetation can be easily incorporated. Ca et al (1998) carried out field observations to determine the influence of a park in Tama New Town, a city to the west of Tokyo Metropolitan Area in Japan. The air temperature measured at 1.2 m above the ground at the grass field inside the park was more than 2°C lower than that measured at the same height in the surrounding commercial and parking areas. A coupled mesoscale urban climate and energy model was used to compute the heat exchange between buildings and the air in the urban canopy layer. In the model, the conduction heat flux to rooms of a building through walls and roof was evaluated by solving the heat conduction equation for the walls and roof with an assumed constant room air temperature; the air-conditioning energy is evaluated based on the cooler's coefficient of performance and the heat gain inside the building which is a function of outdoor air temperature, roof level wind velocity and the internal heat load due to machines, lights and human bodies. Results of computations showed that at noon a difference of 2°C in outside temperature can save almost 15 per cent in cooling energy. Incorporating the area of influence of the park which is 0.5 km, it was found that 4000 kWh of electricity for cooling could be saved, which works out to a saving of US$650.

Alexandri and Jones (2008) used a two-dimensional, prognostic, microscale model to evaluate the thermal effect of covering the building envelope with vegetation for various climates and urban canyon geometries. The differences between the cooling loads when vegetation was placed only on the walls of the buildings compared with when vegetation was placed on both walls and roofs were smaller for humid climates and greater for arid climates due to the different humidity concentrations in the two climatic groups. The authors have also pointed out that the cooling energy demand of buildings in humid climates can be reduced, especially when both walls and roofs are covered with vegetation, reaching up to an 8.4°C maximum temperature decrease for humid Hong Kong. Generally, green walls have a stronger effect than green roofs inside the canyon. However, green roofs have a greater effect at the roof level and consequently at the urban scale. Akbari et al (2001) monitored cooling energy savings from shade trees in two houses in Sacramento, California and it was found that seasonal cooling energy savings of 30 per cent, corresponding to average savings of 3.6 and 4.8 kWh/day, could be achieved.

COOL ROOF

Several studies including field measurements and computer simulations have documented significant energy savings resulting from increasing roof solar reflectance. Measurements conducted at various buildings in a number of cities in the US reported significant savings in cooling energy and peak power by increasing the roof reflectance of buildings.

Akbari (2003) monitored the energy use and environmental parameters of two non-residential buildings in Nevada, US during the summer before and after changing the reflectivity of the roof from 26 per cent to 72 per cent using white paint. The monitored electricity savings were about 0.9 kWh per day (33 Wh/m² per day) and 125 kWh per year (8.4 kWh/m²); at a cost of $0.1/kWh, savings were about $0.86/m²/year. The author agrees that it costs significantly more than this amount to coat the roofs with reflective coatings. However, since the prefabricated roofs are already painted green at the factory, painting them white would not cause additional cost. Hence, a reflective roof saves energy at no extra cost.

Energy simulations conducted by Levinson et al (2005) found that the use of a cool roof on a prototypical non-residential building according to California Title 24 standards yields average annual cooling energy savings of 3.2 kWh/m^2, average annual natural gas deficits of 5.6 MJ/m^2, average annual source energy savings of 30 MJ/m^2 and average peak power demand savings of 2.1 W/m^2. In cities where the cooling degree-days are infrequent, mechanical cooling is used only on the hottest days of the year. In such cases, the installation of a cool roof can potentially obviate the need to operate or even install air-conditioning. He and Hoyano (2008) quantitatively evaluated the thermal improvement effect of sprinkling water on the TiO$_2$-coated external building surfaces using a numerical simulation model. The surface temperature of the walls and roofs with water film was found to be lowered by 2–7°C, the indoor temperature decreased by 2–4°C and the daily building cooling load was reduced by 30–40 per cent compared with a building without water flow. Significant developments have been reported in Athens in the area of cool-coloured coatings (Synnefa et al, 2007). The authors acknowledge the importance of mitigating the heat island effect without compromising on the aesthetics of the building. Ten prototype colour coatings were created using near-infrared reflective colour pigments and tested in comparison with conventionally pigmented colour-matched coatings. It was found that cool-coloured coatings are able to maintain lower surface temperatures than conventionally pigmented colour-matched coatings.

MULTIPLE STRATEGIES

A number of studies have been conducted to evaluate the changes in air temperature and energy consumption due to the installation of various UHI countermeasures. The results of the various scenario analysis is a powerful tool to determine and select the most promising ones for implementation and incorporation in urban design and planning decisions.

A study conducted in Tokyo using a revised urban canopy and building simulation model found that the installation of humidification and albedo-increase technologies as the major UHI countermeasures can reduce the total number of hours for which the outdoor air temperature is more than 30°C during the daytime by more than 60 hours per year (Ihara et al, 2008). For this particular climate, a reduction in the cooling demand and an increase in the heating demand due to albedo increase in the average office building almost cancel each other out. Therefore, there would be no benefit in introducing the humidification and albedo-increase technologies. In addition, humidification and installation of heat sinks were calculated to reduce the energy consumption by 3 and 1 per cent, respectively. Akbari and Konopacki (2005) estimated the potential of heat island reduction strategies such as solar reflective roofs, shade trees, reflective pavements and urban vegetation to reduce cooling energy use in buildings. Savings for direct effect, which means reducing heat gain through the building shell, and indirect effect, which is reducing ambient air temperature, were estimated. It was found that for all building types tested, more than 75 per cent of the total savings were from direct effects of cool roofs and shade trees.

Dhakal and Hanaki (2002) simulated the heat discharge from buildings using a building-scale and mesoscale approach for the representative office, commercial and residential buildings using DOE-2 building energy simulation model. The improvements in the urban thermal environment attained through the various measures were analysed for two types of scenarios, i.e. those related to the management of heat discharge sources and those related to urban surface modifications. The maximum reduction in average temperature for daytime was found to be 0.47°C as a result of greening the areas around the buildings of Tokyo. Similarly, the maximum reduction in average temperature for the evening was found to be 0.11°C by discharging all heat to the ground. Sailor and Dietsch (2007) developed a web-based software tool to assist urban planners and air quality management officials in assessing the potential of urban heat island mitigation strategies. The user of the tool can select from over 170 US cities in which to conduct the analysis and can specify city-wide changes in surface reflectivity and vegetative cover.

DISCUSSION

High-density development and lack of vegetation affect the climate, energy use and liveability of cities. Energy requirements for buildings in the urban area have drawn a great deal of attention in recent years due to the undesirable environmental and economic impacts on the society. The literature demonstrates that energy consumption can be significantly reduced by UHI mitigation strategies. Increasing the albedo of the surface results in a higher percentage of solar radiation being reflected away from the city. Albedo-based mitigation strategies tend to have a significant effect in summer. While cooling in summer is favourable, cooling in winter would result in the increased use of heating energy. Green roofs and green walls cool the microclimate around them which can lead to significant cooling energy saving depending on the climatic type, and the amount and position of vegetation on the building. Increasing the vegetation cover in a city results in a higher level of evaporative cooling. Also, cooling demand can be reduced to zero by covering building surfaces with vegetation in cases where little cooling load is needed. This could lead to successful applications of further passive cooling techniques, especially ones employing ventilation which are not easy to implement in extremely hot urban conditions, in cases of large heat island densities.

CONCLUSION AND FUTURE DIRECTIONS

Researchers around the world have carried out comprehensive assessments to identify the potential causes of urban heat island and the implications for building energy use. The fundamental objective is to provide general guidance to avoid the deterioration of the urban environment. The results are expected to serve as design guidelines for sustainable urban development that will ensure rational energy management and conservation. The contribution of various factors that affect the microclimate of the urban environment was quantitatively evaluated so as to determine and select the most promising ones for implementation and incorporation in urban design and planning decisions. A number of studies indicate that significant reduction in energy consumption can be achieved by urban modifications. Most of the studies related to the energy implications of urban heat islands were based on numerical simulations and lack practical implementation in full-scale buildings. Hence more attention needs to be given to understand what happens in practice by implementing these strategies and conducting long-term measurements. Significant achievements have been reported by the heat island research group at the Lawrence Berkeley National Laboratory at the University of California. Studies are being conducted to rate and label cool materials to be used for building surfaces. The cool materials criteria and standards can be incorporated into building energy performance standards. Significant developments have also been reported in Athens in the area of cool-coloured coatings. When research is carried out on applying cool-coloured coatings to full-scale buildings, we will get a better idea about the practical implementation and real energy savings.

AUTHOR CONTACT DETAILS

Rajagopalan Priyadarsini: School of Architecture and Building, Deakin University, Geelong, Australia; priya@deakin.edu.au

REFERENCES

Akbari, H. (2003) 'Measured energy savings from the application of reflective roofs in two small non-residential buildings', *Energy*, vol 28, no 9, pp953–967

Akbari, H. and Konopacki, S. (2005) 'Calculating energy saving potentials of heat-island reduction strategies', *Energy Policy*, vol 33, no 6, pp721–756

Akbari, H., Pomerantz, M. and Taha, H. (2001) 'Cool surfaces and shade trees to reduce energy use and improve air quality in urban areas', *Solar Energy*, vol 70, no 3, pp295–310

Alexandri, E. and Jones, P. (2008) 'Temperature decrease in an urban canyon due to green walls and green roofs in diverse climate', *Building and Environment*, vol 43, no 4, pp480–493

Ca, V. T., Asaeda, T. and Abu, E. M. (1998) 'Reductions in air-conditioning energy caused by a nearby park', *Energy and Buildings*, vol 29, no 1, pp83–92

Chen, Y. and Wong, N. H. (2006) 'Thermal benefits of city parks', *Energy and Buildings*, vol 38, no 2, pp105–120

Chow, T. T., Lin, Z. and Wang, Q. W. (2000) 'Applying CFD simulation in analyzing split type air-conditioner performance at buildings', *Architectural Science Review*, vol 43, no 3, pp133–140

Dhakal, S. and Hanaki, K. (2002) 'Improvement of urban thermal environment by managing heat discharge sources and surface modification in Tokyo', *Energy and Buildings*, vol 34, no 1, pp13–23

Hassid, S., Santamouris, M., Papinikolaou, N., Linardi, A., Klitsikas, N., Georgakis, C. and Assimakopoulos, D. N. (2000) 'The effect of Athens heat island on air conditioning load', *Energy and Buildings*, vol 32, no 2, pp131–141

He, J. and Hoyano, A. (2008) 'A numerical simulation method for analysing the thermal improvement effect of super-hydrophilic photocatalyst-coated building surface with water film on the urban/built environment', *Energy and Buildings*, vol 40, no 6, pp968–978

Hsieh, C. M., Aramaki, T. and Hanaki, K. (2007) 'The feedback of heat rejection to air conditioning load during the nighttime in subtropical climate', *Energy and Buildings*, vol 39, no 11, pp1175–1182

Ihara, T., Kekegawa, Y., Asahi, K., Genki, Y. and Kondo, H. (2008) 'Changes in year-round temperature and annual energy consumption in office building areas by urban heat island countermeasures and energy saving measures', *Applied Energy*, vol 85, no 1, pp12–25

Kikegawa, Y., Genchi, Y., Yoshikado, H. and Kondo, H. (2003) 'Development of a numerical simulation system toward comprehensive assessment of urban warming countermeasures including their impacts upon the urban buildings' energy demands', *Applied Energy*, vol 76, no 4, pp449–466

Kolokotroni, M., Giannitsaris, I. and Watkins, R. (2006) 'The effect of the London urban heat island on building summer cooling demand and night ventilation strategies', *Solar Energy*, vol 80, no 4, pp383–392

Kolokotroni, M., Zhang, Y. and Watkins, R. (2007) 'The London heat island and building cooling design', *Solar Energy*, vol 81, no 1, pp102–110

Landsberg, H. E. (1981) *The Urban Climate*, Academic Press, New York

Levinson, R., Akbari, H., Konopacki, S. and Bretz, S. (2005) 'Inclusion of cool roofs in non-residential Title 24 prescriptive requirements', *Energy Policy*, vol 33, no 2, pp151–170

Oke, T. R., Johnson, G. T., Steyn, D. G. and Watson, I. D. (1991) 'Simulation of surface urban heat islands under "ideal" conditions at night. Part 2: Diagnosis and causation', *Boundary Layer Meteorology*, vol 56, no 4, pp339–358

Rajagopalan, P., Wong, N. H. and Cheong, K. W. (2008) 'Microclimatic modeling of the urban thermal environment of Singapore to mitigate urban heat island', *Solar Energy*, vol 82, no 8, pp727–745

Sailor, D. J. and Dietsch, N. (2007) 'The urban heat island mitigation impact screening tool (MIST)', *Environmental Modelling and Software*, vol 22, no 10, pp1529–1541

Santamouris, M. (2001) *Energy and Climate in the Urban Built Environment*, James & James Science Publishers, London

Santamouris, M., Papanikolaou, N., Livada, I., Koronakis, I., Georgakis, C., Argiriou, A. and Assimakopoulos, D. N. (2001) 'On the impact of urban climate on the energy consumption of buildings', *Solar Energy*, vol 70, no 3, pp201–216

Synnefa, A., Santamouris, M. and Apostolakis, K. (2007) 'On the development, optical properties and thermal performance of cool colored coatings for the urban environment', *Solar Energy*, vol 81, no 4, pp488–497

Taha, H. (1997) 'Urban climates and heat islands: Albedo, evapotranspiration and anthropogenic heat', *Energy and Buildings*, vol 25, no 2, pp99–103

Wen, Y. and Lian, Z. (2009) 'Influence of air conditioners utilization on urban thermal environment', *Applied Thermal Engineering*, vol 29, no 4, pp645–651

Green Roofs in Buildings: Thermal and Environmental Behaviour

Theodore Theodosiou

Abstract

Although in many countries green roofs still appear to be a novelty, they are in fact a technology that has been known for centuries. During the past decade, urban policies and building regulations in many countries worldwide have led to an increase in the popularity of this technique, mostly due to the belief that green roofs play a key role in environmental restoration in areas where city growth has resulted in significant urban climate deterioration and an increase in the energy consumption of buildings. Until recently, the benefits of green roofs were more qualitative than quantitative in character and they were compared with the benefits of vegetation surfaces. During the past decade, a great number of studies in many different scientific fields have produced important knowledge concerning the actual performance of this green technology and have managed to quantify its performance through a mathematical or experimental approach. Recently, a number of computer models have been introduced that are expected to contribute more to our existing knowledge. The purpose of this study is to present the recent developments in green roof technology. This chapter focuses mainly on the contribution of green roofs to the energy consumption of buildings, but also reports the results of research relating to most of the other environmental benefits that are reported in literature.

■ *Keywords* – green roof; green technology; urban vegetation

INTRODUCTION

Green or planted roofs are a type of roof construction which in their simplest form have existed for thousands of years in many different regions of the world. Their role in improving internal comfort conditions was the main advantage in installing such a roof type during times when natural materials were the only available type of building fabric. In cold climates they contributed to thermal insulation of the roof, while in warm climates they protected the roof from overheating due to their increased solar exposure in summer.

ADVANCES IN BUILDING ENERGY RESEARCH ■ 2009 ■ VOLUME 3 ■ PAGES 271–288

doi:10.3763/aber.2009.0311 ■ © 2009 Earthscan ■ ISSN 1751-2549 (Print), 1756-2201 (Online) ■ www.earthscanjournals.com

Although this climate-dependent behaviour is still one of the reasons for green roof installations, their role as urban ecosystems has been expanded to include other significant characteristics that in many cases seem to be more valuable than the improvement of thermal comfort, such as their environmental benefits.

Until recently, the numerous advantages of green roofs had a qualitative rather than quantitative character, since the scientific research in this field was very limited. In most cases, the existence of plants on the top of a building was regarded as an environmentally friendly construction with a generally positive contribution to the building's energy efficiency. For most building scientists, the extra soil layer on the top of the roof was considered to be simply an additional insulation layer which in the worst case could not increase thermal losses in winter and would shade the conventional construction layers in summer, offering protection from overheating due to solar radiation. For environmental scientists, the benefits of such a construction were regarded as being similar to all the benefits that vegetation can offer.

During the past few years, a lot of research on green roofs has been done or is still in progress. Literature reports a relatively large number of experiments and computer models that aim to investigate the behaviour of green roofs as an integrated building or environmental system. During the same period, a variety of green roof types were constructed in many cities worldwide that gave the opportunity to monitor their behaviour in actual conditions and to evaluate their behaviour in comparison with conventional roof constructions. Recently the findings of this research started to take the form of guidelines and policies in many cities, in order to support the exploitation of green roofs in the most efficient way.

The purpose of this chapter is to review the knowledge gained in recent years in order to present the state of the latest green roof technology.

TECHNICAL DESCRIPTION

Although nowadays complete commercial green roof systems with specific properties of construction and soil layers and vegetation can be found, a typical green roof system is far from being a standard building element regarding the materials used, the layer characteristics and the selection of proper vegetation. This is one of the reasons for the relatively wide range of efficiency levels (in terms of both energy and environmental performance) of green roofs that are reported in the literature, since the additional layers and vegetation that cover the roof have in most cases very few similarities in each examined computer model or experiment.

A typical green roof consists of a lightweight soil mixture and a drainage layer. A fabric filter keeps these layers separated and a special layer under the drainage protects the underlying structure from the vegetation roots (Figure 11.1). High-quality waterproofing is required in order to prevent water leakage. The height of each layer depends on the requirements of the selected vegetation. The role of the drainage layer can be simply to control the moisture of the soil and allow proper drainage, since in many cases saturated soil will permanently damage the roots. In some types of green roof the drainage layer is designed to retain rain or irrigation water in order to keep the soil mixture wet, creating an environment suitable for water-demanding vegetation.

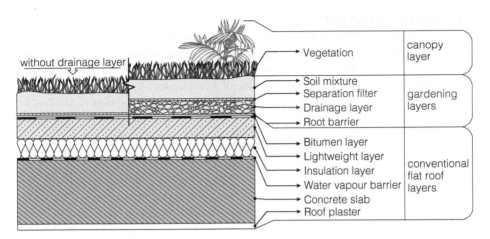

FIGURE 11.1 A typical green roof structure

Regarding the construction properties, the heights of the layers and the maintenance requirements of roof gardens, two main categories are mainly reported: extensive and intensive green roofs (Dunnett and Kingsbury, 2004). However, the limits between these categories are not always clear. The German Landscape Development Research Society (FLL) identifies an additional category, i.e. simple intensive green roofs (FLL, 1995).

Extensive green roofs are the simplest form of such constructions. They are characterized by minimal construction and maintenance costs, soil layer weight and depth. Typical constructions of this type are usually not irrigated and the vegetation consists of self-sustaining and native species of plants that are well adapted to local climatic conditions. Such a green roof type is sometimes referred to as an eco-roof and has the lowest efficiency compared with the other categories, with respect to the building's energy performance. They can be applied to both flat and sloped roofs and to most existing types of building construction since the total additional load is relatively small.

Simple intensive green roofs are similar to the former but are characterized by higher vegetation density. Usually they are irrigated regularly during dry periods to support the vegetation requirements and are designed to be occasionally accessible. They have higher initial and maintenance costs, but their benefits in terms of a building's energy efficiency are better compared with extensive green roofs.

Intensive green roofs are the most demanding category of such constructions. The type of vegetation used is similar to the vegetation planted in gardens at ground level, even including small trees. The soil layers are thick enough to support this type of vegetation, and irrigation and regular maintenance are required. This type of green roof is often referred to as a roof garden and, depending on the total weight of the additional soil layers, can be applied only to buildings that can support the extra weight. It is the most expensive green roof type in terms of both construction and maintenance costs, but it is the most fully accessible, providing an additional area for the users of the building throughout the year.

GREEN ROOF PROPERTIES

Being part of the building shell, a green roof influences thermal flux through the area it occupies. In contrast with every other building element, the gardening layers of the green roof (soil and vegetation) are a living system that interacts both with the building and the environment in a variety of different ways. The most important benefits of green roofs can be summarized as follows (Scholz-Barth and Tanner, 2004; Oberndorfer et al, 2007):

- energy conservation for heating and cooling;
- reduction in the urban heat island effect;
- absorption of air pollutants and dust;
- attenuation of stormwater run-off;
- extension of life for waterproofing layers;
- attractive open space (aesthetic benefits);
- provision of wildlife habitat;
- replacement of vegetation and habitat lost during urban expansion;
- reduction of urban noise;
- social and psychological benefits.

On the other hand, among the major disadvantages of green roofs are the relatively high initial cost and the additional building load that must be supported. In the case of existing buildings, this limits the choice of green roof type to the extensive type, which in many cases does not need additional support.

EFFECTS OF GREEN ROOFS ON HEATING AND COOLING CONSUMPTION OF BUILDINGS

Historically, green roofs were known for their ability to provide thermal insulation in cold climates and to protect from overheating due to the high solar exposure of roofs in warm climates. These properties were much appreciated in times when modern thermal insulation materials and 'cool' materials were not available. In modern times, these properties are still regarded as the main benefits that a green roof can provide to a building, but until recently this knowledge had only a qualitative character since there was no further information to document this position.

The need for more efficient and environmentally friendly buildings, the deterioration of urban climate conditions, continuous urbanization and the limited possibilities that an urban building has to achieve the goal of sustainability, have prompted scientists to investigate green roof technology in order to convert the knowledge from a qualitative into a quantitative form in an effort to exploit and maximize the contribution of green roofs to the energy performance of buildings.

In general, a green roof has two or three additional layers compared with a conventional bare roof and forms a typical green roof construction consisting of:

- the structural layers (concrete slab, insulation layer, waterproofing membrane);
- the drainage layer (optional) that ensures proper removal of excess water or serves as a water reservoir during warm periods;

- the soil-mixture layer where the root system is developed;
- the vegetation layer (or canopy layer) which is the volume covered by the plants.

Every component of a green roof plays an important role in its thermal behaviour (Figure 11.2). The canopy provides solar shading for the soil surface, preventing overheating of the green roof in summer. The efficiency of the shading strongly depends on the vegetation type and its foliage density, expressed by the leaf area index (LAI). Although similar shading can be achieved by the use of a proper shading device, the material of the device would reflect or absorb solar radiation, resulting in solar reflections to the surrounding built environment or thermal transmittance due to its increased temperature. Where foliage is present, this is not the case. Foliage absorbs most of the incoming solar radiation and uses it for its biological functions like evapotranspiration, photosynthesis, etc. The former function is a natural protection measure against leaves overheating, which may destroy the plant. Biologically motivated, forced evaporation from the foliage decreases leaf surface temperature and cools the air in contact with the foliage. As long as there is enough moisture in the soil layer, the intensity of evapotranspiration is analogous to the heat stress, meaning that this biological cooling mechanism is adapted to ambient heat stress and is maximized during times of high solar intensity, when the need for cooling in buildings is also maximized.

Since the air in the canopy layer (the space between the soil surface and the top of the foliage) is kept at a relatively low temperature and dense shading is provided by the foliage, the upper surface of the soil layer is kept relatively cool compared with the ambient temperature. Considering the high thermal capacity of the soil layer, especially in

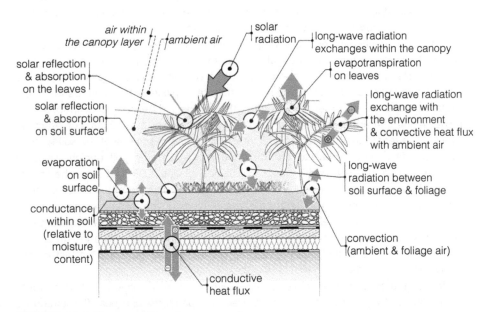

FIGURE 11.2 The energy balance for a green roof

the case of high moisture content, green roofs are in practice a cool 'heat sink' in contact with the roof, which otherwise during summer is highly exposed to solar radiation.

This cooling mechanism is present during the warm part of the year. During colder seasons, evapotranspiration is limited and shading is not a crucial factor for the energy budget due to low solar angles. On the contrary, another biological function – that of root respiration – is present, preventing root freezing and keeping the soil in the root area at temperatures higher than the atmosphere during very low ambient temperatures.

One of the first approaches to the thermal behaviour of green roofs assumes constant thermal properties of the soil layers and calculates thermal transmittance and winter heat flux through different types of green roofs in comparison with conventional flat roof constructions for different conventional insulation layers (Eumorfopoulou and Aravantinos, 1998). In this study, a simplified, steady-state calculation concludes that green roofs can provide a limited improvement in the thermal transmittance of the roof construction, compared with a bare roof with an insulation layer. Despite the fact that thermal transmittance improvement is very limited, the avoidance of extreme temperatures due to the presence of the soil layer results in a reduction in thermal losses, depending on the type of green roof installed. In the case of the extensive green roof type the reduction is small, whereas in the extreme case of a heavy intensive green roof construction with a thick soil layer of 0.5–0.9 m, thermal losses are almost half compared with those of a bare roof. The study concludes that green roofs can lower thermal losses in winter but similar thermal protection can also be provided with a slight increase in the thickness of the thermal insulation layer at greatly reduced cost.

More recent studies confirm these findings in general, but do not always agree on the energy efficiency of green roofs. This can be justified by the fact that the assumptions made in these studies, the climatic conditions examined and the green roof parameters in each case are characterized by significant dissimilarities.

Conclusions concerning energy savings for heating and cooling in different studies are often contradictory in the case of simulation approaches to the performance of a green roof. One reason for this is the large number of parameters used to accurately describe all the energy-related physical phenomena that take place within and beneath the vegetated volume. In order to simplify the model, different assumptions and simplifications that are used in many cases significantly alter the results. For example, some studies regard the thermal conductance and the thermal inertia of the soil layer as a constant value not related to the moisture content of the soil (which varies in accordance to atmospheric and irrigation conditions) (Sailor, 2008) and solve the energy balance of the soil layer like any other conventional opaque element of the building. Other studies assume that the density and the characteristics of the foliage remain constant during the year. In some cases, the drainage layer is filled with water, while in other cases this layer is not even present. Finally, the climatic conditions under which the green roof is investigated in these studies are very different and have a significant effect on the thermal behaviour of the green roof.

Most of the experimental studies dealing with the energy behaviour of green roofs have a comparative nature, focusing on surface temperatures and heat flux through a green roof and a conventional bare roof construction.

Wong et al (2007) conducted temperature measurements on a roof before and after the construction of four extensive green roofs on a multistorey car park in a tropical climate. The examined green roofs had no drainage layer to control moisture and there was no irrigation system installation. The measurements of the bare roof were taken over a period of 22 days in May and June while the same measurements on the green roof constructions were taken over an 18-day period in February and March. They found that a maximum temperature difference of 18°C existed between the bare and the green roof under similar climatic conditions. This temperature difference was reversed during dry periods with no rainfall. They calculated an overall 60 per cent reduction of heat gain in the case of the green roofs, leading to the conclusion that green roofs act as additional thermal insulation, minimizing inward heat flux during warm periods. But in contrast to conventional thermal insulation layers, during the hours when the bare roof was experiencing the highest heat stress, between 2 pm and 6 pm, the heat flux was reversed removing heat from the space beneath the green roof. They concluded that denser vegetation could contribute to a better cooling performance.

Onmura et al (2001) conducted summer field measurements and a wind-tunnel study to investigate the evaporative cooling effect of extensive green roofs with controlled irrigation in Japan. Field measurements were taken in a conditioned building with room air temperatures between 24 and 26°C. Despite the fact that the mean outdoor relative humidity (close to 73 per cent) did not favour evaporation, they showed that the surface temperature on the green roof was lower than that of the ambient air during the daytime and 30°C lower than the surface temperature of a bare roof. They estimated a 50 per cent reduction of heat flux entering the green roof compared with the bare roof. The wind-tunnel study showed that the temperature distribution within the soil layer is greatly influenced by the moisture content.

Experimental measurements and a mathematical approach to the energy behaviour of a green roof installed at a hotel near Athens, Greece, were conducted by Niachou et al (2001). Surface and air temperatures were recorded in August. Temperatures between 26 and 38°C were recorded on foliage by the use of infrared thermography. The room under the green roof showed improved thermal comfort conditions by 2°C compared with a room under a bare roof. The simulation showed that the contribution of the green roof in winter is minor in the case of a well-insulated roof; whereas in the case of buildings with moderate or no insulation, a green roof acts as a thermal barrier. In contrast, during the summer period the indoor air temperatures under a green roof are lower compared with a bare roof and higher during the night, although the more insulated the roof, the less obvious the differences. The greatest total energy conservation was 37 per cent for a non-insulated green roof and 2 per cent for a well-insulated one.

Another simulation study in Greece found a cooling load reduction of 58 per cent for the upper storey under a green roof, while for the entire two-storey building this fluctuated between 15 and 39 per cent (Spala et al, 2008). Regarding the winter period, the heating load conservation ranged between 2 and 8 per cent for the whole building and between 5 and 17 per cent for the highest storey, concluding that green roofs make a major contribution in the summer period and are less efficient during the winter period, although they still make a positive contribution to energy consumption. As the researchers themselves report, this is an encouraging result since most interventions aiming at cooling load reduction usually have a winter penalty.

Sonne (2006) measured surface temperatures on a green roof and on a bare roof and found a maximum average temperature difference of 21°C (54°C for the conventional and 33°C for the vegetated one). During the night, surface temperatures on the green roof were kept higher than the bare roof due to the thermal inertia and the decreased radiative heat losses to the night sky.

Comparing a bare roof and an extensive green one with no irrigation system on two different buildings in Portland in the US, Spolek (2008) calculated (from temperature measurements) reduced heat flux in the case of the green roof by 13 per cent in winter and 72 per cent in summer. Although no passive cooling is reported for the summer period, the temperatures at the top of the concrete slab were lower than that of the bare roof and the ambient air during daytime. Due to the high thermal inertia of the soil layer, the temperature peaks in the case of the green roof are 4–6 hours out of phase compared with the ambient atmosphere and the bare roof.

Measurements on a green roof experiment on a five-storey building in Greece show that a regularly irrigated green roof with a drainage layer that acts as a water reservoir in summer can provide almost constant daytime temperatures under the bitumen layer (Aravantinos et al, 1999). The measurements showed a temperature of 27°C lower than the ambient temperature and 24°C cooler than the temperature at the same point on an adjacent bare roof during daytime. The temperature difference of the external surfaces of both roofs was 31°C during the day and 2°C during the night, with the green roof surface cooler than the exposed reflective surface of the bare roof. Due to the high thermal inertia of both the drainage layer and the wet soil, there was a thermal lag of about 2 days under the soil layer.

Measurements of the dependence of soil thermal properties on its moisture content show that there is a significant difference between dry and wet soils (Sailor et al, 2008). Thermal conductivity in a variety of soil samples increased at a rate of 0.038 W/mK per 10 per cent increase in soil saturation, and the similar rate for specific heat capacity was found to be close to 32 J/kgK. In contrast, in most examined soil mixtures, the albedo of the soil was almost half in mixtures with 60 per cent moisture capacity compared with dry mixtures. In saturated soil, as in the case of a soil surface under irrigation, the albedo can be five times lower compared with that in dry soils. The study reports that the great dependence of thermal properties on moisture and the variation in soil moisture during the day makes it inappropriate to treat the layers of a green roof through a steady-state approach.

Since the total thermal flux that enters or leaves a building through a green roof is related to the size of the green roof cover, high roof–envelope ratios that can be found in single-storey buildings have a stronger effect on the energy conservation of a building, as the study by Martens et al (2008) discusses. In multistorey buildings, the effects of a green roof throughout the year are limited mainly to the upper storey.

The need to predict the thermal behaviour of a planted roof led various scientists to propose transient simulation models that include all the heat fluxes that can be found in a green roof. One of the most referenced models that was validated and exploited in many other studies is that by Palomo (1998). Although Palomo did not integrate this model into building simulation computer software in order to provide information for cooling energy consumption, a detailed parametric sensitivity analysis was presented in order to distinguish the most important factors that can control the thermal behaviour of a green roof. The main conclusion of the study is that a green roof behaves like a thermal insulator that can minimize the inward heat flux from the roof.

Based on this model, Theodosiou (2000) concentrated more on developing the drainage layer module in order to account for water storage during the summer period and used a slightly different module to describe the evapotranspiration process. After validating the model using measurement data for a 2-year period, he also conducted a parametric sensitivity analysis by integrating the model into building simulation software in order to account for the thermal interaction between the rooms beneath the green roof. He concluded that the foliage density is the most important factor when regular irrigation and a drainage layer (also used as a water reservoir) are used to prevent soil from drying and, together with the soil thickness, are the most important design factors for the provision of passive cooling (Theodosiou, 2003a). He also concluded that extensive green roofs behave similarly to an additional thermal insulation layer and do not provide passive cooling like more intensive green roofs can. He also reported that passive cooling is maximized in non-insulated roof constructions, and is noticeable in moderately insulated ones, while in well-insulated roofs the soil layer is actually isolated from the layers beneath and the green roof effects are minimized. Relative humidity is found to be the most important climatic factor. High relative humidity values lower evapotranspiration capabilities and therefore the energy conservation for cooling.

The ability of a green roof to provide passive cooling is also reported by Lazzarin et al (2005). By using field measurements on a green roof installed in a hospital in Italy, and by using an analytical model, the authors found that the heat flux was directed upwards through the wet green roof, providing cooling to the internal space during the summer months. In dry soil conditions, the green roof behaviour was restricted to a role similar to that of additional thermal insulation. During the winter there was slightly higher thermal loss compared with a bare roof.

In another study using a modified version of Palomo's model, Theodosiou (2003b) reported that a green roof in the Mediterranean climate can provide passive cooling only when the indoor air temperature of an air-conditioned building is higher than 23–24°C. In lower internal temperatures, although there is still a reduction in cooling load compared with a bare roof, the heat flux is directed inward. Energy conservation for cooling is reported as 18, 21, 25 and 31 per cent when the room beneath the green roof is kept at temperatures of 20, 22, 24 and 26°C, respectively.

Another green roof model based on Palomo's model was validated and utilized in order to examine the influence of foliage density on the energy performance of a green roof located in India (Kumar and Kaushik, 2005). In this study the authors reported that the green roof in question, in combination with solar thermal shading, can lower indoor conditions by almost 5°C compared with a bare roof, by providing a cooling potential of 3.02 kWh per day.

Recently, a green roof model integrated into a widely available building simulation software programme was proposed by Sailor (2008). The inclusion of a green roof module in a simulation package provides an opportunity for further investigation of the thermal behaviour of green roofs.

From the study of the literature, it is easy to be convinced of the positive contribution that a green roof can make to the energy efficiency of a building. Concerning the behaviour of a green roof in winter conditions, most studies agree that although the additional layers can contribute to an increase in the thermal insulation protection of the roof element, they cannot replace the thermal insulation layer. The actual level of additional insulation provided by the soil and vegetation layers is not clear, since it is closely related to the variable soil-moisture content. In regions with frequent rainfall, the moisture content in winter will be high and will increase thermal conductivity, whereas in dry winter conditions, thermal conductivity will be lower, providing better thermal insulation. Consequently, the additional insulation level is subject to climatic conditions, soil mixture composition and the construction characteristics of the green roof.

This positive contribution is more obvious in the summer period, where all the researchers agree that the presence of a green roof can reduce cooling loads. What is not easy to detect in this period is the ability of a green roof to behave like a passive cooling technique or to simply contribute to enhanced thermal insulation protection. In some studies, although the passive cooling ability is not recognized, the presented measurements or calculated data show that there are times when the heat flux is directed upwards, removing heat from the interior, even in conditioned buildings. This phenomenon which cannot be driven by enhanced thermal insulation (which in extreme cases can only provide zero inward-directed heat flux but not the reverse) still needs to be further investigated in future studies.

Since most of the reported field measurements and simulation studies have been conducted only during the past few years, it is obvious that the study of green roofs is an ongoing procedure that will include more information in the future. This further study is assisted by the availability of more user-friendly green roof simulation modules that are now incorporated into modern whole-building simulation models.

URBAN HEAT ISLAND MITIGATION

In many scientific studies that examine the urban heat island (UHI) effect, 'cool' roofs and vegetation are regarded as the most effective measures that could upgrade the urban climate if they were applied to a large section of a city's surface area (Santamouris, 2007). In all of these studies, vegetation consists mostly of ground-level plants and trees. Although ground-level vegetation shares significant similarities with the vegetation found on green roofs, there are some differences that may alter the efficiency of green roof vegetation in terms of UHI mitigation. The most obvious are the type of plants used in extensive green roofs, which are of low height and have minimum water requirements, and the low height and weight of the soil layer. These differences are minimized in intensive green roofs, where there are fewer restrictions concerning the selection of the plants, which can be similar to the ground-level vegetation.

Nevertheless, even if the direct effects of evapotranspiration to the ambient air are disregarded, the significantly cooler green roof surfaces (even by 30°C cooler) that are reported in the studies mentioned earlier show that green roofs can balance out one of the major factors that causes the UHI effect, that of overheated urban surfaces.

In the literature, there are only a few studies that deal in detail with the impact of green roof vegetation on UHI mitigation. Bass et al (2003), using a mathematical model, calculated that a 50 per cent extensive green roof coverage without installation of irrigation systems in Toronto would cool the ambient air by 1°C in summer. If all these roofs were regularly irrigated, then the cooling effect would be 2°C over the area where the green roofs were installed and 1°C over a larger geographical area.

In a US Department of Energy report, Tanner and Scholz-Barth (2004) noted that because green roofs absorb most of the incident solar radiation, they can provide similar effects to 'cool' roofs without reflecting solar radiation to taller adjacent buildings that could cause an additional cooling load, glare and discomfort to their occupants.

Wong et al (2003a) measured air temperature and mean radiant temperature (MRT) at various heights above a green roof canopy and above a bare roof and found an air temperature decrease of 4.2°C and an MRT decrease of 4.5°C over the green roof. In higher positions the cooling effect was limited, perhaps because of the relatively small size of the green roof.

Takebayashi and Moriyama (2007) measured surface temperatures on bare roofs with high reflective coatings and on a non-irrigated green roof in Japan. They found that the cooler surface was that of the high reflective coating, followed by foliage temperature of the green roof (1–2°C warmer). The temperature of the other surfaces was higher by several degrees. They also reported that the surface temperature of the green roof was expected to be lower in summer.

Surface temperature measurements in two extensive green roofs in Germany showed that wet soil layers are capable of maintaining a 5°C lower temperature compared with dry conditions (Köhler et al, 2003).

Alexandri and Jones (2007) used a two-dimensional model to study the effect of green roofs and walls in the urban climate. They found that due to the redistribution of radiation within the canopy layer, the total radiative heat exchanges were smaller on the green roof compared with a bare roof. If green roofs were installed over a wide urban area, air masses entering the urban canyons would be cooled by the vegetation and there would be a temperature drop even at street level. The authors conclude that the contribution to air cooling is more intense in hot and dry climates like Athens, but even humid regions can benefit from air cooling through green roofs.

All the studies dealing with the green roof effects on UHI agree that the major advantage of green roofs in UHI mitigation is the large amount of space made available for vegetation growth compared with the non-existence of available ground-level areas in the existing urban environment. In every case, the contribution of green roofs is considered to be significant only where there is widespread implementation of green roofs, an action which obviously requires properly planned urban policies. Due to the relatively limited number of studies conducted specifically on the issue of UHI mitigation by green roofs, there are no absolute conclusions concerning the effectiveness of this technology compared with 'cool' roofs, which have been investigated more extensively.

OTHER ENVIRONMENTAL BENEFITS

Stormwater runoff management

Beyond their contribution to a building's energy conservation and to the reduction of the urban heat island effect, green roofs have a significant potential for precipitation and surface water runoff management. The impervious nature of artificial surface materials in the urban environment is the main reason for rapid rainwater runoff and high peak flows after rain events. This is a source of significant problems in urban drainage systems that are sometimes overloaded and in water source contamination by the first flush which contains the highest pollution levels (USEPA, 2003). Both problems are expected to be more intense in the future since global warming is causing an increase in the frequency and intensity of rainfall (Easterling et al, 2000).

By providing permeable surfaces instead of impervious bare roofs, green roofs can retain rainwater in the short term and, depending on the drainage system, can even reduce the amount of rainwater that reaches the urban drainage system in short rain events.

Thomson and Sorvig (2000) measured a total of 18 litres of rainfall reaching the ground after a 10 mm rainstorm, where 200 litres of rain fell onto an 18 m² green roof. Depending on the green roof's construction, the climate and precipitation characteristics, the rainfall retention capability on a yearly basis may range from 75 per cent for intensive green roofs equipped with a drainage layer to 45 per cent for extensive green roofs (Mentens et al, 2006). On an urban scale, the widespread implementation of green roofs could be part of a city's stormwater management policy. As such, many municipal stormwater management companies provide a discount to the utility fee like Portland's Clean River Incentive and Discount programme (CRID) in the US (Liptan, 2003).

Air quality improvement

Vegetation on green roofs also provides air quality improvement as the plants take up NO_x and CO_2 from the urban ambient air (Clark, 2005). Currie and Bass (2008) report that shrubs on green roofs can be as effective as trees in removing PM_{10} (particulate matter 10 μm or less) when green roofs are installed in sufficient quantities. Field measurements on a 4000 m² green roof in Singapore showed that the levels of particles and SO_2 in the air above the green roof were reduced by 6 and 37 per cent, respectively (Tan and Sia, 2005). Yang et al (2008) modelled the air pollutant uptake by 19.8 hectares of green roofs in Chicago and estimated the uptake of four air pollutants (O_3, NO_2, PM_{10}, SO_2) to be 52, 27, 14 and 7 per cent, respectively.

Besides these direct effects of rooftop vegetation, the decrease in surface temperature caused by green roofs can lead to a reduced production of atmospheric ozone (Taha, 1996).

In all cases, the ability of green roofs to improve air quality was found to be site specific and in order to affect the urban air quality, policies that would encourage widespread implementation of green roofs throughout an urban area are required.

Noise reduction

The urban noise reduction following the installation of a green roof was estimated by Renterghem and Botteldooren (2008) to be as high as 10 dB for an extensive green roof.

In their study, the authors found that thicker green roof types do not have a significant additional effect. In other experimental studies of intensive green roofs, the noise reduction inside the building concerned ranged from 8 to 13 dB depending on the frequency (Connelly and Hodgson, 2008), whereas in other studies like Dürr's (1995) a significant reduction close to 46 dB was measured.

ECONOMIC FEASIBILITY OF GREEN ROOFS

Despite the proven energy conservation ability of a green roof, it is worth looking at the economic feasibility of such a technology as presented in the literature. In an effort to evaluate the economic benefits of green roofs solidly as an energy conservation measure for buildings, the main factors that determine the life cycle cost are energy conservation, the life extension of the waterproofing layers, and the initial and maintenance costs, which are related to the type of green roof and to the selection of plants. The beneficial effects on the building are mainly limited to the upper storey of the building and, obviously, in tall buildings, the total energy conservation is limited. By contrast, in single-storey buildings, the energy conservation may be large enough to justify the increased initial cost, assuming that an accurate estimation of the energy conservation can be achieved. In all cases, the tools that could support such a prediction are still at a preliminary stage since they were only recently presented.

One of the greatest barriers for building owners is the initial cost compared with the conventional option of a bare roof, since maintenance costs can be insignificant in the case of extensive green roof types, or even lower than conventional roofs when considering the life extension of the waterproofing layers. A conventional roof will require replacement of the waterproofing layers every 20 years while in the case of a green roof their life is extended up to 90 years (Porsche and Köhler, 2003).

Taking into consideration the life extension of the roof layers, the maintenance costs and the energy conservation profits, a life cycle cost analysis for extensive green roofs in Singapore conducted by Wong et al (2003b) concluded that green roofs do not cost more than conventional flat roofs.

The relatively low penetration of green roofs into the US market and many other countries results in higher construction costs and makes such a construction more expensive than elsewhere like northern Europe or some Asian countries. This is the main reason, according to a study that took into account the economic benefit from stormwater management and air quality improvement, for green roofs to be 10–14 per cent more expensive than bare roofs throughout their lifetime (Carter and Keeler, 2008).

As some of the above studies depict, a major difference of this technology compared with other comparable ones, is the number of environmental and social benefits it brings that are difficult, if not impossible, to quantify and evaluate by means of economic profit. These include their potential for urban heat island mitigation, urban air quality upgrade, aesthetic upgrade and their inherent ecological nature such as increased urban biodiversity. Additionally, increased property value is subject to local market conditions and is difficult to predict. Most of these are public benefits and in many regions of the world, the implementation of green roofs in buildings is supported by municipal authorities by appropriate legislation and financial measures as the following section presents.

GREEN ROOF POLICIES

Since the environmental advantages of green roofs benefit the public and, for many people, are considered to be more important than energy conservation in an individual building, many local authorities worldwide have presented measures to promote the implementation of this technology in urban areas.

In Tokyo, Japan, private buildings larger than 1000 m² and public buildings larger than 250 m² are required to have a green roof that covers at least 20 per cent of the total roof surface. By January 2005, 54.5 ha of green roofs had been installed in the city (TKM, 2007).

In Germany, where the construction of green roofs in many cities is enforced by local building regulations, 13.5 million m² per year is added to the existing green roof coverage (Oberndorfer et al, 2007).

In Toronto, Canada, the city has approved a policy that requires green roof constructions covering 50–75 per cent of a building's footprint, whereas in Portland, Oregon, the requirement is 70 per cent (Carter and Fowler, 2008).

Similar measures are also reported in Switzerland, Austria and elsewhere. Moreover, for buildings requiring an environmental label, a green roof construction can have a contribution as it may count for up to 15 credits under the LEED rating system (Kula, 2005).

DISCUSSION

Regarding the contribution of planted roofs to the energy consumption of buildings, most studies agree that there is a positive effect for the top storey of the buildings. This effect lies mainly in the influence of the thermal flux through the roof element. Researchers agree that although the biological and physical 'mechanisms' that form the thermal effects on a green roof are the same throughout the year, the overall thermal behaviour of a green roof varies considerably between the cold and warm seasons of the year. This is due to the fact that the intensity of the existing energy fluxes varies according to the prevailing climatic conditions.

On the other hand, the extent to which the positive effect of a green roof can provide noticeable energy savings in a building is still not absolutely clear among the researchers, not because of the different scientific approaches, but mainly for the following reasons:

- In the literature, different green roof constructions are investigated in respect of the materials used, the presence of a drainage layer and the thickness of the green roof layers.
- The variety of vegetation type, even within the same experiment, can significantly alter the surface temperature of the soil, affecting the thermal behaviour.
- Each experimental or mathematical study is limited to local climatic conditions, such as relative humidity, to which, as the literature reports, the thermal behaviour is very sensitive, especially during the summer period.
- A crucial factor for the thermal behaviour of a green roof is the moisture content of the soil and the presence of water within the drainage layer. Information regarding these factors is not always clear or similar in the examined studies, leading to significant thermal behaviour differences. This is most important in the case of the use of a green roof for cooling energy conservation, since the evapotranspiration at

the soil surface and in the canopy layer is highly dependent on the water availability in the soil layer during hours of heat stress. Dry conditions in summer, which usually occur in eco-roofs, limit the amount of water that can evaporate in order to provide the cooling effect. In contrast, regular irrigation can provide the necessary water to promote this process and consequently lead to a more intense passive cooling effect.

● The internal air temperature during the experimental study or during the simulation process in conditioned buildings is another factor that is not common in the studies, despite the fact that it is a significant factor in every passive cooling technique. This mainly concerns the conclusions regarding the ability of the green roof to provide passive cooling (heat flow directed upwards through the roof) or simply act as an extra insulation element by limiting the inward-directed heat flux). Mechanical cooling at relatively low air temperatures like 20–22°C can 'deactivate' many passive cooling techniques during the warm hours of the day when the ambient temperature can be more than 10°C warmer than the interior.

From the above, one can easily conclude that the thermal behaviour of a green roof is variable, depending on a relatively large number of factors. These factors are not always possible to control, like the density and other characteristics of the vegetation. Other variables can be designed during the design process such as the characteristics of the soil mixture, the vegetation species, the irrigation programme and the construction pattern.

In all cases during the warm period of the year, the selection of green roof specifications in order to provide energy conservation should take into consideration the climatic conditions, especially the relative humidity since the additional thermal fluxes that can be found on a green roof in comparison with a bare roof are highly dependent upon the humidity of the ambient air, like all the other passive cooling techniques that are based on water evaporation.

On the other hand, despite the fact that green roofs have a positive effect on limiting thermal losses in winter, they cannot replace conventional thermal insulation. A slight increase in the insulation layer thickness can have the same effect at a significantly lower cost. Recent studies concerning the thermal insulation capabilities of additional green roof layers agree that the simplification of increasing the U-value in order to account for the extra soil and vegetation layers is significantly inaccurate due to the variation of the thermophysical properties according to the moisture content. Many experimental and unsteady-state simulation studies have found that there is an insignificant contribution to thermal loss reduction. However, even if the thermal protection in winter is limited compared with insulation materials, a green roof can be regarded as an additional thermal protection measure with a variety of other benefits, such as extending the life of the waterproofing materials, etc.

CONCLUSIONS

This chapter presents the major findings of recent research on the energy performance of green roofs and on their contribution to urban heat island mitigation. Additionally, it reports the results of recent studies related to the other major benefits of green roofs in order to form a more integrated approach to this technology.

Regarding all the benefits that green roofs are reported to provide, both on the scale of individual buildings and on an urban scale, the scientific literature is in agreement on all the positive effects. The quantitative character of these benefits is mostly related to the soil mixture and its moisture content, to the plant selection and climate conditions, with an emphasis on relative humidity which is directly related to evapotranspiration. In cases where the passive cooling potential of green roofs is to be optimized, regular irrigation and a drainage layer fulfilling the role of water reservoir are important. In dry conditions, although there is still noticeable cooling energy conservation, the role of the green roof is similar to enhanced thermal insulation. At this point, someone could argue that since green roofs are regarded as an ecological technology there may be a conflict between the water consumption and the energy conservation. Unfortunately, such a study could not be found although maybe it will form a topic for future study.

Despite the old age of this technology, scientific investigation of green roofs is a relatively new field and more studies are expected to support and to enrich the knowledge gained so far. This can be justified by the fact that most of the literature on green roofs is related to studies conducted during the past few years. The recent introduction of thermal simulation models and the continuing expanding implementation of green roofs in all regions of the world are expected to help scientists and engineers examine this technology more comprehensively.

AUTHOR CONTACT DETAILS

Theodore Theodosiou: Department of Mechanical Engineering, University of Western Macedonia, GR-50100, Kozani, Greece; tgt@uowm.gr

REFERENCES

Alexandri, E. and Jones, P. (2007) 'Developing a one-dimensional heat and mass transfer algorithm for describing the effect of green roofs on the built environment: Comparison with experimental results', *Building and Environment*, vol 42, no 8, pp2835–2849

Aravantinos, D., Theodosiou, T. and Tourtoura, D. (1999) 'The influence of a planted roof on the passive cooling of buildings', in G. Grossman (ed), *Proceedings of ISES Solar World Congress*, vol 2, Jerusalem, Israel, 4–9 July, pp256–364

Bass, B., Krayenhoff, E. F., Martilli, A., Stull, R. B. and Auls, H. (2003) 'The impact of green roofs on Toronto's urban heat island', in *Proceedings of the First North American Green Roof Conference: Greening Rooftops for Sustainable Communities*, Chicago, The Cardinal Group, Toronto, 29–30 May, pp292–304

Carter, T. and Fowler, L. (2008) 'Establishing green roof infrastructure through environmental policy instruments', *Environmental Management*, vol 42, no 1, pp151–164

Carter, T. and Keeler, A. (2008) 'Life-cycle cost-benefit analysis of extensive vegetated roof systems', *Journal of Environmental Management*, vol 87, no 3, pp350–363

Clark, C. (2005) 'Optimization of green roofs for air pollution mitigation', in *Proceedings of 3rd North American Green Roof Conference: Greening Rooftops for Sustainable Communities*, Washington DC, The Cardinal Group, Toronto, 4–6 May, pp482–497

Connelly, M. and Hodgson, M. (2008) 'Sound transmission loss of green roofs', in *Proceedings of 6th Annual Greening Rooftops for Sustainable Communities Conference*, Baltimore, MD, 30 April–2 May, pp1–10

Currie, B. and Bass, B. (2008) 'Estimates of air pollution mitigation with green plants and green roofs using the UFORE model', *Urban Ecosystems*, vol 11, no 4, pp409–422

Dunnett, N. and Kingsbury, N. (2004) *Planting Green Roofs and Living Walls*, Timber Press Inc, Portland, OR

Dürr, A. (1995) *Dachbegrunung: Ein Okologischer Ausgleich Gütersloh*, Bauverlag BV, Gütersloh, Germany

Easterling, D., Meehl, G., Parmesan, C., Changnon, S., Karl, T. and Mearns, L. (2000) 'Climate extremes: Observations, modeling, and impacts', *Science*, vol 289, no 5487, pp 2068–2074

Eumorfopoulou, E. and Aravantinos, D. (1998) 'The contribution of a planted roof to the thermal protection of buildings in Greece', *Energy and Buildings*, vol 27, no 1, pp29–36

FLL (Forschungsgesellschaft Landschaftsentwicklung Landschaftbau) (1995) *Guidelines for the Planning, Execution and Upkeep of Green-roof Sites*, FLL, Bonn

Köhler, M., Schmidt, M. and Laaer, M. (2003) 'Green roofs as a contribution to reduce urban heat island', in *Proceedings of the World Climate and Energy Event*, Rio de Janeiro, Brazil, 1–5 December, pp493–498

Kula, R. (2005) 'Green roofs and the LEED green building rating system', in *Proceedings of 3rd North American Green Roof Conference: Greening Rooftops for Sustainable Communities*, Washington DC, The Cardinal Group, Toronto, 29–30 May, pp113–120

Kumar, R. and Kaushik, S. C. (2005) 'Performance evaluation of green roof and shading for thermal protection of buildings', *Building and Environment*, vol 40, no 11, pp1505–1511

Lazzarin, R. M., Castellotti, F. and Busato, F. (2005) 'Experimental measurements and numerical modeling of a green roof', *Energy and Buildings*, vol 37, no 12, pp1260–1267

Liptan, T. (2003) 'Planning, zoning and financial incentives for ecoroofs in Portland, Oregon', in *Proceedings of 1st North American Green Roof Conference: Greening Rooftops for Sustainable Communities*, Chicago, The Cardinal Group, Toronto, 29–30 May, pp113–120

Martens, R., Bass, B. and Saiz, S. A. (2008) 'Roof-envelope ratio impact on green roof energy performance', *Urban Ecosystems*, vol 11, no 4, pp339–408

Mentens, J., Raes, D. and Hermy, M. (2006) 'Green roofs as a tool for solving the rainwater runoff problem in the urbanized 21st century?', *Landscape and Urban Planning*, vol 77, no 3, pp217–226

Niachou, A., Papakonstantinou, K., Santamouris, M., Tsangrassoulis, A. and Mihalakakou, G. (2001) 'Analysis of the green roof thermal properties and investigation of its energy performance', *Energy and Buildings*, vol 33, no 7, pp719–729

Oberndorfer, E., Lundholm, J., Bass, B., Coffman, R. R., Doshi, H., Dunnett, N., Gaffin, S., Köhler, M., Liu, K. K. Y. and Rowe, B. (2007) 'Green roofs as urban ecosystems: Ecological structures, functions, and services', *BioScience*, vol 57, no 10, pp823–833

Onmura, S., Matsumoto, M. and Hokoi, S. (2001) 'Study on evaporative cooling effects of roof lawn gardens', *Energy and Buildings*, vol 33, no 7, pp653–666

Palomo, E. D. B. (1998) 'Analysis of the green roofs cooling potential in buildings', *Energy and Buildings*, vol 27, no 2, pp179–193

Porsche, U. and Köhler, M. (2003) 'Life cycle costs of green roofs: A comparison of Germany, USA, and Brazil', in *Proceedings of the World Climate and Energy Event*, Rio de Janeiro, Brazil, 1–5 December, pp461–467

Renterghem, V. T. and Botteldooren, D. (2008) 'Numerical evaluation of sound propagating over green roofs', *Journal of Sound and Vibration*, vol 317, no 3–5, pp781–799

Sailor, D. J. (2008) 'A green roof model for building energy simulation programs', *Energy and Buildings*, vol 40, no 8, pp1466–1478

Sailor, D. J., Hutchinson, D. and Bokovoy, L. (2008) 'Thermal property measurements for ecoroof soils common in the western US', *Energy and Buildings*, vol 40, no 7, pp1246–1251

Santamouris, M. (2007) 'Heat island research in Europe: The state of the art', *Advances in Building Energy Research*, vol 1, no 1, pp123–150

Scholz-Barth, K. and Tanner, S. (2004) *Green Roofs: Federal Energy Management (FEMP) Federal Technology Alert*, Report DOE/EE-0298, US Department of Energy

Sonne, J. (2006) 'Evaluating green roof energy performance', *ASHRAE Journal*, vol 48, no 2, pp59–61

Spala, A., Bagiorgas, H. S., Assimakopoulos, M. N., Kalavrouziotis, J., Matthopoulos, D. and Mihalakakou, G. (2008) 'On the green roof system: Selection, state of the art and energy potential investigation of a system installed in an office building in Athens, Greece', *Renewable Energy*, vol 33, no 1, pp173–177

Spolek, G. (2008) 'Performance monitoring of three ecoroofs in Portland, Oregon', *Urban Ecosystems*, vol 11, no 4, pp349–359

Taha, H. (1996) 'Modeling impacts of increased urban vegetation on ozone air quality in the South Coast Air Basin', *Atmospheric Environment*, vol 30, no 20, pp3423–3430

Takebayashi, H. and Moriyama, M. (2007) 'Surface heat budget on green roof and high reflection roof for mitigation of urban heat island', *Building and Environment*, vol 42, no 8, pp2971–2977

Tan, P. Y. and Sia, A. (2005) 'A pilot green roof research project in Singapore', *Proceedings of 3rd North American Green Roof Conference: Greening Rooftops for Sustainable Communities*, Washington DC, The Cardinal Group, Toronto, 4–6 May, pp399–415

Tanner, S. and Scholz-Barth, K. (2004) *Federal Technology Alert: Green Roofs*, DOE Energy Efficiency and Renewable Energy (EERE), www.osti.gov/energycitations/servlets/purl/15009602-KD2isR/native/15009602.pdf (accessed February 2009)

Theodosiou, T. (2000) *Analytical and Experimental Study on the Contribution of Planted Roofs in Passive Cooling of Buildings* (in Greek), dissertation, Aristotle University of Thessaloniki

Theodosiou, T. (2003a) 'Summer period analysis of the performance of a planted roof as a passive cooling technique', *Energy and Buildings*, vol 35, no 9, pp909–917

Theodosiou, T. (2003b) 'Comparative evaluation of a planted roof as a natural cooling technique for buildings', *Ktirio*, vol A–B, no 1, pp43–50 (in Greek)

Thomson, J. W. and Sorvig, K. (2000) *Sustainable Landscape Construction*, Island Press, Washington DC

TKM (Tokyo Metropolitan Government) (2007) *Basic Policies for the 10-Year Project for Green Tokyo*, www.kankyo.metro. tokyo.jp/kouhou/english/

USEPA (United States Environmental Protection Agency) (2003) *Protecting Water Quality from Urban Runoff*, EPA 841-F-03-003

Wong, N. H., Chen, Y., Ong, C. L. and Sia, A. (2003a) 'Investigation of thermal benefits of rooftop garden in the tropical environment', *Building and Environment*, vol 38, no 2, pp261–270

Wong, N. H., Tay, S. F., Wong, R., Ong, C. L. and Sia, A. (2003b) 'Life cycle cost analysis of rooftop gardens in Singapore', *Building and Environment*, vol 38, no 3, pp499–509

Wong, N. H., Yok, T. P. and Yu, C. (2007) 'Study of thermal performance of extensive rooftop greenery systems in the tropical climate', *Building and Environment*, vol 42, no 1, pp25–54

Yang, J., Yu, Q. and Gong, P. (2008) 'Quantifying air pollution removal by green roofs in Chicago', *Atmospheric Environment*, vol 42, no 31, pp7266–7273

publishing for a sustainable future

12

A Review of Building Earth-Contact Heat Transfer

Stamatis Zoras

Abstract
A review and categorization of the methods developed for the prediction of heat transfer in earth-coupled structures is presented. The thermal properties of soils are also briefly discussed. The four categories of tools reviewed are analytical and semi-analytical methods, numerical methods, manual methods and design guides. Numerical methods in three dimensions seem to be the most appropriate ones in building ground-coupled simulation due to their accuracy and ability to handle any complex configuration. Areas of future improvements are also discussed.

■ *Keywords* – earth-contact; heat transfer; conductivity

INTRODUCTION
The reduction of the energy consumption of buildings has been of concern for many years. A combination of limited energy sources and associated environmental issues has led to new techniques for energy saving so as to improve or to conserve the quality of life. In Western Europe, it is reported that 52 per cent of energy used is consumed to maintain acceptable environmental conditions within buildings (The Energy Saver, 1993).

Specifically, this work concerns phenomena related to earth contact. Research on building losses to the ground started in the US during the 1940s (Houghten et al, 1942; Dill et al, 1943, 1945). It has been claimed that in cold climates the heat loss to the ground may be responsible for up to one third or even a half of total heat losses (Adamson et al, 1964, 1973; Adamson, 1973; Claesson and Eftring, 1980; Bahnfleth, 1989a, 1989b; Claesson and Hagentoft, 1991; Hagentoft and Claesson, 1991). The benefits of earth-coupled systems have been of concern in relation to climatic characteristics in China (Anselm, 2008), India (Kumar et al, 2007), Turkey (Aksoy and Inalli, 2006), Finland (Rantala and Leivo, 2006, 2008) and Japan (Iwamae and Matsumoto, 2003). Furthermore, in the US a survey suggested that a waste of about $5–15 billion a year could be attributed to heat transfer to the ground (Claridge, 1987). The increased importance of earth-contact heat transfer against air temperature has been presented by Sobotka et al (1996) in relation to structural design and climatic conditions. In Sweden, ground storage rock construction conserves energy from solar heating (Lundh and Dalenbäck, 2008), while soil may be used as a cooling source in buildings (Givoni, 2007).

Many different tools have been developed for the prediction of heat transfer in earth-coupled structures. Each one of them has its own advantages and limitations. The present review reports on the methods and tools that cope with the calculation of heat transfer to the ground from buildings in contact with the soil.

A variety of modeling and experimental techniques of the soil's temperature calculation have been developed in the past which mostly require past field data for validation and model construction (Jacovides et al, 1996; Mihalakakou, 2002) while based on a one-dimensional transient heat conduction equation (Mihalakakou et al, 1997). With regard to the earth-contact component, it is necessary to recognize that the soil energy exchange process is complex, i.e. multidimensional (Davies et al, 1995; Adjali et al, 2000a; Dos Santos and Mendes, 2004; Rees et al, 2006a). Many alternative simulation methods have been proposed which deal with this complexity with varying degrees of sophistication (for an extensive review see Adjali et al, 1998a).

The complexity of earth-coupled systems inevitably leads to assumptions being made regarding the parameters that influence the processes of heat transfer to the ground. Coupled moisture and heat transfer tools are the ideal ones to simulate energy losses to the ground from buildings (Rock, 2004). However, implications arise due to the non-linear nature of heat transfer concerning moisture transfer and thermal conductivity in relation to temperature variation and other climatic conditions e.g. snow and rain (Adjali et al, 1998b). Recently, it has been stated that coupled simulations could hold for a variety of climates and foundation configurations (Janssen et al, 2004).

The development of numerical methods and the high performance of new computers have provided researchers with tools such that earth-contact heat transfer can be simulated more efficiently. Analytical methods generally give fast and accurate results only for simple cases. Numerical methods can cope with more complex configurations but the computer run time needed is much longer than the run time required by the analytical methods.

THERMAL PROPERTIES OF SOILS

This section reports on the thermal properties of soils that influence the calculation of heat transfer within solids. The heat capacity of a material is required when non-steady solutions are to be determined. In effect, the heat capacity defines the amount of energy stored in a material per unit mass per unit change in temperature ($Jkg^{-1}K^{-1}$). It is given as a function dependent on the different heat capacities and volumes of the constituents (Rees et al, 2000):

$$C_P = v_1\rho_1C_{P1} + v_2\rho_2C_{P2} + \dots\dots\dots \qquad [12.1]$$

where C_p is the total heat capacity of the material and v, ρ and C are respectively the volume fractions, the densities and the heat capacities for each constituent.

The thermal conductivity of a material is the constant of proportionality that relates the rate at which heat is transferred by conduction to the temperature gradient inside the material (Wm^{-1}K^{-1}). In fact, to simulate an earth-coupled system, a problem arises regarding the difficulty in defining the thermal conductivity of the soil. The thermal conductivity of the soil depends on temperature and the different materials (soil grains, water and air) that compose the ground. The soil moisture content influences the temperature of the ground by its transportation and by the latent heat caused from phase changes. Saturated soils transmit heat at a faster rate than unsaturated materials. A good assumption is to neglect the air's thermal conductivity since it is much lower than the conductivities of the other components. It is extremely difficult to simulate the ground to take into account both moisture and temperature changes. So, usually a constant ground temperature and a fixed amount of moisture are utilized, which results in inaccuracies in cases where soil property combinations are of paramount importance. Thus, the most efficient tool would include the calculation of coupled heat and moisture transfer in the ground. Theoretically, for an isotropic material, the thermal conductivity can be expressed as (Carslaw and Jaeger, 1959):

$$\lambda = \frac{Qd}{(T_0 - T_1)St}$$ [12.2]

where λ (Wm^{-1}K^{-1}) is the thermal conductivity, d is the thickness of a plate in the soil (m), Q the heat (J) which flows up through the plate in t seconds from a surface S (m^2) caused by a difference in temperatures $T_o - T_l$ (K). This temperature gradient applies after the soil has reached its steady state with no further significant changes in temperature values. Generally, the calculation of thermal conductivity of soils is not a straightforward and easy procedure due to the generally complex nature of soils (e.g. anisotropy, soils composed of more than one material). This becomes worse if changes in the thermal conductivity are taken into account due to variations in temperature (Givoni and Katz, 1985). Many works have been undertaken aiming to reveal exact formulations for the theoretical or empirical calculation of thermal conductivity. However, the definition of conductivity is always restricted by assumptions being made of homogeneous, isotropic materials and moisture content. Details can be found in an extensive work that has been undertaken by Rees et al (2000).

The thermal resistance R (m^{2}KW^{-1}) of a structural component is the property that characterizes the response of the component to heat transfer fluctuations. For a unit area of a slab of homogeneous material, the thermal resistance is calculated by dividing its thickness l (m) by its thermal conductivity λ (Wm^{-1}K^{-1}) (CIBSE, 1980) and it is expressed as:

$$R = \frac{l}{\lambda}$$ [12.3]

Equation 12.3 describes the thermal resistance within the solid. However, the thermal resistance of a composite solid can be determined by adding the individual resistances of different materials. The heat transfer by radiation and convection at the outer or inner surfaces of building elements or the ground surface can be treated as flow through a thermal resistance R_s (m^{2}KW^{-1}) given by the following equation (CIBSE, 1980):

$$R_S = \frac{1}{Eh_r + h_c}$$ [12.4]

where E is the emissivity factor, h_r (Wm^{-2}K^{-1}) the radiative heat transfer coefficient and h_c (Wm^{-2}K^{-1}) the convective heat transfer coefficient.

The thermal transmittance (U-value) of a structural element gives an indication of the mean rate of heat transfer through the component. It is defined by the reciprocal of the summation of the individual thermal resistances that compose the building element. Therefore, the U-value (Wm^{-2}K^{-1}) is given by the following equation (CIBSE, 1980):

$$U = \frac{1}{R_{SI} + R_1 + R_2 + ... + R_A + R_{SO}}$$ [12.5]

where R_{SI} (m^{2}KW^{-1}) is the inside surface resistance, R_1, R_2 ...... (m^{2}KW^{-1}) the thermal resistances of structural components, R_A (m^{2}KW^{-1}) the airspace resistance and R_{SO} (m^{2}KW^{-1}) the outside surface resistance.

ANALYTICAL AND SEMI-ANALYTICAL METHODS

This section presents a review of the existing analytical and semi-analytical methods for earth-contact heat transfer: their advantages and disadvantages are discussed. The direct application of analytical methods requires some simplifications. However, they are very useful for some simple configurations since they can predict the results in a fast and exact way. Semi-analytical methods require computational processes to perform the solution (e.g. numerical integration) (Adjali et al, 1998a, 1999, 2000b; Rees et al, 2000).

The analytical and semi-analytical methods are reviewed below.

MACEY

Macey's formula addressed a correction factor for the calculation of rectangular floor and ground U-values (Macey, 1949). This was the first attempt to provide an analytical expression for earth-contact heat losses that has taken into account soil thermal conductivity, wall thickness, floor dimensions and the surface air film resistances. For two-dimensional heat flow and with the same floor and soil conductivity, Macey concluded the following equation:

$$U = \frac{2\lambda B}{\pi a} arctanh\left(\frac{2a}{2a + u}\right)$$ [12.6]

where λ (Wm^{-1}K^{-1}) is the soil thermal conductivity, B is a correction factor for rectangular floors, a is half the breadth of the floor and u is the wall thickness. Improvements to this method have been made to make it more realistic. These are:

● Anderson presented an extension to several different shapes (Anderson, 1990, 1991a) and the addition of internal or external insulation (Anderson, 1991b).
● Davies reconsidered the correction applied to the formula, taking into account the finite floor dimensions (Davies, 1993a, 1993b).

LATTA AND BOILEAU

Latta and Boileau used a method based on the steady-state solution (Boileau and Latta, 1968; Latta and Boileau, 1969) and considered measured temperature profiles on the ground to show that the heat flow to the earth's surface follows circular paths. As the thermal conductivity of the soil λ (Wm^{-1}K^{-1}) was already known and calculating the length of these paths (arcs) from a point in the below-ground wall, the heat loss Q_z (Wm^{-2}) at depth z (m) gives:

$$Q_z = \frac{T_i - T_s}{R_w + R_{ins} + \frac{z\pi}{2\lambda}}$$

[12.7]

where $T_i - T_s$ is the temperature difference between outdoors and indoors (°C), and R_w and R_{ins} (Km^{2}W^{-1}) are the thermal resistances of the wall and the insulation, respectively. The authors used the above equation as many times as number of the segments (i.e. 0.3 m length) of the sub-ground into which it has been divided. The basement floor is treated in a similar way. U-values of 0.3 m strips of wall are tabulated for various path lengths, depths and insulation level. This method has been adopted by the ASHRAE guide (ASHRAE, 1993) in 1985 and some modifications were made in 1997. Some limitations should be considered:

● The method ignores heat loss to the deep ground – which is a limitation for predictions during the summer period.
● The real heat flow paths are not concentric as described but a shape somewhere between circular and vertical lines for partially insulated walls.

DELSANTE

Delsante et al (1983) gave a solution for an uninsulated semi-infinite slab-on-ground heat flow, which is a continuation of Fourier analysis (Muncey and Spencer, 1978) where the floor surface is considered to be composed from equally spaced identical rectangular slabs. In fact, Lachenbruch (1967) was the first to consider this analytical solution, but Delsante assumed that the temperature under the wall at the ground level is changing linearly and that the floor is a unique entity to avoid large temperature gradients at the edges and so to produce explicit expressions using Fourier transforms for:

● two-dimensional heat flow under cyclical variations of boundary conditions;
● three-dimensional steady-state heat flow for a rectangular floor. An approximate solution is then derived for the periodic three-dimensional case.

In further work, Landman and Delsante (1987a) presented solutions for partially horizontally insulated slabs-on-ground, and for vertically insulated slabs (Landman and Delsante, 1987b) but without taking into account the water table. Delsante later took into account the indoor and outdoor surface air film resistances with the same floor and soil thermal resistances which were also based on the same approximations, finally producing an expression for the two-dimensional steady-state heat flow (Delsante, 1988). In 1993, Delsante applied the method in relation to water table depth (Delsante, 1993).

KUSUDA

A semi-analytical method developed by Kusuda makes use of Green's functions to solve the three-dimensional dynamic heat flow equation (Kusuda and Bean, 1984). With an approximation that takes into account the effect of floor insulation, the solution for an insulated slab-on-ground floor is then given as:

$$Q = \frac{\frac{\lambda}{l}}{T_R - T_Z} \qquad [12.8]$$

where Q is the heat flux at the floor (Wm^{-2}), λ the thermal conductivity (Wm^{-1}K^{-1}), l the thickness of the floor (m), T_R the mean temperature of the slab (°C) and T_Z is the monthly average sub-floor temperature at a given depth Z (°C) which is calculated by Green's method.

CLAESSON AND HAGENTOFT

Claesson and Hagentoft presented a dynamic analytical solution for the three-dimensional heat flow to the ground from a structure (Claesson and Hagentoft, 1991). By the superposition of three fundamental processes, they derived the expression for the heat loss $Q(t)$ for any outdoor temperature which is dependent on time. These three processes are:

● a steady-state heat transfer;
● a periodic variation of the outdoor temperature;
● a step variation of the outdoor temperature.

The authors assumed a constant indoor temperature and a homogeneous semi-infinite ground with a known temperature at every depth. Further extensions of this work were reported for an insulated rectangular slab-on-ground (Hagentoft and Claesson, 1991) and the effect of ground water flow, at either finite (Hagentoft, 1996a) or infinite rate (Hagentoft, 1996b).

Some years earlier than Claesson's work, Hagentoft proposed a semi-analytical calculation method for predicting the two-dimensional temperature profile and heat flow under a house in steady-state conditions (Hagentoft, 1988). The temperature is given in terms of a Fourier series and the Fourier coefficients are calculated numerically. Several different thickness types can be treated by the method, which assumes a partially horizontally insulated semi-infinite slab, but without taking into account the existence of a water table underneath the slab. More recently, Hagentoft presented an analytical solution for steady-state heat loss from edge-insulated slabs (Hagentoft, 2002a, 2002b).

KRARTI

The method proposed by Krarti is a semi-analytical method for slab-on-ground and basement configurations (Krarti, 1993). It involves the interzone temperature profile estimation (ITPE) technique which is based on both numerical and analytical techniques and consists of dividing the ground and the basement or the slab into several zones of rectangular shape. This method can handle insulation for various geometries. The boundary conditions between the different zones are arbitrarily fixed with physically motivated temperature profiles. Each zone has its solution according to the interzonal functions, for steady-state and periodic conditions in two dimensions (i.e. for all the commonly used insulation configurations) and in three dimensions (i.e. only for uniform insulation configurations), and the transient solution is given only for two-dimensional heat flow. Earth-contact effects are also considered by the ITPE technique for adjacent slab-on-grade floors in the presence of water table effects (Krarti and Piot, 1998).

SHEN AND RAMSEY

A semi-analytical method based on the Patankar and Spalding implicit finite difference technique was presented for two-dimensional periodic steady-state conditions (Patankar and Spalding, 1977). The solution made use of the Fourier series and can handle vertical insulated slabs (Shen and Ramsey, 1983). In further work, Shen et al (1988) complemented the numerical method with the Fourier series technique to reduce storage requirements. Labs et al (1988), based on the latter work, presented a design handbook that was extended to include most of the common basement insulation configurations. The data generated are restricted to certain insulation lengths and R-values. Labs' equations have been applied on earth-contact walls in Kuwait (Al-Temeemi and Harris, 2003).

ACHARD

Achard proposed a method for simple geometries using linking transfer functions to connect the output with the thermal input (Achard et al, 1983). The author used the response factor technique to calculate the transient heat loss from walls in contact with the earth.

SOBOTKA

In 1994, Sobotka suggested an analytical model to calculate the ground temperature based on the seasons (Sobotka, 1994). This method can be useful when it is required to calculate the heat transfer from earth-contact and there are no available measurements to determine the ground temperature field. Various parameters are taken into account in this method, e.g. season, thermal properties of the soil, weather, humidity, surface air convection, site orientation and geographic situation.

CONCLUSION

It is known that the exact behaviour of earth-coupled systems can, theoretically, be determined by analytical solutions. However, difficulties arise in the application of these solutions that have led to simplifications being made in the treatment of the problem. Analytical methods can, however, be applied in a fast and efficient way in certain cases (e.g. simple geometry).

Important limitations are:

● some methods can handle only slabs or only walls (i.e. slabs only: Delsante, Kusuda, Hagentoft, Macey; and walls only: Achard);

- the assumption of semi-infinite geometry;
- the addition of insulation involves extra boundary conditions that are difficult to handle (e.g. Delsante, Kusuda);
- most analytical methods do not deal with the water table level;
- in general, the analytical methods assume constant internal temperature and sinusoidally time-varying outdoor temperature, which does not perform efficiently in some cases.

The most promising method may be the ITPE technique since it can deal with several different insulation and geometry configurations applied to slabs-on-ground and basements, with the effect of the water table level taken into account. The way that the ITPE technique could give optimal underground insulation configurations has also been reported (Krarti and Choi, 1994; Choi and Krarti, 2000; Chuangchid and Krarti, 2001).

NUMERICAL METHODS

A review of numerical methods is presented according to the technique that they use. Numerical methods have been developed to simulate complex systems, where the direct application of analytical solutions is not possible because of the simplifications that are required in order to produce a solution. With a numerical method, most of the variables can be included in one-, two- or three-dimensional solutions. The techniques that they usually make use of are: the finite difference technique which can simulate efficient schemes of an order higher than second, but is adequate only for simple geometries; and finite element and finite volume techniques which can provide flexible modelling when the geometry is complex and without any problems in the definition of dynamic boundary conditions (Ferziger and Peric, 1996).

With a numerical method, the problem has to be solved at discrete locations, which are described by the numerical grid where the solution is given directly from algebraic equations. In a finite difference method, approximations for the derivatives at the grid points have to be selected. In a finite volume method, one has to select the methods of approximating surface and volume integrals. In a finite element method, one has to choose the type of element and associated weighting functions. Additionally, the selected method is valid to solve a specific problem if the solution can satisfy certain properties of the method, which are consistency, stability, convergence, conservation and accuracy (Ferziger and Peric, 1996).

The numerical tools that follow use only one of the above methods or some combination of them.

KUSUDA AND ACHENBACH

One of the first earth-contact numerical simulation tools was the explicit finite difference model in three dimensions developed by Kusuda and Achenbach (1963), which made use of several simplified assumptions. The temperature, humidity and heat transfer were calculated for experimental underground shelters. The programme considered simultaneous heat and vapour transport and took into account solar radiation and convective heat transfer at the earth surface.

DAVIES

Davies undertook the first attempt to couple a ground heat transfer algorithm to a general building simulation code (Davies, 1979). The use of a two-dimensional finite difference technique was considered to predict transient heat transfer from earth-covered buildings. The form of the discretized region (structure and ground) was described as an axisymmetric cylinder. The outdoor and indoor temperature variations were not predefined functions of time but they could be a time series of data and, thus, the programme can handle random external and internal temperature profiles.

WANG

Wang (1981) used a two-dimensional finite element programme to analyse various configurations of slabs and basements. Actually, Wang continued the work of Latta and Boileau (Latta and Boileau, 1969) and showed that the heat flow from a partially insulated basement does not follow circular paths but something between circular and vertical lines. The current version of the ASHRAE *Fundamentals Handbook* suggested the use of the *heat loss factor 'F_2'* which is based on the latter work of Wang and depends on the configuration (i.e. slab or basement, insulation) and the outdoor weekly air temperature.

AMBROSE

A three-dimensional steady-state design solution for the sizing of heating plant was introduced by Billington (1951). However, the design of energy consumption in a building according to weather changes requires three-dimensional unsteady heat transfer to the ground. Ambrose (1981) used a finite difference solution to approach the two-dimensional unsteady-state case and then transformed it to the three-dimensional case according to the required energy consumption. This solution had to be integrated into existing models, many of which were one-dimensional and, thus, the author transformed the solution to an equivalent one-dimensional solution, which yields the same annual energy consumption. The same thinking has been applied by Xie (Xie et al, 2008) by introducing a one-dimensional 'equivalent slab' for each of the recognized heat transfer processes.

SZYDLOWSKI

Szydlowski and Kuehn (1981) determined the thermal effect on different structures (basements, earth-bermed walls and earth-covered shelters) of various insulation configurations (internal, external, level and location) with an implicit finite difference technique.

SAXHOF AND POULSEN

Five Danish test houses were simulated by the development of a two-dimensional finite difference model (Saxhof and Poulsen, 1982). The authors compared the one- and two-dimensional results and revealed that in some known cases (e.g. without thermal bridges), the one-dimensional simulation may be able to predict ground heat transfer while the discrepancy between one- and two-dimensional solutions is 50 per cent for a realistic case.

BLIGH AND WILLARD

The thermal properties of the soil can significantly be changed by phase changes. Bligh was the first to include phase changes in a general finite element thermal analysis model (ADINAT) to calculate the two-dimensional heat transfer from an earth-sheltered building to the ground (Bligh et al, 1980; Bligh and Knoth, 1982; Bligh and Grald, 1983; Bligh et al, 1983; Bligh and Willard, 1985). Sensitivity analysis showed that the selection of soil conductivity must be done very carefully.

WALTON

Walton used two-dimensional results to produce the three-dimensional heat losses from rectangular and cylindrical basements and rectangular slabs to the ground (Walton, 1987). The author used the equivalent R-C electrical analogy to the finite difference equations, which is an approach similar to that of Ambrose.

ROUX

Roux et al (1989) used modal analysis to determine the heat flux between a building and the ground. The author made use of the finite difference technique to solve the discretized equations in one dimension for the walls and two dimensions for the floor. Reduction of the system by choosing an appropriate time constant according to the system's eigenvalues leads to the resulting general matrix, which is solved by modal analysis. This kind of approach can lead to a great saving of computer time.

HASEGAWA AND YOSHINO

A numerical model was developed to consider parametric analysis coupled with the effect of insulation location (Hasegawa et al, 1987; Yoshino et al 1990a, 1990b, 1992). Measurements for a semi-underground house have been taken for a period of 5 years. The results from a two-dimensional finite element model were compared with the experimental data for several different insulation configurations with and without solar gain, transient heating, etc.

BAHNFLETH

A programme that can handle uniform and partial insulation configurations and includes the three-dimensional solution of the diffusion equation was developed to predict heat transfer from slab-on-ground floors with the use of an implicit finite difference scheme (Bahnfleth, 1989a, 1989b; Bahnfleth and Pedersen, 1990). In fact, the method cannot work for configurations that have insulation extended horizontally or vertically from a slab or a wall into the soil. The soil surface temperature is calculated taking into account the solar and infrared radiation, convection and evaporation. A sensitivity analysis showed that the results could be significantly affected by soil conductivity. Additionally, the results were shown to be affected in some cases where the surface moisture transfer was neglected.

SPELTZ

The author presented one-dimensional analysis for the walls of a building and linked it with two- and three-dimensional codes for the soil surrounding the structure

(Speltz, 1980). The two-dimensional model predicted the heat loss to the ground 38 per cent lower than the three-dimensional model because the first one did not take into account corner effects.

CLEAVELAND
In some cases, when a building in a hot climate is to be considered, the performance of both heating and cooling loads must be taken into account. The insulation level can perform efficiently in the minimization of cooling load during summer and heating load during winter. For the consideration of the latter theory, Cleaveland made use of a two-dimensional finite difference model to consider the heat flux between a slab-on-ground floor and the ground (Cleaveland and Akridge, 1990).

RICHARDS
Richards and Mathews (1994) reported that currently available design tools were complicated to use. Thus, the author integrated an existing simplified earth-contact model with the use of a generally applicable numerical thermal programme which made use of the analytical formulation developed by Anderson (1991b), which improves Delsante's (1988) proposal.

REES
Recently, an extensive research programme took place which considered direct measurements of heat transfer from a concrete slab for a period of 2 years (Rees et al, 1995). The results included periodic variations of heat flux through the floor slab of a commercial building, air and ground temperature and moisture content. Numerical analysis of the data was undertaken via a general-purpose finite element heat transfer model. The observations of such a period showed that near the building the temperature of the ground and its moisture content could change significantly while this change is not significant for greater depths underneath the structure. The edge effect near the outer boundary wall and the large thermal storage of the ground were clearly shown in this work. Further exploration of the earth-contact material properties is needed (Rees et al, 2007).

PATANKAR
Modules based on the finite volume technique suggested by Patankar (1980) have been added into general simulating codes for building energy (Davies, 1994; Mihalakakou et al, 1995).

ZORAS AND DAVIES
An improvement of the three-dimensional modelling of earth-coupled structures has been achieved by Zoras and Davies (Zoras, 2001; Zoras et al, 2001, 2002) by the incorporation of three-dimensional generated response factors into a finite volume model. Extrapolation techniques and variable time-stepping schemes were employed according to the scheme solved (i.e. explicit or implicit) to dramatically improve the run times needed by the three-dimensional simulation of the earth-contact domain while retaining the accuracy of the

numerical solution. Implications due to the number of earth-contact rooms were also developed in relation to underground interaction. This method calculates future values of system-dependent variables by the use of weighted coefficients. These coefficients are calculated from linear equations and for unit excitations of other system variables. The response factors technique can only work if the equations that describe the system are linear. The system to be simulated is the earth-contact domain surrounding a building. This can be any combination of soils and structural components. Grid nodes describe this area, with each node defined by a set of linear equations derived from the general finite volume method. Thus, the entire earth-contact domain is linear. Heat transfer response factors time series are generated for the whole area due to any internal or external temperature excitations by the use of a three-dimensional numerical model. The more response factors that are calculated, the better the accuracy for further predictions.

CONCLUSION

Many of the numerical models run with a relatively large time step (e.g. more than one hour) and assume constant internal temperature, and so they fail to account for hourly effects on ground-coupled heat transfer of temperature fluctuations due to thermostat setting, solar gain and lighting schedule. Kusuda et al (1983) found that diurnal fluctuations in the indoor temperature have a large impact on the slab-on-grade floor heat loss or gain. Smaller time steps, however, need more computer run time.

Numerical methods, which are based on two-dimensional solutions, cannot take into account corner effects where the heat flux is three-dimensional. Walton (1987) noted that for an uninsulated rectangular basement, the total annual heat flux obtained by a three-dimensional model was 29 per cent higher than the flux predicted from a two-dimensional model.

Water-level effects can reduce the accuracy of the prediction of models if the water table is neglected. Mitalas (1982) used a three-dimensional code based on the finite element method and observed an underprediction of the heat transfer. He attributed this to the ground water-level effects. Recently, Rees et al (2001) reported that heat flux to the ground could significantly increase with decreasing water table depth.

Numerical methods can deal with 'all' the parameters influencing earth-coupled systems but long computer run times are needed. However, these methods are useful since they provide flexible simulation and accurate results. The addition of numerical methods into general simulation models, e.g. APACHE (Davies, 1994) and TRNSYS (Mihalakakou et al, 1995), for the modelling of the earth coupling yielded tools that can deal with all the aspects influenced in a building (e.g. HVAC components). A two-dimensional solution has been integrated into a whole building simulation software to model the floor heating system (Weitzmann et al, 2005). Moreover, a coupled heat and moisture numerical tool was integrated into building simulation software showing its improved robustness (Dos Santos and Mendes, 2005, 2006). Other ground-integrated whole building simulation tools can be found elsewhere (Krarti and Choi, 1997; Deru et al, 2003).

Generally, earth-contact heat transfer can be most accurately modelled via three-dimensional numerical simulation. However, this kind of approach is inefficient in terms of the run times required. The work by Zoras and Davies has improved the speed of these tools significantly while still retaining a high degree of accuracy. It was claimed that this method is at its most useful when used in repeated simulations (i.e. parametric analysis procedures). The methodology is fast because of the nature of the response factors technique, accurate because the three-dimensional modelling is still used in the initial step of the method and flexible because it can simulate any configuration that a numerical model is able to handle. Other modifications in numerical solutions targeting fast repeated simulations can be found elsewhere (Al-Anzi and Krarti, 2004).

Adjali et al (1998a) reported that the main differences between numerical methods relate to the 'features' offered by each tool, such as:

- ease of use;
- ability to treat moisture transfer at the soil/air surface;
- ability to treat the general issue of heat/moisture coupling;
- user support;
- ability to model in three dimensions;
- a flexible treatment of geometry/insulation placement.

These features effectively describe the most important issues that concern such models.

MANUAL METHODS

In this section, the basic characteristics of several manual methods are reviewed. For the determination of heat transfer between the building and the ground, it is possible to make use of some simplified algorithms with the aid of only a hand calculator. These methods are characterized as manual methods and may be derived from numerical, analytical or experimental techniques. The manual methods have limited applications because of their simplified nature (e.g. reduction of the number of parameters which influence the phenomena or assuming some parameters constant). However, in many cases these methods can yield general conclusions regarding the thermal performance of some buildings.

The methods, which can be characterized as manual methods, are reported below.

SWINTON AND PLATTS

Swinton and Platts established a correlation between heat losses from a basement in Canada versus heating degree-days (Swinton and Platts, 1981). However, this method has limited applications because the correlation was achieved for only three locations. They used a calorimeter box to estimate corner effects for uninsulated basement walls and found that the total heat loss is 10 per cent higher than the loss calculated using the centre wall heat flux that is dominated by two-dimensional heat flow.

SHIPP

This work is based on Patankar and Spalding (1977) and uses a two-dimensional implicit finite difference model to simulate heat flux between earth-sheltered buildings and the ground at the University of Minnesota (Shipp, 1979; Shipp et al, 1981). The model made use of a one-day time step or larger and it calculated annual heating season and cooling season heat flows, Q, for several different cases (e.g. slab-on-grade, basement, insulation, crawl-space configuration, etc.). The following regression equations were derived from the results:

$$Q = B_0 + B_2 HDD/100 + B_3 CDD/100 + B_4 (HDD/100)/R + B_5 (CDD/100)/R + B_6 (HDD/100)(CDD/100) + B_7 R (CDD/100) \qquad [12.9]$$

where B_0–B_7 are tabulated coefficients for various configurations, R is the thermal resistance of the structure and HDD and CDD are heating degree-days (base 12.8°C) and cooling degree-days (base 12.8°C), respectively.

Parker's work was based on the limitation of this method which comes from the single set of soil thermal properties and the fixed indoor temperature used to establish the correlations, and suggested a modification of Shipp's method (Parker, 1986, 1987).

AKRIDGE AND POULOS

The method, known as decrement average ground temperature (DAGT), was developed by Akridge and Poulos and takes into account the ground temperature variations due to the heat losses from the buildings (Akridge and Poulos, 1983). The result from a finite difference model was expressed as follows:

$$Q = F_d U A (T_R - T_S) \qquad [12.10]$$

where F_d is the tabulated decrement factor value, dependent on the soil conductivity and the wall thermal resistance, A is the wall area (m^2), U is the U-value of the wall and $T_R - T_S$ is the temperature difference between the soil and the room (°C). This method is adequate only for entirely underground walls and not for floors.

MITALAS

Mitalas developed the most detailed manual method for predicting the heat loss from basements (Mitalas, 1983), slab-on-grade floors and shallow basements (Mitalas, 1987). Use of two- and three-dimensional finite element models was made to compile tables of heat loss factors for a variety of situations (e.g. insulation, foundation type) and these values were fine-tuned by experimental data to include corner effects. Floor and underground walls were divided into five segments and the total heat flow was estimated by the summation of the heat losses of the different zones Q_n (n = 1–5) using the latter 'linking' factors. Q_n was expressed as:

$$Q_{n,t} = Q_{m,n} + Q_{A,n,t} \sin (2\pi t/12) \qquad [12.11]$$

where Q_m is the annual heat loss (Wm^{-2}), Q_A is the annual amplitude of the heat loss (Wm^{-2}) and t is the time. Mitalas' method was the first to allow the prediction of the heat transfer to the ground at any time of the year. However, Mitalas' correlations are restricted to certain insulation values and limited to particular geometric dimensions.

YARD

In 1984, Yard et al (1984) presented a method called the dimensionless parameter method because of the correlations developed in terms of non-dimensional parameters (e.g. U-values). This method can predict the heat loss from a basement at any time of the year if the ground temperature is a sinusoidal function of time as in Mitalas' theory. The author made use of a two-dimensional finite element analysis to calculate basement heat transfer coefficients in order to generate the dimensionless parameters. The sum of the heat flows from the wall (w) and the floor (f) gives the total heat flow to the ground Q (W):

$$Q = U_f A_f \left(T_b - T_{g,\,f} \right) + U_W A_W \left(T_b - T_{g,w} \right) \qquad [12.12]$$

where A is the area (m²), T_b is the basement indoor temperature (°C) and T_g is the time-dependent effective ground temperature (°C). The $U_{f,\,w}$-values (Wm⁻²K⁻¹) are given as functions of the soil conductivity, insulation configuration and basement depth.

KRARTI

Recently, Krarti used the ITPE method to develop a simplified new tool for heat loss predictions for both slab-on-ground and basement (Krarti and Choi, 1996). The development of the correlations of the heat loss Q from the foundation to the ground was achieved with the use of non-linear regression. Q (Wm⁻²) changes sinusoidally with time and is expressed as:

$$Q = Q_m + Q_a \cos \left(\omega_t - \varphi \right) \qquad [12.13]$$

where Q_m is the annual mean heat loss (Wm⁻²), Q_a is the annual amplitude (Wm⁻²), ω_t is the annual frequency and φ is the phase lag between heat loss and soil surface temperature. The latter coefficients depend on soil thermal properties, insulation R-values and configurations, foundation dimensions, ground surface temperature and indoor temperature. The generated values are tabulated and they were determined using 2000 different slab-on-ground and basement configurations. This was the first work undertaken for such a large number of configurations; by comparison, Mitalas used 100 configurations.

CONCLUSION

Most of the available manual methods for calculating ground heat loss are based on two-dimensional solutions and so they do not take into account corner effects. In fact, Akridge and Poulos' (1983), Yard et al's (1984) and Shipp's (1983) methods were generated from two-dimensional finite difference simulations. The Mitalas method (Mitalas, 1983, 1987) takes into account corner effects. Additionally, Mitalas (1982) investigated corner effects in basements using a three-dimensional code based on the finite element method and found that the three-dimensional corner effects are more significant for partially insulated walls at the floor and uninsulated sections of the wall. As reported in a previous section, Mitalas observed an underprediction of the heat transfer which he attributed to the ground water-level effects. Krarti et al (1988) noted that, for uninsulated floors, the annual average heat loss could be underestimated by as much as 30 per cent if a water table located 3 feet below the ground surface is neglected.

According to the nature of manual methods, it is difficult to include all the parameters mentioned above in such methods. Additionally, limitations arise when climatic data or variable thermal properties must be taken into consideration (e.g. Shipp's coefficients are calculated for a base of 12.8°C heating and cooling degree-days; Swinton's correlations have been developed for typical conditions in Canada for only one set of thermophysical properties).

Mitalas' method can deal with horizontal insulation configurations for a slab-on-ground or basement while Akridge's method can treat only walls. Krarti's method is the most detailed and is able to handle slabs and basements with horizontal or vertical configurations.

DESIGN GUIDES

Design guides are manual methods, but they are reviewed in a different section to give an indication of their importance. These guides are used for the construction of structures because of their ease of use and also because of their link to Building Regulations.

The design guides which are reported below are the ASHRAE (American Society of Heating, Refrigerating and Air-Conditioning Engineers), CIBSE (Chartered Institution of Building Services Engineers), AICVF (Association des Ingenieurs de Climatisation et de Ventilation de France) and CEN (European Committee for Standardization).

ASHRAE

This method has been proposed by ASHRAE (1993, 1997) and is based on the work of Latta and Boileau (1969) and Wang (1981). This guide makes use of tabulated heat losses for basements and for one type of soil conductivity ($1.38 \ Wm^{-1}K^{-1}$) according to depth, path length and insulation configuration. Care should be taken due to edge insulation impacting heat flow path length (Zhou et al, 2002). For slab-on-ground floors, the heat flow Q (W) to the ground is expressed as:

$$Q = F_2 P(T_i - T_0)$$

[12.14]

where F_2 is the tabulated heat loss coefficient ($Wm^{-1}K^{-1}$) which can be found for several different insulation/wall configurations, P is the perimeter of the exposed edge of floor (m) and T_i and T_0 are the indoor and outdoor temperatures, respectively (°C).

As already known, the circular heat flow path theory assumes a single sink at the ground surface, which may be acceptable in the winter when the difference between the interior and the exterior temperature is higher than the difference between the indoor and the deep ground. However, a limitation arises for basements during the summer when the situation is reversed.

CIBSE

The CIBSE guide (CIBSE, 1986) adopted a method based on the steady-state formula of Macey (1949) which is valid only for uninsulated rectangular slab-on-ground floors. Floor U-values are given as:

$$U = \left(2\lambda_e / b\,\pi / 2\right)\,arctanh\,\left((b/2)/(b/2+\omega/2)\right)\exp\,\left((b/2)/l_f\right) \qquad [12.15]$$

where λ_e (Wm^{-1}K^{-1}) is the soil thermal conductivity, b and 1_f are the breadth and the length of floor (m), respectively, and ω is the wall thickness. For insulated floors, the total U-value can be derived by taking the reciprocal of the sum of the thermal resistances for the floor and the insulation.

AICVF

The French guide calculates heat loss Q (W) from floors and walls in contact with the ground (AICVF, 1990) using the equation:

$$Q = K_s L\,(T_i - T_0) \qquad [12.16]$$

where L is the length of the exposed edge (m), T_i and T_0 are the indoor and the outdoor temperatures (°C), respectively, and K_s is a coupling coefficient (Wm^{-1}K^{-1}) which is calculated by a given expression for each case. This coefficient is a function of the type of configuration (floor and wall), the depth, the thickness, and soil and basement thermal conductivities and convective coefficients.

CEN

The most detailed guidance is provided by the CEN (1992) in which U-values are given as:

Slab-on-ground

$$U = U_0 + 2\Delta\psi / B \qquad [12.17]$$

Basement

$$U = \left(AU_{bf} + HPU_{bW}\right)/\left(A + HP\right) \qquad [12.18]$$

where U_0, U_{bf} and U_{bw} are the basic thermal transmittances (Wm^{-2}K^{-1}), $\Delta\psi$ is the edge factor (Wm^{-1}K^{-1}), B is a characteristic dimension of the floor (m), A the basement area (m^2), H the basement depth (m) and P the basement perimeter (m).

The method can handle a large number of configurations where the U-values are given for heated, unheated, etc., including suspended floors. For each situation, the guide provides the monthly, seasonal and annual heat transfer rates with the various coefficients.

CONCLUSION

The ASHRAE method is the most widely used despite the problems associated with the summer basement predictions. The CIBSE method is limited to floors while the AICVF method is more general. The CEN method is the most detailed.

However, the design guides suffer from the same limitations as the analytical and manual methods, as reported previously. A comparative study of design guides can be found in the literature (Adjali et al, 2004).

FINAL CONCLUSIONS

The use of analytical methods gives far more accurate results for the heat losses to the ground from coupled structures. However, these solutions are restricted to simple geometries and linear heat conduction (e.g. dry soil). On the other hand, manual methods and design guides suffer from simplicities and empirical inefficiencies. These days, increased computer power has brought numerical simulation of earth-contact heat transfer closer to engineers through robust coupled tools. Ground domain is integrated in whole building simulation software which improves building thermal envelope modelling.

Inefficiencies in numerical modelling arise due to the lack of initial conditions of the underground domain. This leads to inevitable multi-year simulations that approximate realistic soil temperature fields. The need to pre-condition the simulation for ground field initialization would help three-dimensional numerical simulation of the ground to produce improved results (Rees et al, 2006b). The presence of a new built structure interacts with the surrounding earth-contact domain for a few years until the thermal exchange of the following years remains unchanged year after year. This is also called the time that the earth-contact domain reaches equilibrium due to the presence of a new solid entity. Recently, a progress has been achieved concerning multi-year simulations where the earth-contact domain is simulated rapidly for several years by the use of three-dimensional response factors and variable time step in an implicit scheme (Zoras et al, 2002). Briefly, an idea is addressed for future work that may be able to take advantage of an unconditionally stable implicit scheme, a generally fast explicit scheme and a flexible variable time-stepping scheme. Obviously, this would be a combination of the above three and could be implemented for all finite volume-based numerical models.

The defined unique entities must be described by linear equations in order for most of the above mentioned methods to be implemented. For the ground in particular, a very important issue arises which concerns the variation of conductivity due to temperature changes. This is a non-linear phenomenon and during numerical simulations can only be handled with iterative processes which are extremely time consuming, especially for simulations over long periods. A transformation that converts non-linear effects, such as changes in conductivity due to temperature fluctuations, into linear is of great value. This removes the non-linearity of variable conductivity due to temperature fluctuations through the application of Kirchhoff's transform (Carslaw and Jaeger, 1959). The actual solution has been integrated into a finite element formulation for non-linear heat conduction in the past (Wrobel and Brebbia, 1987). First, justification is needed that the particular formulation leads to a solution for a finite volume method. And second, the particular transformation could be advantageously combined with superposition methods and lead to fast simulations where non-linear phenomena would be taken into account.

In conclusion, a future fully completed tool must deal with variable conductivity, heat and moisture coupling, change of phase, snow cover, convection and evaporation at the earth's surface. Specifically, a very comprehensive review on how to handle soil water content has been carried out by Rees et al (2000). Additionally, convection and evaporation at the earth's surface can be handled according to Krarti's method (Krarti et al, 1995) and some of its applications can be found elsewhere (Mihalakakou et al, 1997).

AUTHOR CONTACT DETAILS

Stamatis Zoras: Laboratory of Environmental and Energy Design, Department of Environmental Engineering, Polytechnic School of Xanthi, Democritus University of Thrace; szoras@airlab.edu.gr

REFERENCES

Achard, G., Allard, F. and Brau, J. (1983) 'Thermal transfer between a building and the surrounding ground', in *Proceedings of the 2nd International Congress, Building Energy Management*, Ames, Iowa, US, pp35–44

Adamson, B. (1973) 'Floor slabs on ground – foundation depth', *Swedish Council for Building Research*, Report R41

Adamson, B., Domner, G. and Ronning, M. (1964) 'Soil temperature under buildings without basements', *Swedish Council for Building Research*, Transaction No 46

Adamson, B., Claesson, J. and Eftring, B. (1973) 'Floor slabs on ground – thermal insulation and floor temperatures', *Swedish Council for Building Research*, Report R40

Adjali, M. H., Davies, M. and Littler, J. (1998a) 'Earth-contact heat flows: Review and application of design guidance predictions', *Building Services Engineering Research Technology*, vol 19, no 3, pp111–121

Adjali, M. H., Davies, M. and Littler, J. (1998b) 'Three-dimensional earth-contact heat flows: A comparison of simulated and measured data for a buried structure', *Renewable Energy*, vol 15, no 14, pp356–359

Adjali, M. H., Davies, M. and Littler, J. (1999) 'Numerical simulation of measured transient heat transfer through the walls, floor and surrounding soil of a buried structure', *International Journal of Numerical Methods for Heat and Fluid Flow*, vol 9, no 4, pp405–422

Adjali, M. H., Davies, M., Rees, S. W. and Littler, J. (2000a) 'Temperatures in and under a slab-on-ground floor: Two- and three-dimensional numerical simulations and comparison with experimental data', *Building and Environment*, vol 35, pp622–655

Adjali, M. H., Davies, M., Riain, C. N. and Littler, J. (2000b) 'In situ measurements and numerical simulation of heat transfer beneath a heated ground floor slab', *Energy and Buildings*, vol 33, pp75–83

Adjali, M. H., Davies, M. and Rees, S.W. (2004) 'A comparative study of design guide calculations and measured heat loss through the ground', *Building and Environment*, vol 39, no 11, pp1301–1311

AICVF (Association des Ingenieurs de Climatisation et de Ventilation de France) (1990) *Chauffagge-Calculs des Deperditions et Charges Thermiques d'Hiver*, Pyc Edition, AICVF, Paris

Akridge, J. M. and Poulos, J. F. J. (1983) 'The decremented average ground temperature method for predicting the thermal performance of underground walls', *ASHRAE Transactions,* vol 89, no 2A, pp49–60

Aksoy, U. T. and Inalli, M. (2006) 'Impacts of some building passive design parameters on heating demand for a cold region', *Building and Environment*, vol 41, no 12, pp1742–1754

Al-Anzi, A. and Krarti, M. (2004) 'Local/global analysis of transient heat transfer from building foundations', *Building and Environment*, vol 39, no 5, pp495–504

Al-Temeemi, A. A. and Harris, D. J. (2003) 'The effect of earth-contact on heat transfer through a wall in Kuwait', *Energy and Buildings*, vol 35, no 4, pp399–404

Ambrose, C. W. (1981) 'Modelling losses from slab floors', *Building and Environment*, vol 16, no 4, 251–258

Anderson, B. R. (1990) *The U-value of Grand Floors: Application to Building Regulations*, IP 3/90, Building Research Establishment, UK

Anderson, B. R. (1991a) 'U-values of uninsulated ground floors: Relationship with floor dimensions', *Building Services Engineering Research Technology*, vol 12, no 3, pp103–105

Anderson, B. R. (1991b) 'Calculation of the steady-state heat transfer through a slab-on-ground floor', *Building and Environment*, vol 26, no 4, pp405–415

Anselm, A. J. (2008) 'Passive annual heat storage principles in earth sheltered housing, a supplementary energy saving system in residential housing', *Energy and Buildings*, vol 40, no 7, pp1214–1219

APACHE (1994) *Design and Simulation Suite*, IES User Manual

ASHRAE (American Society of Heating, Refrigerating and Air-Conditioning Engineers) (1993) *Handbook of Fundamentals*, ASHRAE, Atlanta, GA

ASHRAE (1997) *Handbook of Fundamentals*, ASHRAE, Atlanta, GA

Bahnfleth, W. P. (1989a) *Three-dimensional Modelling of Heat Transfer from Slab Floors*, PhD thesis, University of Illinois at Urbana-Champaign, (also published as USACERL Technical Manuscript E-89/11/ADA210826)

Bahnfleth, W. P. (1989b) *Three-dimensional Modelling of Heat Transfer from Slab Floors*, Technical Manuscript (USA Construction Engineering Research Laboratory)

Bahnfleth, W. P. and Pedersen, C. O. (1990) 'A three dimensional numerical study of slab-on-ground heat transfer', *ASHRAE Transactions*, vol 2, no 2, pp61–72

Billington, N. S. (1951) 'Heat loss through solid ground floor', *Journal of the Institution of Heating and Ventilating Engineers*, pp351–372

Bligh, T. P. and Grald, E. (1983) *A Quantitative Energy Balance on an Earth-sheltered House*, Report to US Department of Energy, Solar Passive Division on Contract No DOE/DE-AC03-80SF-11508, MIT EEBS Report No 21

Bligh, T. P. and Knoth, B. H. (1982) *A Thermal Study of an Earth-sheltered Residence: Instrumentation, Data Progressing Techniques, Soil Temperature, and Heat Flux Data*, Government report under DOE Contract No 17

Bligh, T. P., Meixel, G. P. and Shipp, P. H. (1980) 'The impact of insulation placement on the seasonal heat loss through basement and earth-sheltered walls', *Underground Space*, vol 5, pp41–47

Bligh, T. P. and Willard, T. E. (1985) 'Modelling the thermal performance of earth-contact buildings, including the effect of phase change due to soil freezing', *Computers and Structures*, vol 21, no 1–2, pp291–318

Bligh, T. P., Abtahi, A. and Willard, T. E. (1983) *A New Basement Insulation Method for Retrofit and New Construction to Minimize Winter Heating and Cooling Loads*, Report to Consolidated Edison, New York, MIT EEBS Report No 24

Boileau, G. G. and Latta, J. K. (1968) *Calculation of Basement Heat Loss*, Technical Paper No 292 of the Division of Building Research Council Canada, Ottawa

Carslaw, H. S. and Jaeger, J. C. (1959) *Conduction of Heat in Solids*, Clarendon Press, Oxford

CEN/TC 89 (European Committee for Standardization) (1992) *Thermal Performance of Buildings – Heat Exchange with the Ground-calculation Method*, CEN

Choi, S. and Krarti, M. (2000) 'Thermally optimal insulation distribution for underground structures', *Energy and Buildings*, vol 32, no 3, pp251–265

Chuangchid, P. and Krarti, M. (2001) 'Foundation heat loss from heated concrete slab-on-grade floors', *Building and Environment*, vol 36, no 5, pp637–655

CIBSE (Chartered Institution of Building Services Engineers) (1980) *CIBSE Guide Section A3*, CIBSE, London

CIBSE (1986) *Guide A3: Thermal Properties of Buildings' Structures*, CIBSE, London

Claesson, J. and Eftring, B. (1980) 'Optimal distribution of thermal insulation and ground heat loss', *Swedish Council for Building Research*, vol D33

Claesson, J. and Hagentoft, C. E. (1991) 'Heat loss to the ground from a building – I. General theory', *Building and Environment*, vol 26, no 2, pp195–208

Claridge, D. E. (1987) 'Design methods for earth-contact heat transfer', in K. Boer (ed), *Advances in Solar Energy*, American Solar Energy Society, Boulder, Colorado, pp305–50

Cleaveland, J. P. and Akridge, J. M. (1990) 'Slab-on-grade thermal loss in hot climates', *ASHRAE Transactions*, vol 96, no 1, pp112–119

Davies, G. R. (1979) 'Thermal analysis of earth covered buildings', in *Proceedings of the Fourth National Passive Solar Conference*, Kansas City, pp744–748

Davies, M. G. (1993a) 'Heat loss from a solid floor: New formula', *Building Services Engineering Research Technology*, vol 14, no 2, pp71–75

Davies, M. G. (1993b) 'Heat loss from a solid floor', *Building and Environment*, vol 28, no 3, pp347–359

Davies, M. (1994) *Computational and Experimental Three-dimensional Conductive Heat Flows in and around Buildings*, PhD thesis, University of Westminster

Davies, M., Tindale, A. and Littler, J. (1995) 'Importance of multi-dimensional conductive heat flows in and around buildings', *Building Services Engineering Research Technology*, vol 16, no 2, pp83–90

Delsante, A. E. (1988) 'Theoretical calculations of the steady-state heat loss through a slab-on-ground floor', *Building and Environment*, vol 23, no 1, pp11–17

Delsante, A. E. (1993) 'The effect of water table depth on steady-state heat transfer through a slab-on-ground floor', *Building and Environment*, vol 28, pp369

Delsante, A. E., Stokes, A. N. and Walsh, P. J. (1983) 'Application of Fourier transforms to periodic heat flow into the ground under a building', *International Journal of Heat and Mass Transfer*, vol 26, no 1, pp121–132

Deru, M., Judkoff, R. and Neymark, J. (2003) 'Whole building energy simulation with a three-dimensional ground-coupled heat transfer model', *ASHRAE Transactions*, vol 109, no 1, pp557–565

Dill, R. S., Robinson, W. C. and Robinson, H. E. (1943) *Measurements of Heat Losses from Slab Floors*, National Bureau of Standards, Washington DC, Building Materials and Structures Report BMS 103 9

Dill, R. S., Robinson, W. C. and Robinson, H. E. (1945) *Measurements of Heat Losses from Slab Floors*, US Department of Commerce, National Bureau of Standards, Building Materials and Structures Report BMS 103 3

Dos Santos, G. H. and Mendes, N. (2004) 'Multidimensional effects of ground heat transfer on the dynamics of building thermal performance', *ASHRAE Transactions*, vol 110, Part II, pp345–353

Dos Santos, G. H. and Mendes, N. (2005) 'Unsteady combined heat and moisture transfer in unsaturated porous soils', *Journal of Porous Media*, vol 8, no 5, pp493–510

Dos Santos, G. H. and Mendes, N. (2006) 'Simultaneous heat and moisture transfer in soils combined with building simulation', *Energy and Buildings*, vol 38, no 4, pp303–314

Ferziger, J. H. and Peric, M. (1996) *Computational Methods for Fluid Dynamics*, Springer-Verlag, Berlin, Heidelberg, New York

Givoni, B. (2007) 'Cooled soil as a cooling source for buildings', *Solar Energy*, vol 81, no 3, pp316–328

Givoni, B. and Katz, L. (1985) 'Earth temperatures and underground buildings', *Energy and Buildings*, vol 8, no 1, pp15–25

Hagentoft, C. E. (1988) 'Temperature under a house with variable insulation', *Building and Environment*, vol 23, no 3, pp225–231

Hagentoft, C. E. (1996a) 'Heat losses and temperature in the ground under a building with and without ground water flow – I. Infinite ground water flow rate', *Building and Environment*, vol 31, no 1, pp3–12

Hagentoft, C. E. (1996b) 'Heat losses and temperature in the ground under a building with and without ground water flow – II. Finite ground water flow rate', *Building and Environment*, vol 31, no 1, pp13–19

Hagentoft, C. E. (2002a) 'Steady-state heat loss for an edge-insulated slab: Part I', *Building and Environment*, vol 37, no 1, pp19–25

Hagentoft, C. E. (2002b) 'Periodic heat loss for an edge insulated slab: Part II. A mixed boundary value problem', *Building and Environment*, vol 37, no 1, pp27–34

Hagentoft, C. E. and Claesson, J. (1991) 'Heat loss to the ground from a building – II. Slab on the ground', *Building and Environment*, vol 26, no 4, pp395–403

Hasegawa, F., Yoshino, H. and Matsumoto, S. (1987) 'Optimum use of solar energy techniques in a semi-underground house: First year measurements and computer analysis', *Tunnelling and Underground Space Technology*, vol 2, no 4, pp429–435

Houghten, F. C., Taimuty, S. K., Gutberlet, C. and Brown, C. J. (1942) 'Heat loss through basement walls and floors', *Heating, Piping and Air-Conditioning*, vol 14, pp69–74

Iwamae, A. and Matsumoto, M. (2003) 'The humidity variation in crawl spaces of Japanese houses', *Journal of Thermal Envelope and Building Science*, vol 27, no 2, pp123–133

Jacovides, C. P., Mihalakakou, G., Santamouris, M. and Lewis, J. O. (1996) 'On the ground temperature profile for passive cooling applications in buildings', *Solar Energy*, vol 57, pp167–175

Janssen, H., Carmeliet, J. and Hens, H. (2004) 'The influence of soil moisture transfer on building heat loss via the ground', *Building and Environment*, vol 39, no 7, pp825–836

Krarti, M. (1993) *Energy Calculations for Basements, Slabs and Crawl Spaces*, Final Report ASHRAE TC 4.7 Project 666-TR (University of Colorado)

Krarti, M. and Choi, S. (1994) 'Optimum insulation for rectangular basements', *Energy and Buildings*, vol 22, pp125–131

Krarti, M. and Choi, S. (1996) 'Simplified method for foundation heat loss calculation', *ASHRAE Transactions*, vol 102, no 1, pp140–152

Krarti, M. and Choi, S. (1997) 'A simulation method for fluctuating temperatures in crawlspace foundations', *Energy and Buildings*, vol 26, no 2, pp183–188

Krarti, M. and Piot, O. (1998) 'Time-varying heat transfer from adjacent slab-on-grade floors', *International Journal of Energy Research*, vol 22, no 4, pp289–301

Krarti, M., Claridge, D. E. and Kreider, J. F. (1988) 'The ITPE method applied to time-varying ground-coupling problems', *International Journal of Heat Mass Transfer*, vol 31, pp1899–1911

Krarti, M., Lopez-Alonzo, C., Claridge, D. E. and Kreider, J. F. (1995) 'Analytical model to predict annual soil surface temperature variation', *Journal of Solar Energy Engineering*, vol 117, pp91–99

Kumar, R., Sachdeva, S. and Kaushik, S. C. (2007) 'Dynamic earth-contact building: A sustainable low-energy technology', *Building and Environment*, vol 42, no 6, pp2450–2460

Kusuda, T. and Achenbach, P. R. (1963) 'Numerical analysis of the thermal environment of occupied underground spaces with finite cover using a digital computer', *ASHRAE Transactions*, vol 69, pp439–452

Kusuda, T. and Bean, J. W. (1984) 'Simplified methods for determining seasonal heat loss from uninsulated slab-on-ground floors', *ASHRAE Transactions*, vol 1, no B, 611–632

Kusuda, T., Piet, O. and Bean, J. W. (1983) 'Annual variation of temperature field and heat transfer under heated ground surface: Slab-on-grade floor heat loss calculations', *ASHRAE Transactions*, vol 89

Labs, K., Carmody, J., Sterling, R., Shen, L., Huang, Y. J. and Parker, D. (1988) *Building Foundation Design Handbook*, ORNL Report Sub/86-7214311

Lachenbruch, A. H. (1967) 'Three-dimensional heat conduction in permafrost beneath heated buildings', *Geological Survey Bulletin 1052-B*, US Government Printing Office, Washington DC

Landman, K. A. and Delsante, A. E. (1987a) 'Steady-state heat losses from a building floor slab with horizontal edge insulation', *Building and Environment*, vol 22, pp57–60

Landman, K. A. and Delsante, A. E. (1987b) 'Steady-state heat losses from a building floor slab with vertical edge insulation – II', *Building and Environment*, vol 22, pp49–55

Latta, J. K. and Boileau, G. (1969) 'Heat losses from house basements', *Canadian Builders*, vol 19, no 10, pp39–42

Lundh, M. and Dalenbäck, J. O. (2008) 'Swedish solar heated residential area with seasonal storage in rock: Initial evaluation', *Renewable Energy*, vol 33, no 4, pp703–711

Macey, H. (1949) 'Heat loss through a solid floor', *Journal of the Institute of Fuel*, vol 22, pp369–371

Mihalakakou, G. (2002) 'On estimating soil surface temperature profiles', *Energy and Buildings*, vol 34, pp251–259

Mihalakakou, G., Santamouris, M., Asimakopoulos, D. and Argiriou, A. (1995) 'On the ground temperature below buildings', *Solar Energy*, vol 55, no 2, pp355–362

Mihalakakou, G., Santamouris, M., Lewis, J. O. and Asimakopoulos, D. N. (1997) 'On the application of the energy balance equation to predict ground temperature profiles', *Solar Energy*, vol 60, pp181–190

Mitalas, G. P. (1982) *Basement Heat Loss Studies at DBR/NRC*, DBR Paper No 1045, NRC Canada

Mitalas, G. P. (1983) 'Calculation of basement heat loss', *ASHRAE Transactions*, vol 89, no 1, pp420–438

Mitalas, G. P. (1987) 'Calculation of below-grade residential heat loss: Low rise residential building', *ASHRAE Transactions*, vol 93, no 1, pp743–783

Muncey, R. W. R. and Spencer, J. W. (1978) 'Heat flow into ground under a house', in *Proceedings of Conference on Energy Conservation in Heating, Cooling and Ventilating Buildings*, vol 2, pp649–660

Parker, D. S. (1986) 'Simplified method for determining below grade heat loss', *Proceedings of Congress of International Solar Energy Society*, Montreal, Canada, vol 1, pp254–261

Parker, D. S. (1987) 'F-Factor correlations for determining earth contact heat loads', *ASHRAE Transactions*, vol 93, no 1, pp784–790

Patankar, S. (1980) *Numerical Heat Transfer and Fluid Flow*, Hemisphere, New York

Patankar, S. V. and Spalding, D. B. (1977) *Genmix: A General Computer Program for Two-dimensional Parabolic Phenomena*, Pergamon Press, Oxford

Rantala, J. and Leivo, V. (2006) 'Heat loss into ground from a slab-on-ground structure in a floor heating system', *International Journal of Energy Research*, vol 30, no 12, pp929–938

Rantala, J. and Leivo, V. (2008) 'Thermal, moisture and microbiological boundary conditions of slab-on-ground structures in cold climate', *Building and Environment*, vol 43, no 5, pp736–744

Rees, S. W., Adjali, M. H., Zhou, Z., Davies, M. and Thomas, H. R. (2000) 'Ground heat transfer effects on the thermal performance of earth-contact structures', *Renewable and Sustainable Energy Reviews*, vol 4, pp213–265

Rees, R. W., Lloyds, R. M. and Thomas, H. R. (1995) 'A numerical simulation of measured transient heat transfer through a concrete ground floor slab and underlying substrata', *International Journal of Numerical Methods for Heat and Fluid Flow*, vol 5, no 8, pp669–683

Rees, S. W., Thomas, H. R. and Zhou, Z. (2006b) 'A numerical and experimental investigation of three-dimensional ground heat transfer', *Building Services Engineering Research and Technology*, vol 27, no 3, pp195–208

Rees, S. W., Zhou, Z. and Thomas, H. R. (2001) 'The influence of soil moisture content variations on heat losses from earth-contact structures: An initial assessment', *Building and Environment*, vol 36, no 2, pp157–165

Rees, S. W., Zhou, Z., Thomas, H. R. (2006a) 'Multidimensional simulation of earth-contact heat transfer', *Building Research and Information*, vol 34, no 6, pp565–572

Rees, S. W., Zhou, Z. and Thomas, H. R. (2007) 'Ground heat transfer: A numerical simulation of a full-scale experiment', *Building and Environment*, vol 42, no 3, pp1478–1488

Richards, P. G. and Mathews, E. H. (1994) 'A thermal design tool for building in ground contact', *Building and Environment*, vol 29, no 1, pp73–82

Rock, B. A. (2004) 'Sensitivity study of slab-on-grade transient heat transfer model parameters', *ASHRAE Transactions*, vol 110, Part 1, pp177–184

Roux, J. J., Mokhtari, A. and Achard, G. (1989) 'Modal analysis of thermal transfer between a building and the surrounding ground', in *Proceedings of the 4th Congress, Performance of the Exterior Envelope of Building*, Orlando, US

Saxhof, B. and Poulsen, K. E. (1982) *Foundations for Energy Conservation Houses: A Thermal Analysis based on Examples from Five Low-energy Houses at Hjorteker, Low-Energy House Project*, Report No 130, Danish Ministry of Energy

Shen, L. S. and Ramsey, J. W. (1983) 'A simplified thermal analysis of earth-sheltered building using a Fourier-series boundary method', *ASHRAE Transactions*, vol 89, no 1, pp438–448

Shen, L. S., Poliakova, J. and Huang, Y. J. (1988) 'Calculation of building foundation heat loss using superposition and numerical scaling', *ASHRAE Transactions*, vol 94, no 1

Shipp, P. H. (1979) *The Thermal Characteristics of Large Earth Sheltered Structures*, PhD thesis, University of Minnesota, US

Shipp, P. H. (1983) 'Basement, crawlspace and slab-on-ground thermal performance', in *Proceedings of ASHRAE/DOE Conference – Thermal Performances of Buildings*, Las Vegas, Nevada, pp160–179

Shipp, P. H., Pfender, E. and Blight, T. P. (1981) 'Thermal characteristics of a large earth-sheltered building', *Underground Space*, vol 6, no 1, pp53–64

Sobotka, P. (1994) 'Climatic model for prediction of below ground heat loss: Influence of elevation', *International Journal of Energy Research*, vol 18, no 3, pp401–420

Sobotka, P., Yoshino, H. and Matsumoto, S. I. (1996) 'Thermal comfort in passive solar earth integrated rooms', *Building and Environment*, vol 31, no 2, pp155–166

Speltz, J. J. (1980) 'A numerical simulation of transient heat flow in earth-sheltered buildings for seven selected US cities', MS thesis, Trinity University, US

Swinton, M. C. and Platts, R. E. (1981) 'Engineering method for estimating basement heat loss and insulation performance', *ASHRAE Transactions*, vol 87, no 2, pp343–359

Szydlowski, R. F. and Kuehn, T. H. (1981) 'Analysis of transient heat loss in earth-sheltered structures', *Underground Space*, vol 5, no 4, pp237–246

The Energy Saver (1993) *The Complete Guide to Energy Efficiency*, Pub Gee

Walton, G. (1987) 'Estimating 3D heat loss from rectangular basements and slabs using 2D calculations', *ASHRAE Transactions*, vol 93, no 1, pp791–797

Wang, F. S. (1981) 'Mathematical modelling and computer simulation of insulation systems in below ground applications', in *Proceedings of ASHRAE/DOE Conference – Thermal Performance of the Exterior Envelopes of Buildings*, Orlando, Florida, pp456–471

Weitzmann, P., Kragh, J., Roots, P. and Svendsen, S. (2005) 'Modelling floor heating systems using a validated two-dimensional ground-coupled numerical model', *Building and Environment*, vol 40, no 2, pp153–163

Wrobel, L. C. and Brebbia, C. A. (1987) 'The dual reciprocity boundary element formulation for non-linear diffusion problems', *Computer Methods in Applied Mechanics and Engineering*, vol 65, pp147–164

Xie, X., Jiang, Y. and Xia, J. (2008) 'A new approach to compute heat transfer of ground-coupled envelope in building thermal simulation software', *Energy and Buildings*, vol 40, no 4, pp476–485

Yard, D. C., Morton-Gibson, M. and Mitchell, J. W. (1984) 'Simplified dimensionless relations for heat loss from basements', *ASHRAE Transactions*, vol 90, no 1B, pp633–643

Yoshino, H., Matsumoto, S., Hasegava, F. and Nagatomo, M. (1990a) 'Effects of thermal insulation located in the earth around a semi-underground room: A two-year measurement in a twin type test house without auxiliary heating', *ASHRAE Transactions*, vol 96, pp53–60

Yoshino, H., Matsumoto, S., Hasegava, F. and Nagatomo, M. (1990b) 'Effects of thermal insulation located in the earth around a semi-underground room: Computer analysis by the finite element method', *ASHRAE Transactions*, vol 96, no 1, pp3–111

Yoshino, H., Matsumoto, S., Nagatomo, M. and Sakanishi, T. (1992) 'Five-year measurement of thermal performance for a semi-underground test house', *Tunnelling and Underground Space Technology*, vol 7, no 4, pp339–346

Zhou, Z., Rees, S. W. and Thomas, H. R. (2002) 'A numerical and experimental investigation of ground heat transfer including edge insulation effects', *Building and Environment*, vol 37, no 1, pp67–78

Zoras, S. (2001) *A Novel Tool for the Prediction of Building Earth-contact Heat Transfer*, PhD thesis, Brunel University, UK

Zoras, S., Davies, M. and Adjali, M. H. (2001) 'A novel tool for the prediction of earth-contact heat transfer: A multi-room simulation', in *Proceedings of the Institution of Mechanical Engineers Part C – Journal of Mechanical Engineers*, vol 215, pp1–8

Zoras, S., Davies, M. and Wrobel, L. C. (2002) 'Earth-contact heat transfer: Improvement and application of a novel simulation technique', *Energy and Buildings*, vol 34, pp333–344

Index

US *see* United States of America
user
  behaviour 209–210
  satisfaction 189–191, 193–202, 208, 211–212
utility functions 134
utility trade-offs 206

vapour pressure 223
vegetation 155, 167, 237–260, 265–266, 268
  see also greenery; green roofs
vehicle canopy models 171–174
velocity
  human thermal models 167
  solar chimneys 25, 40
  urban thermal environments 163–164,
    170–171, 173–174
ventilation 24–26, 36–37, 61–62, 263
  see also natural ventilation
vertical channel solar chimneys 32–33
vertical landscaping (VL) 239, 250–258
visual comfort 127, 204, 212
Visual DOE 133
VL *see* vertical landscaping

walk-through surveys 123, 193–194,
  197–201
wall-climbing vertical landscaping
  251–252
walls, green 265–266
Walton method 298
Wang method 297
water heating systems 61, 105–107
water-level effects 300
water/lithium bromide absorption cooling
  47–49
wet cooling towers 55–56, 59–60
wind
  ANNs 97, 103–105
  solar chimneys 36–37, 41
  urban thermal environments 147, 157,
    159–164, 173–174
windows 212
WUFI bio *see* biohygrothermal models

Yard method 303

Zoras and Davies method 299–300